Classical Probability in
the Enlightenment

LORRAINE DASTON

Classical Probability in the Enlightenment

New Edition

With a new preface by the author

PRINCETON UNIVERSITY PRESS
PRINCETON AND OXFORD

First published by Princeton University Press in 1988
New paperback edition, with a new preface by the author, 2023
Paper ISBN 978-0-691-24850-9
Ebook ISBN 978-0-691-24851-6
Library of Congress Control Number: 2023935867

This book has been composed in Linotron Garamond

Cover image: Jean Antoine Pierron. *Les Caprices de la Fortune*, 1786.
Kress Collection (Graphic Materials). CF f9.

TEXT DESIGNED BY LAURY A. EGAN

TO MY FAMILY, OLD AND NEW

CONTENTS

PREFACE TO THE 2023 EDITION ix

PREFACE AND ACKNOWLEDGMENTS xvii

INTRODUCTION xix

CHAPTER ONE

The Prehistory of the Classical Interpretation of Probability:
Expectation and Evidence

1.1 Introduction 3

1.2 Quantitative and Qualitative Probabilities 6

1.3 Expectation, Equity, and Aleatory Contracts 15

1.4 Degrees of Certainty and the Hierarchy of Proofs 33

1.5 Conclusion 47

CHAPTER TWO

Expectation and the Reasonable Man

2.1 Introduction 49

2.2 Expectation as Reasonableness 58

2.3 The Debate over Expectation 68

2.4 Conclusion 108

CHAPTER THREE

The Theory and Practice of Risk

3.1 Introduction 112

3.2 Risk before Probability Theory 116

3.3 The Mathematical Theory of Risk 125

3.4 Risk after Probability Theory 138

3.5 Conclusion 182

CONTENTS

CHAPTER FOUR

Associationism and the Meaning of Probability

4.1	Introduction	188
4.2	Probability, Experience, and Belief	191
4.3	The Dissociation of Objective and Subjective Probabilities	210
4.4	Conclusion	224

CHAPTER FIVE

The Probability of Causes

5.1	Introduction	226
5.2	Bernoulli's Theorem and the Urn Model of Causation	230
5.3	Bayes' Theorem and the Problem of Induction	253
5.4	The Calculus of Induction	267
5.5	Conclusion	293

CHAPTER SIX

Moralizing Mathematics

6.1	Introduction	296
6.2	The Moral Sciences	298
6.3	Testimony and the Probability of Miracles	306
6.4	The Probability of Judgments	342
6.5	Conclusion	368

EPILOGUE

The Decline of the Classical Theory   370

BIBLIOGRAPHY   387

INDEX   413

PREFACE TO THE 2023 EDITION

This is a book about how reason and rationality can sometimes diverge. When the book was first published in 1988, I could not have formulated that short summary because I lacked the conceptual distinction between those two terms, which were then (and still are) used almost universally as synonyms, at least in English.[1] Yet this is a book about the power of just such conceptual distinctions: between objective and subjective probabilities; between mathematics and mathematical models; between long-term and short-term risks; between normative and descriptive uses of the calculus of probabilities. Even more, it is about how historical circumstances can obscure or reveal such distinctions, and why the efforts of the best and the brightest of one epoch can appear absurd or downright benighted to their successors only a few generations later. Once a conceptual distinction becomes thinkable, even self-evident, the ideas of those who did not share it appear at best muddled and at worst reckless. Branches of probability theory dealing, for example, with the probability of witness testimony or the probability that an event that has happened just once will happen again, both perfectly respectable pursuits within the framework of classical probability, struck later probabilists and philosophers as mad, bad, and perhaps dangerous—even though they had been advanced by mathematicians of the caliber of Pierre-Simon Laplace and Siméon-Denis Poisson. The work of the intellectual historian was to reconstruct the reasonableness of those classical probabilists who understood their mathematics as a reasonable calculus, and that is what this book attempted to do.

---

[1] Other languages are more fastidious: German, for example, distinguishes between *Vernunft* and *Rationalität*, the first a capacious faculty that embraces all manner of thought and deliberation, the second more narrowly confined to the calculation of the best means to achieve the ends sought.

What was classical probability, what did it set out to do, and why should we still care about it? Between about 1650 and 1840, some of the finest mathematicians of the era—among them Blaise Pascal, Pierre Fermat, Christiaan Huygens, Gottfried Wilhelm Leibniz, and a whole bevy of Bernoullis (Jakob, Nicholas, and Daniel)—tried to create a quantitative way of thinking about uncertainty: the uncertainty of a game of chance, a voyage to a distant port on stormy seas, the price of next year's wheat crop, the credibility of a witness in court, whether an inoculation for smallpox would protect or kill, and any of the other myriad events whose outcome was at best a matter of educated guesswork. From the outset, there was something intrinsically paradoxical about applying mathematics, the most certain of the sciences, to uncertainty. There was also something paradoxical about the way in which the classical probabilists defined their task: to create a mathematical model faithful to the way enlightened people already did make decisions under conditions of uncertainty, but yet to correct that reasoning whenever it diverged from the conclusions drawn from the mathematical models.

These two fault lines ran straight through the work of the classical probabilists, and the book follows their struggles to rescue their reasonable calculus from repeated reproaches that it was fundamentally unreasonable. If the mathematical calculation dictated that gamblers pay an infinite amount to play a game (named after the city of St. Petersburg) that no same person would spend more than pocket change on, who was right? If in the aggregate the probability of dying from a smallpox inoculation was about one in two hundred as compared to a one in seven probability of dying from the disease, was it nonetheless reasonable to resist having one's only child inoculated? Did it make sense to try to quantify the probability that a witness was telling the truth or that the judges would reach the correct conclusion? Even had reliable statistics been available, was it sound business policy to standardize the costs of insuring a voyage to a given destination on the basis of past experience, given that circumstances were constantly changing? These were the sort of practical problems that the classical probabilists attempted to solve mathematically. Later generations of mathematicians and philosophers would condemn many of their efforts as egregiously unreasonable and eventually enforced a strict separation between the axiomatized version of probability theory and its applications. But for the classical probabilists from Pascal and Jakob Bernoulli through Laplace and Poisson, mathematical probability was identical with its applications, and together mathematics and applications mapped out how to be reasonable in an unstable, unpredictable world. The book is therefore a history of reason in the Age of Reason.

Although the problems addressed by classical probability during the Enlightenment are no longer the essence of the mathematical theory of probability, they are still alive and well in every realm where risk must be assessed, from insurance to rational choice theory to stock market investment to the psychology of judgment under uncertainty. Many of the same puzzles that taxed the ingenuity of the classical probabilists still kindle lively debate among economists, psychologists, and game theorists. At the heart of almost all of these conundrums over what constitutes rational behavior under conditions of uncertainty is the same clash between the results of a formal calculus of rationality, be it Bayes' theorem or game theory, and what most people would deem to be the reasonable course of action in a given situation. Confronted with just such a tension between the rational and the reasonable in the famous case of the Prisoner's Dilemma problem, psychologist R. Duncan Luce and economist Howard Raiffa concluded in their influential 1957 introduction to game theory for social scientists that the rational strategy would be for both prisoners to defect, even though the outcome would be better for both if they instead opted to cooperate: "Of course, it is slightly uncomfortable that the two so-called irrational players will both fare much better than two so-called rational ones. . . . No, there appears to be no way around this dilemma."[2]

Such conundrums pit reason against rationality, a conceptual distinction that only slowly became clear and consequential to me in light of subsequent work on the history of objectivity, algorithms, and rules.[3] All

---

[2] R. Duncan Luce and Howard Raiffa, *Games and Decisions: Introduction and Critical Survey* (1957) (New York: Dover, 1985), 96. On the history of the various versions of the Prisoner's Dilemma, see Mary S. Morgan, "The Curious Case of the Prisoner's Dilemma," in *Science without Laws: Model Systems, Cases, Exemplary Narratives*, ed. Angela N. H. Creager, Elizabeth Lunbeck, and M. Norton Wise (Durham, NC: Duke University Press, 2007), 157–185; and Paul Erikson, *The World the Game Theorists Made* (Chicago: University of Chicago Press, 2015).

[3] Lorraine Daston and Peter Galison, *Objectivity* (New York: Zone Books, 2007); Paul Erikson, Judy L. Klein, Lorraine Daston, Rebecca Lemov, Thomas Sturm, and Michael D. Gordin, *How Reason Almost Lost Its Mind: The Strange Career of Cold War Rationality* (Chicago: University of Chicago Press, 2013); and Lorraine Daston, *Rules: A Short History of What We Live By* (Princeton, NJ: Princeton University Press, 2022).

three of these terms cluster around the pole of rationality: explicit, often formal protocols for providing unambiguous, unique, and replicable answers to well-posed questions in controlled contexts. Examples might include the use of statistical tests of significance to decide whether the results of an experiment are due merely to chance or manifest a stable regularity, or the use of a computer algorithm to determine the length of sentences for prisoners convicted of a given crime, or the use of Bayes' theorem to ascertain whether the positive result of a medical test is likely to be correct, given the base rates for the disease in the relevant population.

In none of these cases need rationality necessarily part ways with reason. But since the generalized, mechanical procedures of rationality do necessarily (and usually deliberately) abstract from the particulars of an individual case—whether in the name of objectivity, impartiality, or fairness—they are vulnerable to at least three kinds of failures. First, they may include faulty assumptions: the wrong features of the situation may have been built into the procedures. This has been shown to be the case in many machine-learning algorithms of the judicial sentencing type. Because they are generated from data about past sentencing, any biases in past practice will creep willy-nilly into the algorithm, despite the best intentions of those who introduce such algorithms to reduce bias. Second, the implicit preconditions for applying the procedures may have been overlooked: in the case of the medical test, it makes a vast difference as to whether the patient is selected from the population at large for random screening or has come in with symptoms of the disease. Third, all mechanical procedures can and usually will be gamed, as the replication crisis in some branches of science has shown in the case of statistical significance testing of experimental results.

This is where reason steps in—and also why traditional accounts of reason have been so difficult to square with later efforts to formalize its workings, whether in objective methods in science, formal decision-making procedures in law and medicine, or the rigid rules that have made modern bureaucracies (and, more recently, computer algorithms) notorious. Both reason and rationality strive for universality, but reason is context-sensitive whereas rationality is not. Rationality presupposes a domain of application that is stable, uniform, and in which a few big causes swamp the effects of many small ones: think about how the big cause of gravitation swamps the small causes of air perturbations and shape in calculating the trajectory of a falling object. This is the world assumed by the classical probabilists when they calculated the risks involved in insurance, smallpox inoculation, or gambling. Yet time and time again their conclusions seemed to diverge from those of exactly the enlightened people whose good sense the calculus of probabilities sought to capture in mathematical form. The objections to the mathematical solutions pointed to ornery particulars and unstable circumstances that might suddenly capsize expectations based on what happens most of the time. Merchant ships voyaging between Marseille and Venice during the fair-weather summer months usually sailed into port unscathed and could therefore be insured at a nominal rate. But what if news of pirates or plague en route reached the insurers at the last moment? Even the minute chances of winning a lottery might look very different from the perspectives of a prosperous burgher and an impoverished worker buying a ticket: for the one, the ticket was a long shot indulged as a whim; for the other, an extravagance, but the only possibility of escaping lifelong misery by legal means. We are still all too familiar with the difficulties of applying medical recommendations, however well-founded on statistical regularities, to individual cases in which there are grounds to suspect special circumstances.

Reason intervenes to adjust rationality to context. When Daniel Bernoulli attempted to resolve the apparent paradox of the St. Petersburg game by positing that the satisfaction players derived from the game was not a linear function, or when Jean d'Alembert modeled how the state and parents might differently calculate the risks of smallpox inoculation for children, these mathematicians were using reason to tweak rationality—to make the reasonable calculus live up to its name. In contrast to more modern versions of formal models of judgment under uncertainty, which assume that any deviation from the dictates of, for example, Bayes' theorem signals a cognitive defect, the classical probabilists treated such deviations as defects of their formal models. Theirs was a mathematical theory that straddled the line between the prescriptive—how reasonable people ought to behave— and the descriptive—how they actually did behave. The contrast with their modern counterparts, who sometimes proudly trace their lineage back to the classical probabilists, is striking. Had d'Alembert, Bernoulli, Condorcet, or Laplace been confronted with the conundrums of Prisoner's Dilemma in game theory or the Allais paradoxes in economics, it is most unlikely that they would have dismissed these apparent exceptions to formal models of rationality out of hand.[4]

They would also have puzzled over the latest candidates to crystallize rationality out of formal procedures, machine-learning algorithms. In contrast to earlier versions of rationality reduced to transparent rules, such as logic or even earlier Artificial Intelligence programs, these algorithms evolve mechanically but inscrutably. Not even the programmers who developed the algorithms understand how they develop as they are fed thousands and thousands of examples with feedback. The accuracy and validity of the algorithms so derived depends crucially on the examples, both their representativeness and whether or not past examples can be relied on in the future. Facial recognition algorithms trained on only white male faces are examples of what can go wrong with this particular form of rationality, as are just-in-time supply chains suddenly capsized by a global pandemic. In both cases, rational procedures that might work well in carefully circumscribed contexts founder when context changes. And in both cases, the opacity of the processes by which the algorithms work baffle the efforts of reason to correct for the narrowness of rationality.

The classical probabilists were no enemies of rationality—quite the contrary. But they did not yet live in a world that was so orderly that rationality need no longer be adjusted by reason, though they dreamed of such a world. This would be a world so uniform, so stable, so predictable that particulars would always coincide with universals, statistical regularities with the individual case, the rule with its every application, without exception. But they knew they did not yet live in such a world and were therefore willing to correct their calculus—a sterling example of would-be rationality and still revered as such—by reason when one contradicted the other. It is a measure of how much more closely at least certain parts of the modern world have come to approximate their dream that rationality has turned the tables on reason in many branches of inquiry, from economics to the psychology

---

[4] On the Prisoner's Dilemma, see Anatol Rapoport, "The Use and Misuse of Game Theory," *Scientific American* 207 (1962): 108–119; on the Allais paradoxes, Maurice Allais, "Le comportement de l'homme rationnel devant le risque: Critique des postulats et axiomes de l'Ecole Américaine," *Econometrica* 21 (1953): 503–546.

of decision making.[5] When context-sensitive reason clashes with context-insensitive rationality, it is usually the latter that wins out in these fields. Only when contexts widen, shift, or suddenly collapse altogether do the limitations of rationality without reason become glaring, just as an earthquake can topple buildings that seemed rock-solid only minutes before.

Rationality is a part of reason, the part that can be safely formalized because certain assumptions about an orderly world hold. But "irrationality" seems to be our only word for the whole of unreason, a sign of the extent to which we not only conflate reason and rationality but also grant rationality pride of place. What might a more capacious definition of unreason look like, one as capacious as reason itself? It would certainly include violations of rationality where rationality can be safely applied: for example, where reliable, up-to-date information sets Bayesian priors or where alternative risks can be accurately weighed against each other. But it might also include violations of the good sense enshrined by the classical probabilists, when reason reins in rationality in a still-messy world.

---

[5] For an overview of these approaches, see Alfred R. Miele and Piers Rawling, eds., *The Oxford Handbook of Rationality* (Oxford: Oxford University Press, 2004).

PREFACE AND
ACKNOWLEDGMENTS

This book is a joint product of interests in the history of mathematics, in the elusive concept of rationality, and in that best of all possible centuries, the Enlightenment. It originally took shape as a doctoral dissertation submitted to the Department of the History of Science at Harvard University, and the present version still records my debt to my fine teachers there, in particular Erwin Hiebert, I. Bernard Cohen and Dirk Struik.

Since then the study has evolved from dissertation to book with the help of colleagues both near and far. I am particularly grateful to Lorenz Krüger and the other members of the research group at the Zentrum für interdisziplinäre Forschung (1982–83), Universität Bielefeld, for many stimulating discussions on all matters probabilistic. The richness and range of these discussions are echoed in the proceedings of the group: Lorenz Krüger et al., eds., *The Probabilistic Revolution*. My colleagues in the Program in the History of Science and the Department of History at Princeton University taught me to ask a whole new set of questions about the relationship between theory and practice; I hope Chapter Three of this book does justice to that lesson. Joan L. Richards has been an indefatigable and perceptive critic of draft after draft; Gerd Gigerenzer has sharpened my understanding of many issues in classical probability theory by comparisons with kindred themes in contemporary psychology. So many kind people have done me the friendly and collegial service of listening to and discussing my ideas on this subject that I cannot list them here without trying the patience of printer and reader, but all have my hearty thanks.

It is a pleasure to thank the people and institutions that made my research possible and pleasurable. The generous support of the Zentrum für interdisziplinäre Forschung at the Universität Bielefeld, the National Endowment for the Humanities, and the Howard Foundation gave me the means and the leisure to read, travel, and write. I would like to thank the staffs of the many libraries and archives that I consulted for their graciousness and access to materials: the Harvard College Library, the Library of Congress, the Archives de L'Académie des Sciences, the Bibliothèque de l'Institut, the Bibliothèque Nationale, the Archives Nationales, the Royal Society of London, the Equitable Assurance Company, the Salle des Cartes et Plans (Fonds Quetelet) at the Bibliothèque Royale in Brussels, and the AMEV library in Utrecht.

The jacket illustration, "Les Caprices de la Fortune" by Jean-Antoine Pierron (Paris, 1786), is courtesy of the Kress Collection, Baker Library, Harvard Business School.

Books in the becoming require more than an author. Mrs. Vanessa Noya, Ms. Melissa Daston, and Mrs. Marie Daston typed the final manuscript with patience and care. My family, both old and new, has been my refuge and my comfort throughout, and I dedicate this book to them with all my thanks.

INTRODUCTION

There are truths which are not for all men, nor for all times.
—VOLTAIRE, *Letter to Cardinal de Bernis, 23 April 1761*

What does it mean to be rational? This book is about a two-hundred-year attempt to answer this question, the classical theory of probability. Between roughly 1650 and 1840 mathematicians of the caliber of Blaise Pascal, Jakob Bernoulli, and Pierre Simon Laplace labored over a model of rational decision, action, and belief under conditions of uncertainty. Almost all of the problems they addressed were couched in these terms: When is it rational to buy a lottery ticket? accept a scientific hypothesis? sell off an expectation of a future inheritance? invest in an annuity? believe in God? Mathematical probability theory was to be the codification of a new brand of rationality that emerged at approximately the same time as the theory itself, or rather of a more modest reasonableness that solved everyday dilemmas on the basis of incomplete knowledge, in contrast to the traditional rationality of demonstrative certainty. As the probabilists never tired of repeating, their theory was intended to reduce this prosaic good sense to a calculus.

Classical probability theory thus came to be viewed as a "reasonable calculus" within an intellectual milieu in which older notions of rationality were being challenged and redefined. The sixteenth-century revival of Greek skepticism, combined with the blistering polemics of the Reformation and Counter-Reformation, had subverted traditional criteria for belief in religion, philosophy, and the sciences. To counter this erosion of belief, a group of seventeenth-century thinkers, including Marin Mersenne, Pierre Gassendi, Hugo Grotius, John Tillotson, Robert Boyle, and John Locke, advanced a new philosophy of rational belief which Richard Popkin has called "constructive skepticism." While conceding the skeptical claim that absolute certainty lay beyond human grasp in all but a very few areas, these writers asserted that the conduct of daily life furnished sufficient, if imperfect standards for moral certainty. It was rational to believe or act in religious or philosophical matters if comparable evidence would persuade a "reasonable man" to adopt a course of action in his daily affairs. This pragmatic rationality of partial certainty contrasted sharply with the demonstrative certainty demanded by the Scholastics and later by the Cartesians. Between the poles of absolute certainty and total doubt, the reasonable man interpolated and compared degrees of certainty. These were the probabilities that the mathematicians sought to quantify in their theory.

To talk about the classical *theory* of probability is somewhat misleading, for it suggests a separation between theory and applications that would have been foreign to these early probabilists. They conceived of their field as more akin to, say, celestial mechanics than to algebra—that is, as a mathematical model of a certain set of phenomena, rather than as an abstract theory independent of its applications. The link between model and subject matter is considerably more intimate than that between theory and applications: if Euclidean geometry does not happen to describe the space we live in, it is not thereby discredited as a mathematical theory; but if a model of lunar perturbations fails to tally with astronomical observations, it is grounds for revising or discarding the model. The classical probabilists aimed at a model of rationality under uncertainty—of the intuitions of an elite of reasonable men—and when their results did not square with those intuitions, they tinkered with the model to bring about a better fit.

If the term "theory" in the current sense would have jarred the sensibility of an eighteenth-century probabilist, the term "classical" in this context grates upon twentieth-century ears. We believe that prosody or architecture may evolve through classical, romantic, or modernist styles, but not mathematics. Mathematical theories are indeed said to "unfold" or "progress." However, the image is one of continuous expansion, not of development in stages, coherent in themselves and distinct from one another. Contrary to this view, I shall argue that the classical theory of probability was coherent and distinct in just this stylistic sense. A characteristic interpretation of probability and an equally characteristic objective stamped the applications of the calculus of probability—and it was then little more than the sum total of its applications—almost from its inception until the mid-nineteenth century. Interpretation and objectives together not only dictated the right answers to problems; they also determined the *kind* of problem the calculus of probabilities could solve. When probabilists abandoned both the interpretation and the objective of the classical theory around 1840, they therefore abandoned whole areas of application as well. Projects that Bernoulli and Laplace had considered respectable and even central to their theory, like the probabilities of causes and testimony, struck later probabilists as patently absurd, for they no longer subscribed to the classical interpretation and objectives. They did not reject their predecessors' work as wrong in any technical sense, but rather as nonsensical: probability theory was no longer about that kind of problem. Metaphors of expansion or accumulation hardly do justice to this state of affairs. I concede that the adjective "classical" is a somewhat arbitrary choice to describe this mathematical style, but since it is commonly applied to the literature and art of the same period, I shall use it, *faute de mieux*.

Both the classical interpretation and objective often elude the modern student of probability, to whom the works of the early probabilists may seem more muddled than coherent. Bernoulli and Laplace sometimes read like the latest journal article, and at other times like the bizarre speculations of a crank. This double vision, which makes the classical probabilists seem at once familiar and fanciful, is an artifact of conceptual distinctions that are the stock-in-trade of every modern probabilist, but which were rarely made by the classical probabilists—indeed, the emergence of such clear-cut distinctions spelled the end of the classical interpretation. It is now a matter of heated debate whether mathematical probabilities measure objective frequencies of events or subjective degrees of belief, but classical probabilists were able to assume that the two were equivalent, thanks largely to early Enlightenment theories of associationist psychology. They therefore could slide between objective and subjective senses with an ease that confounds their successors, who have been weaned on the distinction.

Other conceptual categories fundamental to the classical interpretation also gradually fissioned into oppositions. I have already mentioned how classifying the calculus of probabilities as part of what was then called "mixed mathematics," along with hydrodynamics and celestial mechanics, blurred the distinction between theory and application. Shorn of its applications, the classical "theory" of probability would have ceased to exist. This is why classical probabilists counted problems of the St. Petersburg sort as paradoxes—not because they revealed any mathematical contradiction, but rather because the straightforward mathematical results were at odds with good sense. In short, the model did not match the phenomena. D'Alembert's was an extreme but not hysterical voice when he argued that such worrisome problems were grounds for scrapping the calculus of probabilities altogether.

Within the model itself a gap gradually opened up between the prescriptive and descriptive senses of rationality. Originally the probabilists set themselves the task of describing the intuitions of an elite of reasonable men in order to prescribe those of the hoi polloi in the form of mathematical rules. It was a project wholly in keeping with the Enlightenment campaign against "vulgar errors" and shared many of the same assumptions. In particular, the mathematicians took it for granted that reasonable men agreed in their intuitions about action and belief in the face of uncertainty. But cracks inevitably emerged in this hypothetical consensus: the good sense of the learned judge was not, for example, that of the prudent merchant, a discrepancy that joined the probabilists in a century-long battle over the proper definition of expectation. A similar conflict arose when smallpox inoculation was introduced into Europe in the mid-eighteenth century. Many of the enlightened preferred the large, long-term risks of the disease to the small, short-term risks of the inoculation itself, drawing Daniel Bernoulli and d'Alembert into a debate over whether the calculus of probabilities should correct or follow these intuitions. Yet despite these difficulties, the probabilists clung to the ideal of a unitary, universal rationality until the trauma of the French Revolution and its aftermath finally persuaded them otherwise. This ideal allowed them to straddle the line between prescriptive and descriptive accounts in a way that later became intolerable.

These slow separations of conceptual units into distinctions between objective and subjective probabilities, between theory and application, and between normative and descriptive aims did not come about exclusively or even largely because of technical mathematical developments. Rather, the background assumptions that had held these units together changed, unraveling connections that had once seemed self-evident. Contemporary ideas about fair contracts, the economic role of luxury, the ability of the mind to translate experience into belief, and the nature of thought itself all played a part in making the classical theory of probability plausible. Against the much altered intellectual background of the early nineteenth century, however, the classical theory became more an "aberration of the intellect" than "good sense reduced to a calculus." As heirs to these distinctions, we are hard put to understand the classical theory and its decline without reconstructing this shifting background. Enlightenment jurisprudence, political economy, psychology, and philosophy contributed to the notion of rationality that the probabilists hoped to mathematize, and are therefore part of this history of the classical interpretation.

The conviction that rationality could indeed be reduced to a calculus lay at the very heart of the classical program. The recurring Enlightenment dream of a calculus that would convert judgment and inference into a set of rules stemmed from the seventeenth-century fascination with methods of right reasoning, with the further embellishment that the rules were mathematical. A calculus was deemed superior to a mere method in that its rules yielded unique, unambiguous solutions that brooked no further argument. As Leibniz remarked of his own plan for such a calculus, the Universal Characteristic, the mathematical form coerced assent once—and here was the catch—everyone accepted the rules. Accepting the rules was not only a matter of the specifics, but also of general preconditions: were sound judgment and correct inference really the implicit application of rules? And was the world really stable, simple, and regular enough for such general rules to hold in most cases? Probabilists had to embrace both a psychology and a metaphysics in order to embark upon the quantification of rationality.

The history of the classical theory of probability is in fact an extremely instructive case study in the preconditions for quantification. Out of the crowd of qualitative notions of probability current in the mid-seventeenth century, mathematicians seized upon only a few. Their choices appear in part to have been governed by what might be called proto-quantification. Some kinds of probability—legal evidence, odds on a wager, the value of a future event—were already cast in quantitative terms of sorts, albeit rather crude and arbitrary ones, and these were just the ones that became part of the earliest formulations of the calculus of probabilities. Quantification depends on some perceived analogy between subject matter and the available mathematics, and protoquantification thrusts these analogies to the fore. In the case of classical probability theory, Enlightenment theories of associationism, induction, political economy, and testimony suggested further analogies that led to new domains of application.

However, analogy alone is not sufficient for quantification: the phenomena must also be conceived as constant and orderly enough to yield to mathematical treatment. The phenomena of the classical probabilists were, from our vantage point, extraordinarily diverse, ranging from the credibility of witnesses to the pattern of human mortality, but all were assumed to exhibit at least the long-term regularities compatible with Bernoulli's and later Bayes' theorems. Here the mathematical tools shaped ideas about the subject matter by imposing a greater determinacy upon phenomena previously thought to be unruly. Prevailing ideas about the subject matter also shaped the mathematical tools, as in the protracted attempts of the probabilists to tailor the definition of expectation to the specifications of one or another sense of rationality. Quantification is thus a process of mutual accommodation between mathematics and subject matter to create and sustain the analogies that make applications possible. The process is not necessarily irreversible, for analogies sometimes decay: as the later history of the classical theory of probability shows, mathematical theories can lose as well as gain domains of application.

In this study I examine the rise and fall of the classical theory of probability with an eye toward what it can tell us about quantification and about changing notions of rationality. I hope that it will complement and supplement the several fine studies on mathematical and philosophical probability that have appeared in recent years. Chronologically, it covers the period roughly between those treated in Ian Hacking's *Emergence of Probability* and Theodore Porter's *The Rise of Statistical Thinking, 1820–1900*. Thematically, it brings together mathematical and philosophical developments dealt with separately in the work of Ivo Schneider, Oskar Sheynin, and Stephen Stigler on the one hand, and that of Barbara Shapiro and Keith Baker on the other. In Chapter One we will look at the prehistory and origins of mathematical probability to discover which of the ambient qualitative notions of probability became quantitative, and why; I conclude that legal influences were particularly strong at this early stage. Legal models give way to economic ones in Chapter Two, which traces how the definition of probabilistic expectation reflected changes and conflicts in the Enlightenment idea of rationality, placing ever greater strain on the prescriptive/descriptive dualism of the classical theory. In Chapter Three, I turn from theory to practice in order to assess to what degree mathematical probability changed attitudes and actions concerning real risks, using gambling and insurance as examples. In Chapter Four I explain how the doctrine of the association of ideas first joined and then sundered objective and subjective senses of probability in the classical theory, paving the way for the applications to causal reasoning in the natural sciences described in Chapter Five. Applications to the moral sciences, particularly jurisprudence and history, are the topic of Chapter Six. The Epilogue deals briefly with the reasons for the decline of the classical theory in the first half of the nineteenth century.

The main characters in my story are intellectuals and their ideas. They were never very far removed from the practical matters of the day, as Chapters Three and Six make particularly clear, but their approach to the problems of, for example, pricing insurance policies or reforming the judicial system was characteristically abstract. This does not mean that their approach was narrow: the classical probabilists drew upon a very broad range of ideas from disciplines outside of mathematics to construct models of rational belief and action. However, it was an approach based on general principles rather than piecemeal solutions; on idealization rather than concrete particulars—in short, an intellectual's approach. Therefore I have written their story as intellectual history, with an emphasis on the history of science. Even Chapter Three, in which I discuss the divergence of popular and learned beliefs about chance, is primarily about ideas, although not primarily those of intellectuals. Ideas certainly do not exhaust the whole of history, but they are as "real" as any other part thereof.

This is a study of a mathematical theory, but of a mathematical theory about rationality in an age intoxicated by reason. It is consequently intended as a contribution to Enlightenment history as well as to the history of science and mathematics. Philosophers, economists, and psychologists may also take an interest in the work of the classical probabilists, for many of the classical problems later migrated to other fields, along with the classical assumptions. With this varied audience in mind, I have tried to keep technical material to a minimum. (Readers interested in pursuing the mathematical developments in greater detail are directed to the work of Ivo Schneider and Oskar Sheynin listed in the bibliography, and to Stephen Stigler's excellent study, *The History of Statistics. The Measurement of Uncertainty before 1900*, which appeared after the manuscript of this book was completed.) However, some grasp of mathematics is necessary to appreciate the extent to which the applications treated in Chapters Five and Six incorporated assumptions that did *not* follow strictly from the mathematics per se. The point of including the equations is not to bar the nonspecialist reader, but rather to show the extent to which extra-mathematical assumptions permeated the mathematical formulations themselves. In all cases I have striven to make the ideas behind the symbols clear.

My intention throughout has been to reconstruct and explain, not to vindicate or criticize the classical theory of probability. I have tried to show that if by 1840 the "calculus of good sense" had become antithetical to good sense, it was good sense rather than the calculus of probabilities that had changed. If the writings of Jakob and Nicholas Bernoulli, Abraham De Moivre, Laplace, and other classical probabilists strike modern commentators as incomprehensible or simply ludicrous, it is largely because we navigate by a different map of the intellectual terrain, and by a different pole star of rationality. Once the context in which the classical theory of probability had originated and flourished disappeared, so did that peculiar conception of rationality that had celebrated the theory as the reasonable calculus.

Classical Probability in
the Enlightenment

The Prehistory of the Classical Interpretation of Probability: Expectation and Evidence

1.1 Introduction

Although the famous correspondence between Blaise Pascal and Pierre Fermat first cast the calculus of probabilities in mathematical form in 1654, many mathematicians would argue that the theory achieved full status as a branch of mathematics only in 1933 with the publication of A. N. Kolmogorov's *Grundbegriffe der Wahrscheinlichkeitsrechnung*. Taking David Hilbert's *Foundations of Geometry* as his model, Kolmogorov advanced an axiomatic formulation of probability based on Lebesgue integrals and measure set theory. Like Hilbert, Kolmogorov insisted that any axiomatic system admitted "an unlimited number of concrete interpretations besides those from which it was derived," and that once the axioms for probability theory had been established, "all further exposition must be based exclusively on these axioms, independent of the usual concrete meaning of these elements and their relations."[1] Although philosophers, probabilists, and statisticians have since vigorously debated the relative merits of subjectivist (or Bayesian), frequentist, and logical interpretations as means of applying probability theory to actual situations, all accept the formal integrity of the axiomatic system as their departure point.[2] The mathematical

---

[1] Andrei Kolmogorov, *Foundations of the Theory of Probability*, trans. Nathan Morrison (New York: Chelsea Publishing Company, 1950), p. 1.

[2] Ernest Nagel, *Principles of the Theory of Probability*, in *International Encyclopedia of Unified Science*, vol. 1, part 2, Otto Neurath, Charles Morris, Rudolf Carnap, eds. (Chicago: University of Chicago Press, 1955), pp. 368–369.

theory itself preserves full conceptual independence from these interpretations, however successful any or all may prove as descriptions of reality.

This logical schism between the formal axiomatic system and its concrete interpretations is not unique to probability theory: geometry, algebra, and the calculus have also been translated into purely formal systems and explicitly divorced from the contexts from which they emerged historically. For modern mathematicians, the very existence of a discipline of applied mathematics is a continuous miracle—a kind of prearranged harmony between the "free creations of the mind" which constitute pure mathematics and the external world.

Although these are pressing issues for the philosopher of mathematics, they tend to blur historical vision. While innumerable interpretations may logically satisfy the axioms of the mathematical theory of probability, in point of fact the historical development of the theory was dominated almost from its inception until the mid-nineteenth century by a single interpretation, the so-called "classical" viewpoint. Throughout the eighteenth and nineteenth centuries, probabilists understood the classical interpretation and the mathematical formalism underlying it to be inextricable—indeed, to be one and the same entity. If any distinction between the levels of application, interpretation, and formalism existed in the minds of the classical probabilists, the hierarchy in which these levels were arranged reversed the modern order: the mathematical formalisms of probability theory were justified to the extent that they matched the prevailing interpretation and field of application, rather than the interpretation and its ensuing applications being sanctioned to the degree that they satisfied the formal axioms.

Where did the classical interpretation come from? Seventeenth-century texts—literary, religious, philosophical, medical, scientific, legal—abound with references to "probability" of one sort or another, and two recent works have studied these proliferating, mutating usages in fascinating detail.[3] My question about the origins of the classical interpretation cuts at right angles to these concerns: out of the swarm of probabilistic notions abroad at mid-century, which ones supplied the first mathematical probabilists with concepts and problems—and why? Posed in this way, it is a question about quantification. Recasting ideas in mathematical form is a

---

[3] Ian Hacking, *The Emergence of Probability* (Cambridge, Eng.: Cambridge University Press, 1975), and Barbara J. Shapiro, *Probability and Certainty in Seventeenth-Century England* (Princeton: Princeton University Press, 1983).

selective and not always faithful act of translation. In the seventeenth-century geometrization of mechanics, only local motion survived from that cluster of phenomena Aristotle had called change: a falling body, a growing oak, a wavering mood. Similarly, only some of the ambient seventeenth-century views about what probability meant passed through the filter of the mathematical methods invented by Pascal, Fermat, Christiaan Huygens, and Jakob Bernoulli. Those that did changed their meaning as well as their form. John Wilkins's philosophical certainty, envisioned as three ascending stages of moral, physical, and metaphysical assurance, was not identical to Jakob Bernoulli's full continuum of degrees of certainty ranging from zero to one, any more than Galileo's description of rest as an infinite degree of slowness was identical to scholastic distinctions between the states of rest and motion. Quantification was not neutral translation. This chapter is about how certain qualitative probabilities became quantitative ones in the latter half of the seventeenth century, and created the classical interpretation in the process.

Fitting numbers to the world changes the world—or at least the concepts we use to catch hold of the world. A world of continua spanning rest and motion, certainty and ignorance does not look like a world of sharp either/or oppositions. But the world can change the numbers as well. To be more precise: if we want our mathematics to match a set of phenomena with reasonable accuracy, we may have to alter (or invent) the mathematics to do so. The tandem development of mechanics and the calculus in the seventeenth century is full of examples of new mathematical techniques that mimicked motion: Giles Roberval's velocity method of finding tangents, or Isaac Newton's machinery of fluxions and fluents. The case of classical probability theory is less dramatic in a mathematical sense, for probabilists had few new techniques to call their own until the end of the eighteenth century. Yet this very lack of new mathematical content bound mathematical probability all the more firmly to its applications. Since it belonged wholly to what we would now call applied mathematics, probability theory stood or fell upon its success in modeling the domain of phenomena that the classical interpretation had mapped out for it. Failure threatened not just this or that field of application, but the mathematical standing of the theory itself. Hence classical probabilists bent and hammered their definitions and postulates to fit the contours of the designated phenomena with unusual care. I shall deal at length with examples of their handiwork in Chapter Two; here I only wish to point out that quantification is a two-way street. Neither the original subject matter nor the mathematics emerges entirely unchanged from the encounter.

The classical interpretation of probability was the result of such an encounter between a tangle of qualitative notions about credibility, physical symmetry, indifference, certainty, frequency, belief, evidence, opinion, and authority on the one hand, and algebra and combinatorics on the other. By looking closely at the problems posed by the early mathematical probabilists, and the concepts they used to solve them, we can locate the point of intersection between the quantitative and the qualitative. Of all the then available meanings of probability, which were grist for the mathematicians' mill, and why? Once the mathematicians had made their choice, to what kind of program of applications did it commit them? I shall argue that seventeenth-century legal practices and theories shaped the first expressions of mathematical probability and stamped the classical theory with two of its most distinctive and enduring features: the "epistemic" interpretation of probabilities as degrees of certainty; and the primacy of the concept of expectation. Moreover, legal problems provided the principle applications for the classical theory of probability from the outset. Even the earliest problems concerning games of chance and annuities were framed in legal terms drawn from contract law, and, as will be seen in subsequent chapters, classical probabilists of the eighteenth and nineteenth centuries continued to include other sorts of legal problems, such as the credibility of testimony and the design of tribunals, within the compass of their theory.

1.2 Quantitative and Qualitative Probabilities

No monistic explanation can satisfactorily account for so complex an intellectual phenomenon as the advent of mathematical probability, and I do not intend to put forward any such here. However, I do claim that more than any other single factor, legal doctrines molded the conceptual and practical orientation of the classical theory of probability at the levels of application, specific concepts, and general interpretation. Although some historians have noted in passing the legalistic tone of the writings of the early probabilists, they have tended to regard the more explicitly juridical formulations, such as that of Gottfried Wilhelm Leibniz, as idiosyncratic. Ernest Nagel mentioned the medieval arithmetic of proof in a survey of premodern notions of probability;[4] Alexandre Koyré commented upon the lawyerly approach of the Pascal/Fermat correspondence;[5] Ian Hacking discussed Leibniz's probabilistic proposal to settle conflicting property claims.[6] Ernest Coumet has systematically pursued these allusions in his illuminating discussion of the relationship between Jesuit casuistry, seventeenth-century contract law, and mathematical probability, but only with respect to the Pascal/Fermat correspondence.[7]

Yet the works of the early probabilists are full of legal references. Pascal, in a 1654 address to the Académie de Paris on his current scientific projects, described his research on the "géométrie du hasard" as a means of determining equity: "The uncertainty of fortune is so well ruled by the rigor of the calculus that two players will always each be given exactly what equitably [*en justice*] belongs to him."[8] Huygens and Johann De Witt presented the fundamental propositions of the calculus of probabilities in terms of contracts and equitable exchanges; Part IV of Jakob Bernoulli's *Ars conjectandi* bristled with legal examples; Nicholas Bernoulli wrote an entire dissertation on the applications of mathematical probability to the law. As A. A. Cournot observed in 1843, the early probabilists had for the most part little idea of how their new calculus might be applied to "the economy of natural facts," being primarily concerned with the "rules of equity."[9] The spirit, if not the letter, of Leibniz's views on the close connection between the calculus of probabilities and jurisprudence was widely shared by his contemporaries.

Before going on to argue this claim in detail, however, we must take some account of the alternative theories put forward by historians about the roots of mathematical probability. My survey of this large and growing literature will be necessarily brief, and directed principally toward the adequacy of these explanations for understanding why *mathematical* proba-

[4] Nagel, *Principles*, p. 348.

[5] Alexandre Koyré, "Pascal savant," in *Blaise Pascal, l'homme et l'oeuvre*, Cahiers de Royaumont (Paris: Éditions de Minuit, 1956), p. 291.

[6] Hacking, *Emergence*, chapter 10.

[7] Ernest Coumet, "La théorie du hasard est-elle née par hasard?" *Annales: Économies, Sociétés, Civilisations* 25 (May-June 1970): 574–598.

[8] Blaise Pascal, "Celeberrimas Matheseos Academiae Parisiensi" (1654), in *Oeuvres complètes de Pascal*, Jean Mesnard, ed. (Paris: Bibliothèque Européene-Desclès de Brouwer, 1970), vol. 1, part 2, p. 1034.

[9] Antoine Augustin Cournot, *Exposition de la théorie des chances et des probabilités* (Paris, 1843), pp. 86–87.

bility emerged when and how it did. I do not contest the value of these accounts for understanding the increasing complexity and importance in the early modern period of probabilistic notions more broadly construed: they are rich in insights that will, I believe, eventually help rewrite the intellectual history of the era. But because I am here interested in which specific kinds of probability passed through the strait and narrow gate of quantification, my criteria for an adequate explanation will be correspondingly narrow and specific.

The prehistory of mathematical probability has excited much interest among historians of mathematics, perhaps because the rise of a mathematical approach to chance in the seventeenth century seems at first glance long overdue. The origins have been sought in astronomy, fine arts, gambling, medicine, alchemy, and the insurance trade. The quest for antecedents has been a frustrating one, uncovering proto-probabilistic thinking everywhere and nowhere. Certain passages of Aristotle, for example, could be construed as an embryonic version of statistical correlation or a scale of subjective probabilities; with an even greater effort of the imagination, Bayes' theorem may be discovered in medieval Talmudic exegesis.[10] However, these philosophical discourses on the nature of chance and rules of thumb for dealing with situations fraught with uncertainty (e.g., an astrological prediction or a medical prognosis) not only fall short of a mathematical treatment of probability considered in and of themselves, but they also manifestly failed to generate such a theory.

More clear-cut elements of mathematical probability, such as an enumeration of all possible outcomes for the throw of several dice, surface as early as the tenth century,[11] but, like the promising hints scattered through the classical and medieval philosophical texts, evidently bore no mathematical fruit. Plausible practical sources of mathematical probability prove equally sterile upon investigation. Despite the popularity of gambling since time immemorial, games of chance apparently did not

---

[10] O. B. Sheynin, "On the prehistory of the theory of probability," *Archive for History of Exact Sciences* 12 (1974): 97–141, especially pp. 101, 119; Nachum L. Rabinovitch, *Probability and Statistical Inference in Ancient and Medieval Jewish Literature* (Toronto and Buffalo: University of Toronto Press, 1973), pp. 58–60; S. Sambursky, "On the possible and the probable in ancient Greece," *Osiris* 12 (1956): 35–48.

[11] M. G. Kendall, "The beginnings of a probability calculus," *Biometrika* 43 (1956): 1–14, reprinted in *Studies in the History of Statistics and Probability*, E. S. Pearson and M. G. Kendall, eds. (Darien, Conn.: Hafner, 1970), vol. 1, pp. 19–34.

suggest notions of stable statistical frequencies or combinatorial derivations of probabilities until the sixteenth century. The sale of maritime insurance and annuities, both known since ancient times and revived on an impressive scale by fourteenth-century Italian entrepreneurs, also failed to spark a mathematics of chance or even a compilation of statistics, as we will see in Chapter Three. Even the Problem of Points—the division of stakes in an interrupted game of chance which prompted the seminal Pascal/Fermat correspondence of 1654—had been posed in a mathematical context as early as 1494, in Luca Pacioli's *Summa de arithmetica, geometrica, proportioni et proportionalita*.[12]

These attempts to trace the ancestry of mathematical probability usually founder on the issue of timing. Although chance figured in philosophical speculation and practical dealings since ancient times, mathematical probability did not emerge until the middle of the seventeenth century. What was the intellectual seed crystal introduced during this critical period that permitted ambient and often ancient ways of thinking about chance to coalesce in mathematical form?

The catalyst does not appear to have been mathematical. Mathematical prerequisites posed no obstacle. In its original form, probability theory presupposed only elementary combinatorics, and although the work of Pascal, John Wallis, Leibniz, F. van Schooten, and lesser-known figures such as Jean Prestet on this subject kindled mathematical interest during the latter half of the seventeenth century,[13] combinatorial thinking appears to have been more stimulated by nascent probability theory than the reverse. Almost all of the major works on combinatorics were published after the first treatise on mathematical probability, Huygens's *De ratiociniis in aleae ludo* (1657), appeared. Wallis's *Discourse on Combinations, Alternations, and Aliquot Parts* was published as an appendix to the English edition of his *Treatise on Algebra* (1685). Pascal's *Traité du triangle arithmétique* was apparently printed in 1654 (though circulated in 1665),[14] the same year as his exchange with Fermat. However, Pascal's original solution to the Problem of Points reveals that he recognized the relevance of the arith-

---

[12] Kendall, "Beginnings," p. 27.

[13] Eberhard Knobloch, "Musurgia universalis: Unknown combinatorial studies in the age of Baroque absolutism," *History of Science* 17 (1979): 258–275; also his "The mathematical studies of G. W. Leibniz on combinatorics," *Historia Mathematica* 1 (1974): 409–430.

[14] Pascal, *Oeuvres*, vol. 1, part 2, pp. 33–37.

metic triangle only belatedly, and initially shied away from the combinatorial method it embodied. Leibniz's *Dissertatio de arte combinatoria* was published in 1666; Prestet's *Elemens des mathématiques* in 1675 (Book II discussed combinations and permutations). Van Schooten's comments on combinations appeared in his *Exercitationum mathematicarum* of 1657, to which Huygens's *De ratiociniis* was appended. While van Schooten's work must have influenced his student Huygens's approach to probability, his own remarks on the subject were brief and schematic, serving as the basis for a discussion of prime factorization rather than possible outcomes.

Thus, extended mathematical treatments of combinations postdated the earliest published treatise on probability theory. Indeed, some of the most comprehensive treatments of combinations and permutations appeared as supplements to works on probability, such as Part II of Jakob Bernoulli's *Ars conjectandi* (1713) and Part I of the second edition of Pierre de Montmort's *Essai d'analyse sur les jeux de hazard* (1713). The two subjects developed in tandem.

Nor did any new philosophy of chance develop during this period, although the protracted religious controversies that wracked Europe during the sixteenth and seventeenth centuries did persuade an increasing number of thinkers of the vanity of human pretensions to certainty.[15] Classical probabilists from Jakob Bernoulli through Laplace followed the Thomist line:[16] from the perspective of an omniscient God (or later Laplace's secularized supercalculator), the events of the universe were fully determined. Chance was merely apparent, the figment of human ignorance. Until the nineteenth century, no mathematician, scientist, or philosopher appears to have contemplated the possibility of genuinely random phenomena except to dismiss the idea as nonsensical: causeless events were unthinkable. Indeed, from Hobbes and Spinoza through d'Holbach, the philosophical climate of opinion during the seventeenth and eighteenth centuries grew ever more resolutely deterministic. Let Abraham De Moivre speak for these deterministic probabilists. True chance, he claimed, "imports no determination to any *mode of Existence*; nor indeed to *Existence* itself, more than to non-existence; it can neither be defined nor understood: nor can any Proposition concerning it be either affirmed or denied, excepting this

[15] See Richard Popkin, *The History of Scepticism from Erasmus to Descartes* (Assen, Netherlands: Van Gorcum, 1964), chapter 1; Shapiro, *Probability*, chapters 1–3.

[16] Edmund F. Byrne, *Probability and Opinion* (The Hague: Martinus Nijhoff, 1968), pp. 293–296.

one, 'That it is a mere word.' "[17] The random was simply unintelligible. Although some historians have attributed the tardy development of mathematical probability to the "absence of a notion of chance," the writings of the classical probabilists do not remedy this dearth. On the contrary, they strenuously denied both the subjective and objective existence of real chance. However, failure to articulate a concept of randomness evidently did not hinder the birth and growth of mathematical probability from the seventeenth through the mid-nineteenth centuries.[18]

These and other aspects of the literature on the prehistory of probability have been treated at greater length recently by Ian Hacking in the most sophisticated and stimulating treatment of the subject to date.[19] Concerned with the emergence of a concept of probability in the broadest sense, Hacking ranges over philosophical, practical, and legal, as well as mathematical themes. However, he argues that all of these kinds of probability (1) share a common feature that stamps our understanding of probability to this day, namely a dual aleatory and epistemic aspect; and (2) could not fully emerge in any form before a "mutation" in the concept of evidence prepared the way around the turn of the seventeenth century. According to Hacking, astrologers, physicians, alchemists, and other sixteenth-century practitioners of the nondemonstrative "low" sciences evolved a new concept of diagnosis that linked overt "signs" to hidden properties, and at the same time associated these natural signs with an authoritative text, the "book of nature." Thus the old, epistemic meaning of probability as belief or opinion warranted by authority merged with the new, aleatory idea of observed (if unexplained) correlations between events (e.g., between fever and disease, comets and the death of kings) to create the concept we still recognize as probability.

This, much telescoped, is Hacking's thesis, and it has provoked considerable controversy among historians and philosophers. Like any important and original claim, it is vulnerable to challenges at several levels. Do all significant seventeenth-century (not to mention later) usages of "probability" really reduce to the epistemic ("opinion derived from authority") and aleatory ("natural signs correlated by experience") elements Hacking believes to constitute probability? Using only the English literature of the period, Barbara Shapiro has documented many other shades of probability,

[17] Abraham De Moivre, *Doctrine of Chances*, 3rd edition (London, 1756), p. 253.

[18] Sheynin, "Prehistory," p. 141.

[19] Hacking, *Emergence*, chapters 1–5.

11

including degrees of certainty or assurance; reasonable doubt; verisimilitude; worthiness to be believed (credibility); epistemological modesty.[20]
In fact, seventeenth-century probability had more than Janus's two faces;
it was more a group of visages loosely assembled in a family portrait. Conversely, several key seventeenth-century instances of concepts and methods
using one or another of Hacking's two constituents do not mention the
word "probability": for example, John Graunt's analysis of the London
bills of mortality. Only with the benefit of hindsight can we exclude some
of the ideas that seventeenth-century writers did call "probability" and
include some they did not.

Hacking's dissection of probability, now and then, into two and only
two constituents has the great advantages of conceptual clarity and of setting the standards for the solution of a knotty historical problem, namely,
when and why did the concept of probability emerge? In essence, Hacking
reasons that the concept X has components a, b, c; if we find a, b, c, then
X has entered the realm of the thinkable. Alas for the clear-thinking
historian, the contexts in which a, b, c occur may be so disparate from one
another, or so alien to current sensibilities, that we can hardly glue these
bits and pieces together to form any single notion at all, much less a
familiar one.

However, even those who accept Hacking's account of the two constituents of probability might question his explanation of how they came to
be fused together just when they did. For Hacking, the aleatory element—
probabilities as observed frequencies—derives from the sixteenth-century
doctrine of natural signs, which created a new kind of "internal" evidence
of things rather than of testimony. This is the "diagnosis," the inference
from one particular to another, which Hacking claims achieved "a new
conceptualization" in the works of Renaissance practitioners of the low
sciences like Paracelsus.[21] A great deal depends on the novelty of the "diagnosis," for the epistemic element of "opinion" had been the standard
meaning of "probability" for centuries: a new kind of nondemonstrative
knowledge—and a link between new and old—is needed to resolve the
problem of timing that bedevils all historians of probability. Daniel Garber and Sandy Zabell have collected instances from the medieval handbooks of law and rhetoric that show that the idea of internal evidence was
firmly established in the Latin West after the twelfth century, with a dis-

---

[20] Shapiro, *Probability*. [21] Hacking, *Emergence*, pp. 34–37.

tinguished classical pedigree. They also point out that the link between such natural signs and the writ of God—Hacking's bridge between "old" epistemic and "new" aleatory elements—is a venerable one.[22]

My aim here is not to develop yet another account of the rise or emergence of probabilistic notions in the early modern period. In order to settle even the prior question of whether there was indeed such a rise (Hacking and Shapiro claim there was; Garber and Zabell are dubious) would involve a thorough canvassing of most of the classical, medieval, and Renaissance learned corpus to establish a baseline. Probability could surface almost anywhere, and did. Unlike Hacking and the majority of his predecessors and critics, I do not believe that the origins of mathematical probability were identical to those of conceptual probability. But I do maintain that some concepts are more readily quantified than others, and that this was the case within the conceptual field available to the early probabilists. That they had a choice in the matter is evident not only from the several sorts of probability concepts available in the mid-seventeenth century, but also from the way in which the domain of applications for probability theory later shifted. As we shall see in later chapters, eighteenth- and nineteenth-century probabilists sometimes diverged sharply in their views about what kinds of problems their theory could solve. Subjects to which classical probabilists from Jakob Bernoulli through Laplace had devoted much attention—such as the probability of testimony or the probability of causes—were rejected out of hand by their successors: probability theory was no longer "about" those matters.

What was probability theory about in the second half of the seventeenth century? The time-honored answer is: games of chance.[23] This answer has much to be said for it, for the pioneers of mathematical probability— Gerolamo Cardano, Galileo, Pascal, Fermat, Huygens—all solved gambling problems. But it is also incomplete and misleading: incomplete, because it omits the other important applications concerning evidence, demography, and annuities that very soon accreted to the theory in the work of De Witt, Edmund Halley, John Craig, Jakob and Nicholas Bernoulli; misleading, because it suggests that gambling provided the early probabilists with the conceptual framework in which they posed and

[22] Daniel Garber and Sandy Zabell, "On the emergence of probability," *Archive for History of Exact Sciences* 21 (1979): 33–53.

[23] See, for example, Isaac Todhunter, *A History of the Mathematical Theory of Probability from the Time of Pascal to That of Laplace* (London and Cambridge, Eng., 1865).

solved their problems. Upon closer examination, the works of the early probabilists turn out to be more about equity than about chances, and more about expectations than about probabilities. These ideas and the applications they stimulated—for example, to games of chance and annuities—came, as we shall see, largely from the law. The next generation of probabilists owed a second debt to jurists, this time for the interpretation of mathematical probability as a degree of certainty. This new interpretation in turn spawned a new set of applications concerning evidence both in and out of the courtroom. Both these legal borrowings share a common feature: while neither the doctrine of contract equity nor that of evidence were truly quantified, practice in one case and theory in the other had given them a proto-quantitative form that made them seem ripe for a thoroughgoing mathematical treatment. Equally or more familiar seventeenth-century senses of "probability," like verisimilitude, were never so conceived, and hence never made it into the mathematicians' repertoire of applications.

The principal contributions of jurisprudence to early mathematical probability were thus twofold. First, early probabilists like Jakob Bernoulli and Huygens drew upon legal doctrines concerning aleatory contracts—that is, those involving some element of chance, such as games of chance and annuities—as sources not only of problems, but also of fundamental concepts and definitions. Aleatory contracts, like all contract law, centered upon considerations of equity and fair exchange among partners. Classical probabilists quite explicitly translated the legal terms for an equitable contract into mathematical expectation—that is, the value of an uncertain prospect—and made expectation, rather than probability per se, the departure point for the first expositions of mathematical probability. Second, legal theories of evidence supplied probabilists with a model for ordered and even roughly quantified degrees of subjective probability. The hierarchy of proofs within Roman and canon law led mathematicians to conceive of degrees of probability as degrees of certainty along a graduated spectrum of belief, ranging from total ignorance or uncertainty to firm conviction or "moral" certainty.

Thus jurisprudence furnished two striking features of the classical interpretation of probability: the subjective understanding of probability as a "degree of certainty"; and the prominence of the concept of probabilistic expectation. Classical probability theory retained these legal elements, albeit in modified form, throughout its career.

The remainder of this chapter is divided into two parts. Section 1.3 explains how probabilistic expectation derived from the doctrine of aleatory contracts and examines its critical role in the first formulations of mathematical probability. Section 1.4 describes the relationship between the hierarchy of proof in Roman and canon jurisprudence and the subjective or epistemological orientation of classical probability theory.

1.3 Expectation, Equity, and Aleatory Contracts

Between July and October of 1654 the mathematicians Blaise Pascal and Pierre Fermat exchanged a number of letters that tradition recognizes as the beginning of mathematical probability theory. There were of course precursors, chief among them Gerolamo Cardano's manuscript *Liber de ludo aleae* (composed circa 1530, but first published in 1663). Historians have hesitated to count Cardano's work as the origin of probability theory for a number of reasons: only a part of the brief treatise actually deals with the computation of chances, and like most of what Cardano wrote, the treatment now seems odd, peppered as it is with personal anecdotes, philosophical reflections, classical allusions, and much hardheaded advice on cheating, strategies, and the psychology of competition. However, the book is very revealing of some of the early conceptual difficulties facing the mathematical theory, and we shall return to it in Section 1.4 to illuminate later developments. But while Cardano's work was without influence, the Pascal/Fermat correspondence created a research tradition, complete with problems and concepts, that dominated the field for over fifty years. On these grounds alone it deserves its traditional place in the history of mathematical probability, and I shall not break with that tradition.

Apparently at the instigation of the mathematical dabbler and man-about-town, the Chevalier de Méré,[24] Pascal posed the following "Problem of Points" to Fermat: Two players, A and B, each stake thirty-two pistoles on a three-point game. When A has two points and B has one, the game is interrupted. How should the stakes be divided? Fermat's solution, as it can be pieced together from the extant correspondence (particularly Pascal's reply of 24 August 1654), seems to rest upon a full enumeration of all possible outcomes. Pascal's approach, which has been described as "re-

---

[24] See Hacking, *Emergence*, chapter 7, for the circumstances.

15

cursive,"[25] rejected Fermat's combinatorial method as unwieldy and potentially liable to error.[26] Pascal's solution, which fortunately survives in full,
was based on expectations rather than combinations. With the aid of the
definition of expectation formulated by later probabilists, player A's expectation would be

$$(1/2)(32) + (1/2)(64) = 48 \text{ pistoles.}$$

Pascal got the same answer, but his line of reasoning was different. The
later definition breaks expectation down into the product of probabilities
and their associated outcome values, both of which are assumed to be
known a priori. In contrast, Pascal made expectation and equality of condition the primitive concepts of his analysis. Since player A is assured
thirty-two pistoles no matter what the outcome of the next round, Pascal
contended that it is only the remaining thirty-two pistoles that are at issue. Because "le hasard est égal" for both A and B in the upcoming round,
Pascal decided they should halve the remaining thirty-two pistoles. In
modern notation, A's expectation would be

$$(1)(32) + (1/2)(32) = 48 \text{ pistoles.}$$

However, the modern notation is misleading in its suggestion that the
two conceptualizations of the problem are symmetric, even though they
are equivalent. In fact, only one term of Pascal's solution dissected expectation into distinct probability and outcome factors, and even then the
terms must be used advisedly: the 1/2 factor derived from the equality of
condition between the two players; also, the thirty-two pistoles did *not*
represent the outcome value for A's winning the next round of play. Although Pascal clearly knew the outcome values of A's winning or losing
the next round, and understood Fermat's combinatorial solution, he chose
to analyze the problem in terms of certain gain and a remainder subject to
equitable distribution. Only after this fundamental expectation has been
established do probabilities of any description enter the argument, and
then only to endorse halving the residual amount as fair. Unlike Fermat,
Pascal's strategy consisted in eliminating explicit considerations of probability from as much of the problem as possible, substituting certain gain

[25] Kokiti Hara, "Pascal et l'induction mathématique," *Revue d'Histoire des Sciences* 15
(1962): 287–302.

[26] Pascal, *Oeuvres*, vol. 1, pp. 1147–1153.

and equity in their place. Fermat's solution took equiprobable combinations as fundamental; Pascal's approach was built upon expectation. Both mathematicians viewed the problem as one of determining expectations rather than probabilities.

Pascal's distrust of combinations stemmed largely from his belief that they were both cumbersome to manipulate and ambiguous to enumerate (an objection voiced more strongly by Roberval and later by Jean d'Alembert), rather than from any suspicion that Fermat's methods were invalid. Once Pascal realized that combinations (i.e., coefficients of the terms of the binomial expansion) could be systematically read off from the arithmetic triangle, he himself favored this approach to the mathematical analysis of games of chance. (When Pascal claimed that Fermat's method "has nothing in common with my own," he apparently meant that Fermat had suggested no mechanical means of finding combinations such as the arithmetic triangle provided.) Nonetheless, expectation remained fundamental in the treatments of Huygens and De Witt, and continued to play an important role in classical probability theory even after probabilities came to be defined explicitly in terms of ratios of combinations.

Why expectations instead of probabilities? The answer lies in the Problem of Points itself, which had tested mathematical mettle long before the Chevalier de Méré posed it to Pascal,[27] and arose out of a primarily legal context which made equity the paramount consideration. Consider another of the earliest discussions of quantitative probabilities, in the concluding chapter of Antoine Arnauld and Pierre Nicole's famous *La logique, ou l'Art de penser* (1662), better known as the Port Royal *Logique*. The authors[28] criticize those who err on either the side of excessive caution or recklessness in the conduct of their daily affairs. Readers are advised to consider not only "the good and the bad in itself, but also the probability that it will or will not happen, and to consider mathematically [*géometriquement*] the proportion that all of these things have together." In other words, decisions should be based on the expectation of the outcomes.

The example given to illustrate this counsel drove home the association with probabilistic expectation. Ten players each contribute one unit coin to the pot; each has the possibility of losing one or gaining nine, but the

---

[27] Kendall, "Beginnings," pp. 26–27.

[28] There is some speculation about the authorship of Part IV; see Hacking, *Emergence*, pp. 73–74.

game is so designed that it is "nine times more probable" that any given player will lose one coin rather than gain the other nine: "Thus each hopes for nine écus, has one écu to lose, nine degrees of probability of losing one écu, and only one to win the nine écus: this makes for perfect equality."[29] This then, is what was meant by the injunction to consider the proportion "mathematically": the ratio of "degrees of probability" for gain or loss is inversely proportional to the ratio of the gain or loss itself.

Although this dictum yielded results equivalent to those given by the later definition of expectation as the product of the probability and the outcome value, the conceptual slant differed significantly. The Port Royal version of expectation offered no general means of reckoning probabilities beyond the conventional estimation of odds for equiprobable cases. There was no mention of combinations. Like other early expositions of mathematical probability, the Port Royal *Logique* made expectation rather than probability the central concept, in order to ascertain the conditions that made risk "equitable." These expositions concentrated on problems of rational decision in the face of uncertainty and of the terms of a fair game or just division of stakes, as in the Problem of Points. The Problem of Points antedated the Pascal/Fermat correspondence by at least a century; Pacioli and Nicolo Tartaglia were among the mathematicians who had made unsuccessful attempts to solve it.[30] All of these solutions, including that proposed by Pascal and Fermat which laid the foundations for mathematical probability, grappled with the issue of a "fair" distribution based on a true measure of expectation.

Ernest Coumet has situated the Pascal/Fermat correspondence against the background of late sixteenth- and seventeenth-century legal and theological discussions that debated whether risk taking in trade should be exempted from church prohibitions against gambling and usury.[31] This controversy focused attention on the class of contract law known as aleatory, because it dealt with agreements involving an element of chance: insurance, games of chance, annuities, and so forth. I would like to pursue Coumet's insight beyond the immediate origins of mathematical probability, into the works of the first generation of mathematical probabilists, in order to show how the legal doctrine of aleatory contracts continued to

[29] Antoine Arnauld and Pierre Nicole, *La logique, ou l'Art de penser* (1662), Pierre Clair and François Girbal, eds. (Paris: Presses Universitaires de France, 1965), p. 353.
[30] Kendall, "Beginnings," pp. 26–27.
[31] Coumet, "Théorie," pp. 579–582.

exert a strong influence on the calculus of probabilities at the level of definitions and proofs as well as that of applications. This persistent legal slant guaranteed the concept of expectation a prominent place in the classical theory of probability.

Like all contract law, treatises on aleatory contracts sought to specify conditions of equity and rules for exchanging goods in-hand for the more or less likely prospect of other, more valuable goods. By the sixteenth century, aleatory contracts were an established category in civil—that is, Roman—law, although the types of situations covered by this designation varied from jurist to jurist. Charles Du Moulin, for example, distinguished between the licit practices of purchasing annuities and wheat futures and the reprehensible pastime of gambling, although he admitted that all involved "uncertainty and danger." The prohibition against gambling was primarily a moral one, and did not prevent Du Moulin from treating all such aleatory cases jointly in his *Summaire du livre analytique des contractz usures* . . . (1554). In general, aleatory contracts included any formal agreement in which chance might figure, including not only insurance and games of chance, but also inheritance expectations and even risky business investments. The legal discussions all revolved around the same issue: as contracts, such agreements must assure all parties of maximum "reciprocity" or equality of terms. How should the (certain) price of an uncertain gain be assessed in order to preserve the rule of equity?

Although the answers to this question were largely qualitative, they display attempts to "proportion" risk to gain in a way that provided the prototype for probabilistic expectation. The discussions of risk sharing among business partners are particularly revealing on this point. Many seventeenth-century jurists hoped to override church proscriptions against usury by equating interest reaped on investments in, for example, a merchant-shipping expedition, with the legitimate earnings paid for work done or services rendered. Investors, it was argued, deserved a share of the profit for having shared the risks. "Mixed" partnerships in which some partners supplied capital and others labor dated from Roman times,[32] and by the sixteenth century it had become common practice for one partner to assume the "péril des deniers"—a kind of insurance policy—as their contribution to the venture.

[32] Eli F. Heckshaw, *Mercantilism*, trans. Muriel Shapiro (London: George Allen & Unwin, 1935), vol. 1, p. 332.

19

Jurists defended the right of such risk-bearing partners, who essentially functioned as insurers, to a share of the profit, known as the "price of the peril."[33] This practice, derived from the Roman *foenus nauticum* (bottomry), had been linked to usury by some medieval canon lawyers. In this type of arrangement, the shipowner need not repay the loan if the goods are lost at sea. By the sixteenth century, however, the risk was widely accepted as a title to profit even in the so-called triple contract, which involved a second, separate insurance contract as well as the original contract of partnership. The third element was a contract "by which an uncertain future gain is sold for a lesser certain gain."[34]

By the early seventeenth century, Hugo Grotius had extended this equation of risk with earnings to exonerate bankers from usury. Dutch financiers were justified in charging merchants a 12 percent interest rate on loans, as opposed to the standard 8 percent rate, "because the hazard was greater. The justice and reasonableness indeed of all these regulations must be measured by the hazard or inconvenience of lending." Already in 1645, the Sacred Congregation of Propaganda, in a reply to a Jesuit request that Chinese converts who lent at interest be granted a dispensation from usury strictures, spoke for many jurists in approving such loans "provided that there is considered the equality and probability of danger, and provided that there is kept a proportion between the danger and what is received."[35]

Rules for translating risk into compensation remained for the most part qualitative, but were guided in spirit by the so-called Rule of Fellowship (i.e., distributive proportion), which specified that the profit of each partner should be proportional to his investment. Every sixteenth-century text on practical arithmetic included a section on the Rule of Fellowship, illustrated with numerous examples and problems. Some also discussed the "double" Rule of Fellowship, which took into account the duration as well as the amount of the investment.[36] Probabilities, or rather expectations,

[33] François Grimaudet, *Paraphrase des droicts des usures pignoratifs* (Paris, 1583), p. 92.

[34] John T. Noonan, Jr., *The Scholastic Analysis of Usury* (Cambridge, Mass.: Harvard University Press, 1957), pp. 137–151, 209.

[35] Noonan, *Usury*, pp. 281–283, 289.

[36] See, for example, Estienne De La Roche, *L'Arismethique* (Lyons, 1520); Thomas Masterson, *His First Books of Arithmeticke* (London, 1652); Pierre Forcadel, *L'Arithmeticque* (Paris, 1557); Simon Stevin, *L'Arithmetique* (Leyden, 1585); P. Taillefer, ed., *Methodiques institutions de la vraye et parfaite arithmetique de Iacques Chauvet* (Paris, 1615); also David Murray, *Chapters in the History of Bookkeeping, Accountancy and Commercial Arithmetic* (Glasgow: Jackson, Wylie, & Co., 1930), pp. 144, 437–445.

were thus familiarly, if qualitatively, conceived in terms of proportions by the late sixteenth century, and this is the format in which early probabilists like Huygens expressed their mathematical versions of expectation.

Jurists and theologians concerned with accommodating usury prohibitions to commercial practices posed the type of questions the probabilists addressed: "What is the price one should offer to those who undergo the perils, and other fortuitous events, to which everything is subject in commerce, and especially money; what is the sum proportioned to the indefinite and uncertain gain which pledges as backing for a Society of Merchants?"[37] Other aleatory contracts posed analogous problems for jurists. Although the jurists who attempted to specify conditions of equity for contracts involving uncertain outcomes, such as insurance policies on seabound cargoes, were no more interested in quantifying risks on a statistical basis than were their clients, they did argue that profits should be scaled according to risk. This precept led to a qualitative conception of expectation as a compound of the magnitude of the risk and the value of the outcome, one very similar to the Port Royal *Logique*'s dictum that both probability and contingent advantage should be considered "in proportion." Like the Port Royal authors, the jurists were primarily concerned with the equality of expectation as a precondition for a fair game, insurance policy, division of profits, price of a lottery ticket, and so on. The determination of the component probabilities that conditioned the outcome values was of secondary interest. It was equal expectations, not equal probabilities, which in most cases guaranteed equitable terms, and it was equity which interested the jurists:

> And these sorts of agreements have their justice in that one prefers a certain and known portion, either of profit or of loss, to the uncertain expectation of events; and the other on the contrary finds it to his advantage to hope for a better condition. Thus there is a kind of equality in their portions, which renders their agreement just.[38]

Contracts were the backbone of the natural law school of jurisprudence of the late sixteenth and seventeenth centuries,[39] since they cemented con-

---

[37] R.P.E. Bauny, *Somme des pechez qui se commettent en tous les etats* (Lyon, 1646), p. 227.

[38] Jean Domat, *Les loix civiles dans leur ordre naturel* (1689–94), nouvelle édition . . . augmentée des Troisième et Quatrième Livres du Droit Public, par M. de Héricourt (Paris, 1777), p. 30.

[39] See Otto Gierke, *Natural Law and the Theory of Society 1500 to 1800*, trans. with an introduction by Ernest Barker (Cambridge, Eng.: Cambridge University Press, 1934), vol.

senting individuals together to form a society, and even transcended the social bonds: "For the words of agreement between contracting parties are even stronger than those, on which society is founded." Grotius's influential *De jure belli ac pacis* (1625) stipulated that "In all contracts, natural justice requires there should be an equality of terms,"[40] and examined the degrees of equity pertaining to various sorts of contracts, from sales of goods to international treaties, in great detail. Jean Domat, the prominent seventeenth-century French jurist and friend of Pascal, also maintained that "The use of contracts [*conventions*] is a natural consequence of the order of civil society, and of the bonds which God creates among men."[41] Domat summarized the general rules of equity as rendering to all their just expectations, honoring promises and obligations, and taking care to "do hurt to no man." Charles Du Moulin claimed that even the etymology of the word "contract" denoted "mutual attraction and reciprocity."[42]

Jurists considered contracts that worked even implicitly to the disadvantage of one of the parties to be just cause for legal action, and aleatory contracts, including games of chance, were no exception. In his discussion of contracts involving chance, Samuel Pufendorf contended that players' risk of winning or losing must be in "just proportion" to the stake, and that all must share "equally the risk of winning or losing."[43] Domat also made equality of condition the essential guarantee of equality in aleatory agreements. For example, a partnership of as yet childless men might legally arrange to provide their daughters' marriage portions from joint stock, even though only some of the partners might ultimately be able to take advantage of the provision: "The state in which all of them share, with the same uncertainty of the event and with the same right, having rendered their condition equal, also makes their agreement just."[44] Domat's guarantee of equality stemmed from the shared (and therefore equal) subjective

---

1, pp. 76–78; Leonard Krieger, *The Politics of Discretion: Pufendorf and the Acceptance of Natural Law* (Chicago and London: University of Chicago Press, 1965), pp. 99–118.

[40] Hugo Grotius, *The Rights of War and Peace* (1625), trans. A. C. Campbell (Washington and London: M. Walter Dunne, 1901), p. 147.

[41] Domat, *Loix civiles*, p. 19.

[42] Charles Du Moulin, *Summaire du livre analytique des contractz usures, rentes constituées, interestz & monnoyes* (Paris, 1554), f. 15v.

[43] Samuel Pufendorf, *Le droit de la nature et des gens* (1682), trans. Jean Barbeyrac (London, 1740), vol. 2, pp. 503–504.

[44] Domat, *Loix civiles*, p. 99.

uncertainty of the partners, as well as their equal claims to the dowries. Scholastic theologians even mounted a moral defense of gambling in cases where there was "equality of uncertainty, peril, or chance."[45]

This strict legal insistence upon absolute equality among the parties to risk found its way into the early literature of mathematical probability in a sometimes exaggerated form. Cardano, for example, makes equality of condition—equality of opponents in rank and skill, of bystanders, of money, and of situation, as well as of the dice—his "Principale fundamentum in Alea," and warns that those who deviate from this cardinal rule are "unjust."[46] The Port Royal *Logique* is more narrowly concerned with equality of the chance setup rather than the social status of the players, but even here it is an idea of legal rather than mathematical equality that demands that the situation of all the players be absolutely identical. One could easily invent situations in which the condition of the players was equalized by a balancing of odds and stakes, and we know from Cardano that dice games of the sort were well known.[47] But the jurists held to a more rigid standard as a further guarantee of equity.[48]

The doctrine of aleatory contracts thus furnished the late seventeenth-century probabilists with a set of concepts and problems. Jurists seeking the fair price for an annuity, a lottery ticket, or partnership share thought in terms of expectation, rather than risk per se, and the first mathematical probabilists did as well. This is why Pascal described the results of the new mathematics of chance as rendering to each of the players what was due to him *en justice*. Expectation had the advantage of being already quantified in legal practice, for contracts specified the purchase price of an uncertain gain. If the means for reckoning that price in any given case were nebulous, the price itself was exact.

These influences are palpable in the work of Christiaan Huygens, author of the first published work in mathematical probability, and of Johann De Witt, who applied Huygens's precepts to the problem of pricing annuities. Although Pascal and Fermat invented the calculus of probabilities in their 1654 correspondence, their letters remained unpublished until 1679. Christiaan Huygens, visiting Paris in 1655, heard about the problems addressed in this exchange from Giles Roberval and Claude Mylon,

[45] Du Moulin, *Summaire*, f. 186v.

[46] Hieronymus Cardanus, *Liber de ludo aleae*, in *Opera Omnia* (Lyons, 1663), facsimile reprint (Stuttgart-Bad Cannstatt: Friedrich Fromann Verlag, 1966), vol. 1, p. 263.

[47] Cardanus, *Ludo*, chapter 14.   [48] Pufendorf, *Droit*, p. 504.

a friend of Carcavy, who was the intermediary between Pascal and Fermat. Although Huygens met neither of the principals, he worked out his own solutions to the Problem of Points, and after ascertaining that his answers tallied with those of the French mathematicians, composed a brief treatise, which he sent to his former teacher Frans van Schooten on 20 April 1656.[49] Huygen's *De ratiociniis in aleae ludo* was published as an appendix to van Schooten's *Exercitationum mathematicarum libri quinque* (1657), and was subsequently translated into Dutch, English, and French.[50] It was later reprinted, with commentary, as Book I of Jakob Bernoulli's *Ars conjectandi*.

Huygens's treatise set forth, strictly speaking, a calculus of expectations rather than of probabilities. Huygens posed problems on the fair division of stakes or the "reasonable" price for a player's place in an ongoing game, rather than questions about the probabilities of events themselves. Considered by itself, Huygens's fundamental principle—his definition of expectation—sounds suspiciously circular:

> I begin with the hypothesis that in a game the chance one has to win something has a value such that if one possessed this value, one could procure the same chance in an equitable game [*rechtmatigh spel*], that is in a game which works to no one's disadvantage.[51]

Since later probabilists *defined* an equal or fair game as one in which the players' expectations equaled the price of playing the game (i.e., the stake), Huygens's explanation of expectation in terms of a fair game seems to lead nowhere. However, Huygens here assumed that the notion of an equal game was a self-evident one for his readers. The alternative definition, which gained currency in the eighteenth century, derived expectation and the criterion for a fair game from the definition of probability, expressed as the ratio of the number of combinations favorable to the event to the total number of combinations. This route remained closed to Huygens. Instead, he appealed to an intuitive, or at least nonmathematical, notion of equity: in this case, the equitable exchange of expectations and the conditions of a fair game.

[49] See Henri Brugmans, *Le séjour de Christian Huygens à Paris et ses relations avec les milieux scientifiques français* (Paris: Librairie E. Droz, 1935), p. 40; also "Avertissement," in *Oeuvres complètes de Christian Huygens*, Société Hollandaise des Sciences (The Hague: Martinus Nijhoff, 1920), vol. 14, pp. 1–30.

[50] "Avertissement," in *Oeuvres*, pp. 4–5.

[51] Huygens, *De ratiociniis in ludo aleae*, in *Oeuvres*, vol. 14, p. 60.

Huygens's formulation of expectation was drawn from contemporary doctrines of contract law. Huygens could presume the self-evidence of a fair game or exchange because these were the staples of seventeenth-century legal theory and practice. These legal discussions (recall especially Pufendorf's stipulation that players must all have an equal chance of winning or losing), along with Huygens's definition of equal expectation, may have presumed equiprobability by requiring "equal conditions" among players, but they did so only tacitly. Seventeenth-century jurists did assess trade-offs between various risks and stakes, quantitatively if unmathematically, in the cases of fluctuating insurance premiums, partnerships formed with mixed contributions of capital, labor, and risk bearing, and other probabilistic situations. A well-honed sense of an equitable contract, even one hinging on uncertain outcomes, could be assumed, as could the legal conception of expectation. Hence, Huygens's definition of equal expectations in terms of fair exchange or game, one which worked to the "disadvantage" of no one, would not have struck a contemporary reader as tautologous. The conditions of equity and the legal paradigm of a just contract had been firmly and independently established in legal usage and daily practice. Later probabilists such as Nicholas Bernoulli reversed this order by defining equity in terms of equal expectations,[52] but throughout the eighteenth century probabilists returned to the model of an equitable exchange.

Huygens's propositions and examples made frequent use of this legal device of a fair exchange. In order to prove that the expectation of each of two players in an equal game that awards a sum a to the winner and b to the loser is $(a + b)/2$ Huygens argued in what again appears to be a closed circle. Both players stake an amount x, and agree to offer the loser a consolation prize of a, so that the possible outcome values will be a or $2x - a$. Because "this game is equitable, and thus I have an equal chance" at either outcome,[53] Huygens defines b as equal to $2x - a$, and concludes that he could bet $(a + b)/2$ with another player and make the same arrangement for a consolation prize a. Later probabilists would summarize this argument by asserting that the equation between the stake x and the expectation $(a + b)/2$ guarantees a fair game. Huygens, however, assumed $x = (a + b)/2$ (by setting $b = 2x - a$) *because* the game is, by hypothesis,

[52] Nicholas Bernoulli, *De usu artis conjectandi in iure* (1709), chapter 4, in *Die Werke von Jakob Bernoulli*, Basel Naturforschende Gesellschaft (Basel: Birkhäuser, 1975), vol. 3.

[53] Huygens, *Ludo*, p. 62.

a fair one. He insured that the game would be fair a priori by constructing completely symmetric conditions for all players, and by arranging a series of deals, each certified as self-evidently equitable, among the players to convert one mathematical expression of expectation into another. Returning to his initial hypothesis, Huygens asserted that expectations were equal when they could be fairly traded for one another.

This method of circumventing any explicit statement of equiprobability became more involved as the outcomes (or "chances") proliferated, for Huygens had to posit as many players, bound in as many subcontracts, as there were possible outcomes. Once again, the modern order of reasoning regarding expectations was inverted: instead of the game being fair because the probabilities (and therefore the expectations in a symmetric game) are equal for all players, the probabilities are (implicitly) equal because the game is assumed fair—and the game is fair because the conditions of the players are indistinguishable, as shown by their willingness to exchange expectations in a series of "equitable" subcontracts. For Huygens, expectation represented a mathematical version of equity.

Expectation later came to be defined as a composite notion, the product of the more fundamental components of probability and outcome value:

> In all cases, the Expectation of obtaining any Sum is estimated by multiplying the value of the Sum expected by the Fraction which represents the Probability of obtaining it.[54]

For Huygens and the first generation of probabilists, however, expectation was the irreducible concept from which probability could be in theory derived if the outcome value were known. I have suggested that this order of precedence made sense in the context of legal theories that estimated expectations rather than risks, and that aimed at equalizing these expectations in partnerships and other contracts. Except in extremely simple cases, such as coin tossing and dice throwing, combinatorial arguments were not feasible, and the data required for statistical evaluations were generally unavailable. Although mathematicians like Leibniz, Huygens, and Jakob Bernoulli were quick to perceive the relevance of Graunt's political arithmetic to probability theory, the two disciplines emerged independently of one another. The first attempt to apply probability to annuities made no direct use of statistics and adopted Huygens's methods of expectations.

[54] De Moivre, *Doctrine*, p. 3.

De Witt's *Waerdye van Lyf-Renten* (*Treatise on Life Annuities*), originally written as a series of letters to the Estates-General of the United Provinces of Holland and West Friesland in 1671, was one of the earliest attempts to extend the new mathematics of probability to other sorts of aleatory contracts besides games of chance. Although the sale of annuities dated back to Roman times, there is little evidence that rates were computed on the basis of mortality statistics. That annuities, as well as compound interest, were fixtures of finance and trade by the late sixteenth century can be seen from the tables on annuities and compound interest for various rates and periods regularly appended to late sixteenth- and seventeenth-century treatises of practical arithmetic.[55] De Witt's originality lay in his attempt to estimate the probability of death as a correlate of age, and in his extension of Huygens's calculus of expectations to a new class of problems. De Witt skirted the principal obstacle to such generalizations, *viz.*, the need to deal with what are apparently nonequiprobable outcomes such as age at death, by simply assuming equiprobability for the risk of dying between the ages of three and fifty-three, and assigning proportional probabilities for earlier and later ages on the basis of educated guesswork.

Although De Witt welcomed Johannes Hudde's empirical confirmation of his guesswork with data culled from the records of past holders of Dutch annuities,[56] the initial lack of mortality statistics did not undermine his confidence in his original conclusions, which were no more grounded in statistics than the rules of thumb of the Roman jurists had been. This insouciance is more easily understood within the context of the established practice of gauging the value of the expectation of an insurance policy, an annuity, or other aleatory contract "by eye." De Witt had, after all, been trained in the law. Huygens's mathematics provided him with a more precise method of treating concepts already certified by long use. Although the even greater quantitative precision to be achieved through statistics would have been—and was—immediately appreciated, it was not consid-

---

[55] See, for example, William Purser, *Compound Interest and Annuities* (London, 1634); John Kersey, ed., *Mr. Wingate's Arithmetick*, 5th edition (London, 1670).

[56] See Société Générale Néerlandaise d'Assurances sur la Vie et de Rentes Viagères, *Mémoires pour servir à l'histoire des assurances sur la vie et des rentes viagères au Pays-Bas* (Amsterdam, 1898), pp. 24–33, for a French translation of the correspondence between Hudde and De Witt on this subject. The original correspondence is preserved in the National Archives in Amsterdam; the AMEV Library in Utrecht holds Hudde's manuscript reckoning sheets, *Stads-finatie geredresfeert in den jaare 1679 . . . Balansenenz: Betreffende de lofen lijfrenten*.

ered a prerequisite for an accurate analysis of the relative financial advantages of redeemable and life annuities.

De Witt's brief treatise is therefore instructive as an early mathematical codification of concepts and practices previously implemented by rules of thumb and seasoned judgment. As in Huygens's treatise, from which De Witt borrowed liberally, games of chance furnished many of the illustrations. However, De Witt hoped to branch out to other sorts of aleatory contracts. De Witt's vocabulary is even more legalistic than Huygens's, rephrasing Huygens's hypothesis explicitly in terms of equitable contracts:

> I presuppose that the real value of certain expectations or chances of objects, of different value, must be estimated by that which we can obtain from equal expectations or chances, dependent on one or several equal contracts.[57]

As in Huygens's definition of equal expectations, only the presumption of an independent notion of "equal contract" rescued De Witt's definition from tautology. Although De Witt's examples of such equal contracts are all fair games with equiprobable outcomes, he did not single out either "equiprobability" or "probability" as distinct concepts requiring definition: these notions were subsumed within the definition of an equal contract, one which balanced the advantages and disadvantages of all parties as precisely as possible. De Witt's demonstration of the proposition (corresponding to Huygens's Proposition III) that the value of several expectations or "chances" is to be computed by summing the value represented by the chances, and by then dividing this sum by the number of chances,[58] relied on an exchange of equal contracts among partners in completely symmetric situations. The symmetry both insured the legality of the contract—in the words of Pufendorf, all run "equal risks"—and obviated the need for explicit discussion of probabilities per se. De Witt could use "expectation" and "chance" as synonyms because the number of outcomes in each example was designed to equal the number of partners, which in turn

[57] De Witt's rare treatise is reprinted in Jakob Bernoulli, *Werke*, vol. 3, pp. 327–350. It was already hard to come by in Bernoulli's time, and he importuned Leibniz in vain for a copy. An English translation by F. Hendriks is reprinted in Robert G. Barnwell, *A Sketch of the Life and Times of John De Witt* (New York, 1856). All quoted passages are taken from this translation; see De Witt, *Waerdye van Lyf-Renten* (1671), in *Werke*, vol. 3, p. 329; Barnwell, *Sketch*, pp. 82–83.

[58] De Witt, *Waerdye*, in *Werke*, vol. 3, pp. 331–332; Barnwell, *Sketch*, p. 86.

equaled the number of chances. Of course, this condition held only if the chances were assumed to be equiprobable, and the chances were equiprobable according to De Witt, because the contracts were fair. By inverting the last claim, as later probabilists would, the proof collapses into circularity. Without equal contracts and the concomitant notion of the symmetric status of the partners, there would be no grounds for asserting equal expectations or (implied) equal probabilities.[59]

Twentieth-century critics of classical probability theory have commented at length on the circular assumption of equiprobable outcomes built into the classical definition of probability, and the Principle of Indifference invoked to defend that assumption. Laplace's definition rests on both:

> The theory of chance consists in reducing all the events of the same kind to a certain number of cases equally possible, that is to say, to such as we may be equally undecided about in regard to their existence.[60]

Ernest Nagel objects that "if 'equipossible' is synonymous with 'equiprobable,' " then the classical definition in terms of a ratio of favorable to total number of equipossible alternatives "is circular, unless 'equiprobable' can be defined independently of 'probable.' "[61] The Principle of Indifference cannot be used to salvage the classical definition, because it does not yield unique probability values.[62]

I have argued that the first formulations of mathematical probability were heavily indebted to seventeenth-century legal notions of contract. These granted late seventeenth-century probabilists a reprieve from the difficult task of justifying the useful assumption of equiprobable outcomes. Equal expectations, rather than equiprobable outcomes, were the departure point for the earliest mathematical treatments. By defining

[59] For further early examples of the various sorts of aleatory contracts treated mathematically, see Jakob Bernoulli, *Meditationes*, nos. 159, 162, 169, in *Werke*, vol. 3, pp. 42–48, 66, 71; and Nicholas Bernoulli, *De usu*, also in *Werke*, vol. 3, pp. 287–326.

[60] Pierre Simon de Laplace, *Essai philosophique sur les probabilités* (1814), in *Oeuvres complètes*, Académie des Sciences (Paris, 1878–1912), vol. 7, p. viii; all cited passages are taken from the English translation of the 6th edition by Frederick Wilson Truscott and Frederick Lincoln Emory, *A Philosophical Essay on Probabilities* (New York: Dover, 1951), p. 6.

[61] Nagel, *Principles*, p. 388.

[62] See John Maynard Keynes, *A Treatise on Probability* (London: Macmillan, 1943), chapter 4.

equal expectations in terms of equitable contracts, which legal usage permitted probabilists to take as self-evident, these expositions could avoid transparent circularity, although the definitions and demonstrations always threatened to close in upon themselves. In games of chance, the physical symmetry of the die or coin lent credence to the legal contention that players enjoyed equal prospects and therefore were not in violation of equity. Hence the somewhat vague allusions to the "equal ease" or "facility" with which certain events could occur: as Huygens phrased it in the original Dutch version of his treatise, "each chance can come about equally easily."[63]

Historians and philosophers of probability theory have viewed the equiprobability assumption from the perspective of the current dichotomy between subjective, epistemological versus objective, frequentist interpretations of probability. From the epistemological standpoint, the a priori assumption of equiprobable outcomes rests on the Principle of Indifference, as in Laplace's formulation. The objectivists argue that the physical symmetry of, for example, a fair coin, validates the assumption that in the long run both sides will turn up with equal frequency.[64] When this opposition is superimposed on the discussions of the classical probabilists, they appear to vacillate, sometimes siding with the subjectivists and sometimes with the objectivists. Hacking has suggested that early probabilists were able to tolerate such ambiguity by taking refuge in the parallel ambiguities in the usage of "possibility," alternately favoring its epistemological and objectivist nuances: "By explaining probability in terms of possibility writers of an earlier period could usefully equivocate."[65] Equity played a similar role in the first expositions of mathematical probability via the fundamental hypothesis concerning equal expectations. As long as probabilists could take equity as an irreducible, undefined concept, the vexing issue of equiprobability could be dodged.

However, in the more complicated situations that involved unequal risks, the problems of ascertaining probabilities were all but insoluble without more extensive statistical information. Roman-canon law offered some rules of thumb, like those of Ulpian, but in general jurists left such uncertain matters to the discretion of an experienced judge, to be arbi-

[63] Huygens, *Ludo*, p. 65. See also Cardanus, *Ludo*, p. 64, for a similar expression.

[64] Hacking, "Jacques Bernoulli's *Art of Conjecturing*," *British Journal for the Philosophy of Science* 22 (1971): 209–229, on p. 210.

[65] Hacking, "Equipossibility theories of probability," *British Journal for the Philosophy of Science* 22 (1971): 339–355, on p. 341.

trated on a case-by-case basis.[66] Like the business of setting insurance premiums, the preferred method brought wide experience and discretion to bear on each individual case, considered on its particular merits.[67] The statistical approach of John Graunt heralded a new way of thinking, but assumptions of equiprobability lingered. De Witt was obliged to assume equiprobable chances of dying in any six-month period between childhood and old age; Graunt also presumed that the same proportion (about 3/8) of the population died every ten years.[68] Until Jakob Bernoulli's limit theorem legitimated the practice of equating statistical frequencies and probabilities in at least some cases, and fact-gathering projects like Edmund Halley's Breslau table of mortality[69] furnished those frequencies, an independent notion of probability applied to anything other than games of chance would have been superfluous. Even in games of chance, the enumeration of the combinations of equiprobable outcomes quickly became unmanageable, as Pascal had complained to Fermat. The problem was further complicated by the mixture of elements of chance and skill in the games analyzed by the early probabilists.[70]

Perhaps this is why Thomas Bayes, whose method of finding a posteriori probabilities seemed to free probabilists from considering only a priori equiprobable cases, chose to return to the expectation-centered approach of Huygens to probability, although Abraham De Moivre's direct estimation of probability as "a Fraction whereof the Numerator be the number of Chances whereby an Event may happen, and the Denominator the number of all Chances whereby it may happen or fail"[71] would have presumably been known to him. Bayes began his posthumous (1763) essay with an

---

[66] S. P. Scott, trans., *The Civil Law, including the Twelve Tables, the Institutes of Gaius, the Rules of Ulpian, the Opinions of Paulus, the Enactments of Justinian, and the Constitution of Leo* (Cincinnati: Central Trust, 1932). See also Du Moulin, *Summaire*, ff. 187r.-v., for rough methods of estimation. To judge from manuals on annuities, the important temporal variable was interest rather than age; see the tables in William Purser, *Compound Interest*.

[67] See Section 3.4.2.

[68] John Graunt, *Natural and Political Observations Mentioned in a Following Index and Made Upon the Bills of Mortality* (London, 1662).

[69] Edmund Halley, "An estimate on the degrees of mortality of mankind, drawn from curious tables of the birth and funerals at the city of Breslaw; with an attempt to ascertain the price of annuities upon lives," *Philosophical Transactions of the Royal Society of London* 17 (1693): 596–610.

[70] See, for example, Jakob Bernoulli, *Meditationes*, no. 160, in *Werke*, vol. 3, pp. 48–64.

[71] De Moivre, *Doctrine*, p. i.

exposition of the basic principles of the calculus of probabilities, because according to his literary executor and editor Richard Price, Bayes knew of no source which gave a "clear demonstration of them."[72] Price gave no indication of when Bayes composed the memoir, and it is barely possible that at that time Bayes knew neither Montmort's *Essai d'analyse sur les jeux de hazard* (1708, 1713), Jakob Bernoulli's *Ars conjectandi* (1713), nor any of the three editions or supplements of De Moivre's *Doctrine of Chances* (1718, 1730, 1756), although the very title of his essay suggests the contrary. It seems more likely that he found these treatments in some way unsatisfactory.

In any case, Bayes' own exposition of the "general laws of chance" fell squarely within the original expectation-based treatments of probability. Like Huygens, Bayes built his proofs around the reasonable trade of expectations. For example, if the payoff depends on both events A and B happening, Bayes argues that news that B had occurred did not alter the initial expectation:

> For if I have reason to think it less, it would be reasonable for me to give something to be re-instated in my former circumstances, and this over and over again as often as I should be informed that the second event had happened, which is evidently absurd. And the like absurdity plainly follows if you say I ought to set a greater value on my expectation than before, for then it would be reasonable for me to refuse something if offered me upon condition I would relinquish it, and be re-instated in my former circumstances.[73]

If it is "unreasonable" to either buy or sell at a higher price, Bayes concludes that the expectations must therefore be equal.

By 1763 (the third edition of De Moivre's *Doctrine of Chances* had appeared in 1756), Bayes' definition of probability as "the ratio between the value at which an expectation depending on the happening ought to be computed, and the value of the thing expected on its happening" might well have struck knowledgeable readers as outdated. Price felt constrained to explain that his friend had chosen to overlook "the proper sense of the word *probability*" because whatever confusion surrounded the meanings of

---

[72] Thomas Bayes, "An essay towards solving a problem in the doctrine of chances," *Philosophical Transactions of the Royal Society of London* 53 (1763): 370–418, on p. 375.

[73] Bayes, "Essay," p. 380.

"chance" or "probability," "all will allow [expectation] to be its proper measure in every case where the word is used" since "expectations ought to be estimated so much the more valuable as the fact is more likely to be true, or the event more likely to happen."[74] By defining probability as the quotient of the expectation divided by the outcome value, Bayes of course assumed that expectation was the fundamental concept (one which could be appreciated, as Price suggested, directly by common sense), rather than the product of probability and outcome value. Moreover, "reasonable" barters might set a price upon—or at least an equivalence between—expectations, just as equitable contracts certified an equivalence of condition.

1.4 Degrees of Certainty and the Hierarchy of Proofs

Bayes was the last great representative of the expectation tradition in mathematical probability until the twentieth century.[75] By 1750, the mathematical theory was truly about probabilities rather than expectations, although its primary applications—the fair price of lottery tickets, insurance premiums, annuities—would still have been familiar to a lawyer specializing in aleatory contracts. The pitched battle over the proper definition of expectation,[76] as well as the slowly increasing store of statistics, helped unseat the concept in the mathematical literature. However, mathematical probability theory owed a second debt to the law, this one more enduring if in the end no less controversial. This was the interpretation of probability as a degree of certainty. Although this or closely related notions were ambient in the philosophical literature of much of the seventeenth century, it surfaced in mathematical form comparatively late. There are hints in Leibniz's writing, in the Port Royal *Logique*, and in an article in the *Philosophical Transactions* on the erosion of witness credibility, but the first full-fledged expression of the idea is Jakob Bernoulli's, in the posthumous *Ars conjectandi* (1713). Once introduced, however, neither the idea nor the set of applications that went with it ever fully disappeared from the mathematical theory, despite occasionally fierce attacks. In the

[74] Bayes, "Essay," p. 375.

[75] Hacking, *Emergence*, pp. 96–97.

[76] Lorraine J. Daston, "Prudence and equity: Expectation in classical probability theory," *Historia Mathematica* 7 (1980): 234–260; see also Section 2.2–3, below.

remainder of this chapter, I shall trace its immediate ancestry, once again taking the works of the early probabilists as my departure point and concentrating on the problem of quantifying the qualitative. There is no better place to begin with than Jakob Bernoulli's famous treatise.

Jakob Bernoulli's *Ars conjectandi* (1713) was the most important mathematical work on probability until Laplace's treatise on the subject a century later. The very format of the treatise revealed both the present state of the theory and its future direction: Part I is a reprint of Huygens's tract, with annotations; Part II treats the doctrine of combinations and permutations which became the mathematical backbone of the theory in the eighteenth century; Part III applies the results of Part II to further gambling problems; and Part IV applies the theory to matters of "civil, moral and economic life," crowned with the statement and proof of the limit theorem that bears Bernoulli's name. It was the reflections, applications, and mathematics contained in Part IV that impressed the classical theory of probability with its distinctive character, and although few of the elements were wholly original to Bernoulli, it was he who gave them their sharpest and most influential formulation until Laplace. And it is arguable that even Laplace's views on the nature and scope of probability theory (though certainly not the mathematics with which he enriched it) deviate very little from those of Bernoulli. We owe to Bernoulli the classical interpretation of probability as a state of mind rather than as a state of the world, and the first outright definition of a probability (*probabilitas*) in a mathematical context: "*Probability* is a degree of certainty [*gradus certitudinis*] and differs from it as a part from the whole."[77] This shift of emphasis from expectations to probabilities, and from equiprobable outcomes to measures of certainty, brought mathematical probability theory a new set of applications to belief and credibility, inaugurated by Bernoulli himself. Before turning to these applications and to the conceptual background that made *degrees* of certainty thinkable, we must first briefly examine what Bernoulli meant by "certainty."

Some historians of probability have argued that determinism was an obstacle to the emergence of a mathematical theory of chance.[78] Such views are hard to reconcile with the long list of self-confessed determinists who either contributed to mathematical probability or took a keen interest in it: as Hacking remarks, the plain historical record would indeed suggest

[77] Jakob Bernoulli, *Ars conjectandi* (1713), in *Werke*, vol. 3, p. 239.
[78] Kendall, "Beginnings," pp. 31–32.

just the contrary.[79] Jakob Bernoulli is unequivocal on the subject in his introduction to Part IV: "All things which exist or are acted upon under the sun—past, present, or future things—always have the greatest certainty."[80] We have already heard De Moivre's similar opinion. Laplace echoed them in a celebrated passage:

All events, even those which on account of their insignificance do not seem to follow the great laws of nature, are as a result of it just as necessary as the revolutions of the sun. In the ignorance of the ties which unite such events to the entire system of the universe, they have been made to depend upon final causes or upon hazard, accordingly as they occur and are repeated with regularity, or appear without regard to order; but these imaginary causes have gradually receded with the widening bounds of knowledge and disappear entirely before sound philosophy, which sees in them only the ignorance of the true causes.[81]

Bernoulli insists that the throw of a die is "no less necessary [*non minus necessario*]" than an eclipse. The only difference between the gambler and the astronomer lies in the relative completeness of their respective knowledge of dice and eclipses, and a people ignorant of astronomy might well gamble on the occurrence of eclipses. "Contingent" events only exist relative to our ignorance, so that one man's contingency might easily be another's necessity.[82]

This is the creed of the classical probabilist. From an enlightened (or perhaps superhuman) viewpoint, all events are necessary, so probabilities measure the partial certainty upon which we ill-informed mortals must ground rational belief and action. A perfected science would no longer need probability theory. It proved an extraordinarily tenacious creed, for only in the latter part of the nineteenth century did some radical spirits suggest that probabilities might measure genuine randomness and variability, and even then this alternative interpretation made very slow headway.[83] Determinism, old-fashioned necessitarian or new-fangled mechan-

[79] Hacking, *Emergence*, p. 3.

[80] Bernoulli, *Ars conjectandi*, p. 239.

[81] Laplace, *Essay*, p. 3; *Oeuvres*, vol. 7, p. vi.

[82] Bernoulli, *Ars conjectandi*, p. 240.

[83] See Michael Heidelberger, "Fechner's indeterminism: From freedom to laws of chance," in *The Probabilistic Revolution*, vol. I, *Ideas in History*, Lorenz Krüger, Lorraine J. Daston, and Michael Heidelberger, eds. (Cambridge, Mass.: MIT Press, 1987), pp. 117–156.

ical, lay at the very heart of this classical creed. Why did the classical probabilists insist so paradoxically on ridding the "calcul du hazard" of all real "hazard"?

Cardano's long-unpublished treatise on gaming provides an important hint. Cardano based his calculations of the chances of various dice throws on the idea of a "circuit," or the number of throws necessary to realize all possibilities. For a fair die, Cardano claims that a full circuit is completed in six throws, since all outcomes can happen equally easily. Of course, this is literally false, and Cardano readily acknowledged that in practice more than six throws may be required to turn up all six faces. In this and all other similar cases, he explained the discrepancy between calculation and actual outcome by "luck." Only in the case of many trials do the calculations come close to the fact of the matter, and even then they remain an "approximation" subject to the perturbations of fortune.[84] "Luck" for Cardano was not merely a *façon de parler*; he was convinced of its reality. Would-be gamblers were warned to avoid more fortunate opponents, for good fortune is as fixed a trait as skill. After an anecdote about a spectacular win of his own, he argued that such protracted winning or losing streaks have to do with fortune rather than the fluctuations of mere chance.[85]

Cardano was no determinist, and he was not greatly troubled by the interventions of chance (small, temporary fluctuations) or luck (systematic trends or streaks). Indeed, since he believed in a daimon at his own side, they added more spice to the game. However, they all but vitiated the practical importance of his calculations.[86] Given Cardano's metaphysics, this was only to be expected: mathematical calculations concern the necessary, but cannot govern the contingent. Inscrutable but real forces like luck would from time to time throw off the match between calculated and actual outcomes. (In the short run, chance could also introduce discrepancies, but these were not systematic and therefore not so serious.) At one point Cardano suggested that an infinite number of throws would render a calculated outcome "almost necessary [*proximé necesse*],"[87] but this is an unexplored aside, patently without practical ramifications, like his flirtation with the idea that his own good luck might follow unknown rational principles.[88]

[84] Cardanus, *Ludo*, chapters 9–11.

[85] Cardanus, *Ludo*, chapters 8, 20.

[86] Cardanus, *Ludo*, chapter 10.

[87] Cardanus, *Ludo*, p. 267.

[88] Cardanus, *Ludo*, p. 271.

However, later probabilists like Bernoulli and De Moivre were more ambitious for their calculations, and their unrelenting determinism licensed them to be so. Once the new metaphysics had banished luck and chance as figments of human ignorance, there was every reason to expect a match of necessary calculations with (newly) necessary events. Many seventeenth-century thinkers believed that at least some kind of local determinism was a precondition to applying the necessitarian machinery of mathematics: Galileo, for example, despaired of a true science of air currents, which he believed to be irreducibly contingent.[89] The probabilists still faced Cardano's problem—observed frequencies deviated more or less from calculated probabilities—but since they had eliminated the contingent in their metaphysics, they were confident that it could be solved. By their own lights, the chief achievements of Bernoulli and De Moivre lay in scaling down Cardano's infinite number of throws to a finite, calculable number needed to guarantee that calculations and fact would very likely match. This is why determinism, far from stifling mathematical probability theory, actually promoted it. A calculus saddled with Cardano's daimon had no future.

Determinism also made certainty of the fallible human sort the subject matter of probability theory, for necessary events per se left little room for objective probabilities. But determinism did not require that certainty take the form of a graduated continuum of fractional degrees that Bernoulli gave it. Whereas at least parts of Bernoulli's interpretation of probability, such as the opposition of divine omniscience and human uncertainty, might be traced back as far as Aquinas, his understanding of certainty as a continuum was quite novel by classical and medieval standards. The opposed categories of "probability" (or "opinion") and "certainty" long antedate the mathematical discussions of the seventeenth century as important components of medieval moral theology. Knowledge was either apodictic, supported by demonstration, or persuasive, endorsed by a more or less compelling body of opinion and argument. The moral doctrines of probabiliorism, equiprobabiliorism, and probabilism assigned relative weightings to law, expert theological opinion, and individual liberty in ambiguous situations.[90] Within the field of probable knowledge,

[89] Galileo Galilei, *Discorsi* . . . (1638), in *Opere*, Antonio Favaro, ed., Nuova Ristampa della Edizione Nazionale (Florence: G. Barbéra, 1968), vol. 8, p. 277.

[90] Thomas Demain, "Probabilisme," in *Dictionnaire de théologie catholique*, A. Vacant and E. Mangenot, eds. (Paris: Letouzey et Ané, 1935), vol. 13, part 1, cols. 417–619. .

medieval philosophers followed Aquinas in ordering rival positions by their degree of probabililty according to their plausibility or reasonableness in disputation. Despite such attempts to rank probable assertions by the comparative force of the arguments mustered in their defense, medieval philosophers regarded probability and certainty as a sort of two-valued logic.[91] In contrast, the mathematical theory of probability supposed a continuous spectrum of probabilities, spanning the extremes of impossibility and certainty, which required a method of quantifying, or at least ordering *degrees* of certainty.

How did such a method arise? How did seventeenth-century mathematicians come to view the relationship of probabilities to certainty as that of parts to the whole, in Bernoulli's phrase, rather than as incommensurable categories like "true" and "false"?

Moral theology seems a dubious source, despite its suggestive vocabulary: the rough Thomist equation of "probable" and "rhetorically defensible" admitted many alternative dimensions along which opinions might be ordered, none of them easily measured. The classical rhetorical tradition that derived from Aristotle and Cicero was also a qualitative one. For Aristotle, probabilities (ἔνδοχα; "resting on opinion") concern that which generally but not invariably happens, and along with signs and examples form the material of rhetorical enthymemes.[92] (It should be noted that Aristotle sharply distinguishes chance events, which are purposeless and exceptional, from probable ones, which are the rule.)[93] The Roman and medieval rhetoricians followed suit, parroting the classical definitions.[94] Nor should the ordered stages of certainty—moral, physical, and metaphysical—of Glanvill, Wilkins, Boyle, and other early Royal Society luminaries be confused with a full continuum of degrees.[95] As Keynes has shown, not everything that can be rank-ordered can be meas-

---

[91] Edmund F. Byrne, *Probability and Opinion: A Study in the Medieval Presuppositions of Post-Medieval Theories of Probability* (The Hague: Martinus Nijhoff, 1968), pp. 139–258, 293.

[92] Aristotle, *Rhetoric*, I.ii.14–15, 1357a24–35; II.xxv.8, 1402b13–25.

[93] Richard Sorabji, *Necessity, Cause and Blame: Perspectives on Aristotle's Theory* (London: Duckworth, 1980), pp. 151–152.

[94] Garber and Zabell, "Emergence," pp. 44–47; James J. Murphy, *Rhetoric in the Middle Ages* (Berkeley, Los Angeles, and London: University of California Press, 1947), pp. 5–6, 13–14.

[95] Shapiro, *Probability*, provides numerous examples, although she tends to assimilate them to a full continuum of degrees. See also Section 2.2, below.

ured,[96] and indeed, the Royal Society writers rarely attempted even a rough quantification. Rather, they regard the various levels (particularly moral certainty) as thresholds of belief, and as an expansion of Aristotle's view that not every subject admits the same degree or kind of certainty.[97]

One of the first attempts to actually quantify the ancient dictum "most of the time" notably occurs in a work sprinkled with other examples of the new mathematical approach to probability, the Port Royal *Logique*. The Port Royal discussion is as remarkable for its restrictions on quantitative reasoning as for its use of it, and marks a transitional point between the qualitative tradition of the orators and Jakob Bernoulli's thoroughgoing (if not wholly convincing) mathematical treatment of the same topic composed some twenty-five years later. The context of the Port Royal example, evaluating the reliability of testimony, also supplies a clue to another source of semiquantified probabilities.

The relevant Port Royal section arises as a response to a familiar legal and historical perplexity: when does testimony warrant belief? The Port Royal authors recommend two criteria: the intrinsic credibility of the fact itself (*circonstances intérieures*); and the extrinsic credibility of the witnesses (*circonstances extérieures*).[98] For example, there are historical witnesses to both affirm and deny the story of Constantine's baptism by St. Sylvester. Since the fact of the baptism itself is quite possible, the *Logique* devotes the bulk of its discussion to extrinsic credibility—how the ulterior motives and probity of both camps of historians should bear on the assessment of their testimony. In general, we should be guided by "la plus grande probabilité," or what happens "incomparably more often" in our evaluation of these circumstances. To the late seventeenth-century eye, the "incomparably more often" begged to be quantified, however roughly, and in their next example, the Port Royal authors do just that. But the example is a cautionary one, meant to warn of the pitfalls of blindly following the usual without attending to the vitiating particulars of the case at hand. Out of a thousand notarized documents, we are told, 999 are correctly dated. Therefore it is "incomparably more probable" that this particular one has not been predated. But should we hear that the notaries involved have a bad reputation, and moreover stand to profit from such a falsification, we should suspend our belief based on what happens "incom-

---

[96] Keynes, *Treatise*, chapter 3.

[97] Aristotle, *Nicomachean Ethics*, I.i, 1094b25.

[98] Arnauld and Nicole, *Logique*, p. 340.

parably more often." And if further witness reports reach us that the document was indeed antedated, we must reverse our initial belief in the probable.[99]

The steady drift in this example from quantitative probability (999 cases out of 1,000) to qualitative countervailing circumstances highlights the difficulty of quantifying the rhetoricians' probabilities. Since the special circumstances are by their nature exceptional, and since their relative weighting is bound to the individual case, they hardly invite quantification so readily as the case that occurs most of the time. Just when these qualitative considerations begin to erode the plausibility of the quantitative commonplace was a matter of judgment, not calculation, even for the Port Royal authors bent on providing their readers with rules.

If, however, we turn to Jakob Bernoulli's treatment of a similar topic, the evaluation of courtroom evidence, we see no such hesitation. (He even tackled the notary problem.) Bernoulli adopts the Port Royal vocabulary of intrinsic (concerning a "thing's cause, effect, subject, accessory, or sign") and extrinsic ("derived from the authority and testimony of men") proofs. (The *Ars conjectandi* was intended as a companion piece to the *Logique*, known in Latin as the *Ars cogitandi*.) Intrinsic proofs include the accused's pallor under interrogation and the blood-stained sword found in his house; extrinsic confirmation comes from a witness's report of a quarrel between the victim and the accused on the night of the murder.[100] Moreover, Bernoulli's distinction between "pure" and "mixed" proofs corresponds to the Port Royal distinction between certain and probable signs:[101] pure proofs implied a conclusion necessarily; mixed proofs, only contingently. In Bernoulli's example, a brother fails to write. He may be dead, a cause which entails his failure to write necessarily; or he may be preoccupied with business, a state which may or may not prevent him from writing. The equivocal evidence of mixed proofs comprised the bulk of legal practice, and Bernoulli's examples show that he had such applications in mind. A man is stabbed in an unruly crowd; witnesses are able to testify only that the murderer wore a dark cloak. Four members of the crowd are so attired, but this constitutes only mixed proof that any one of them in particular committed the crime: the proof (dark cloak) only par-

[99] Arnauld and Nicole, *Logique*, pp. 349–350.

[100] Bernoulli, *Ars conjectandi*, pp. 241–243.

[101] Arnauld and Nicole, *Logique*, p. 53; see also Aristotle, *Rhetoric*, I.ii.14–15, 1357a24–35; Garber and Zabell, "Emergence," pp. 42–44.

tially implicates any individual suspect among them. If the suspect turns pale under interrogation, this further sign would, Bernoulli contends, constitute pure proof only if the pallor were caused by a stricken conscience.

Bernoulli attempted to formulate the analysis of pure and mixed proofs mathematically, along two dimensions: a proof might or might not exist; and extant proof might indicate a particular conclusion, or instead indicate others (or even its opposite). If

b = number of cases in which a given proof exists

c = number of cases in which a given proof does not exist, and

$a = b + c;$

β = number of cases in which the given proof did indicate a certain conclusion

γ = number of cases in which it did not, and

$\alpha = \beta + \gamma,$

then the probability that such a proof both exists and tends toward the specified conclusion will be

$$b\beta/a\alpha = b/a(\beta/\alpha) + c/a(0).$$

(If the proof does not exist, Bernoulli reasoned that it could neither prove nor disprove any conclusion; hence the zero term.)[102]

Bernoulli's examples have a distinct legal flavor, and his conceptualization of proofs owed as much to the jurists as to the Port Royal authors. I shall argue that the Roman-canon system of evaluating proofs also supplied a precedent for thinking about such matters quantitatively—that is, for conceiving certainty in terms of fractions of a whole.

During the sixteenth and seventeenth centuries Continental jurists refined and consolidated the theory of judicial proof inherited from Roman and canon law. They did not invent this elaborate system of proofs, which was already entrenched in Italy and France by the thirteenth century. (The famous *Constitutio Criminalis Carolina* of 1532 established Roman-canon doctrines of "Indizienlehre" in Germany; England never accepted these in criminal procedures, but followed them in most civil cases.) In contrast to the persuasive techniques of the rhetoricians, the Roman-canon theory of proof was intended as "a set of mechanical rules . . . to assign arithmetic

[102] Bernoulli, *Ars conjectandi*, pp. 243–247.

values of evidence" according to an elaborate hierarchy of indices, suspicions, conjectures, and presumptions hammered out by generations of doctors of law.[103]Confessions, oaths, written documents, and witnesses all supplied these indices in varying degrees of probative force, to be summed by the judge.

This system of so-called "legal" proofs had evolved from a twelfth-century opposition between a true or demonstrative proof (*probatio vera*) and a presumptive one (*probatio ficto*), and in response to the abolition of proof by ordeal by the Fourth Lateran Council in 1215. Legal proofs combined presumptions of varying degrees of strength to attain the "complete" proof (*probatio plena*) required for conviction in capital cases. The resulting "arithmetic of proof" assigned fractional measures to each type of evidence, natural or testimonial; for example, the corroborative testimony of two unimpeachable eyewitnesses constituted a complete proof. In the absence of two such witnesses, a full proof might be constructed out of a sum of half- and even quarter-proofs.[104] Legal treatises instructed jurists in the delicate weighting and balancing of both internal and external evidence required by such addition.[105] Judges were advised to consider the reputation of the witness; his age, sex, and social standing; his relationship to the accused; any private interest in the case; and comportment under interrogation (paleness, vacillation, and timorous manner all reduced credibility). Inferences regarding internal evidence offered by the facts themselves were similarly shaded. Presumptions were inferences drawn from known facts to doubtful ones, and ranged from "violent" (e.g., the accused fleeing the scene of a murder with an unsheathed bloody sword) to "remote" (e.g., pallor as a sign of pregnancy.)[106]

As the language of "complete" and "half" proofs suggests, jurists attempted to roughly quantify, or at least to order, the strength of evidence relative to a full proof: a "close" or conclusive index might count one-half

[103] John H. Langbein, *Prosecuting Crime in the Renaissance: England, Germany, France* (Cambridge, Mass.: Harvard University Press, 1974), p. 238.

[104] John Gilissen, "La preuve en Europe du XVIe au debut du XIXe siècle. Rapport de synthèse," in *La preuve. Deuxième partie: Moyen Age et temps modernes. Recueils de la Société Jean Bodin pour l'histoire comparative des institutions*, vol. 17 (1965), pp. 755–833.

[105] See, for example, Jacobus Menochius, *De praesumptionibus, conjecturis, signis et indicis commentaria, in VI distincta libros* (Cologne, 1595); Alberic Allard, *Histoire de la justice criminelle au seizième siècle* (Ghent, Paris, and Leipzig: Hoste/Durand & Laurier/Durr, 1868), Part III, *titre* IV, chapter 3, contains further references.

[106] Domat, *Loix civiles*, pp. 274–275.

of the total measure of proof required; the testimony of a witness impeached by various debilities (youth, previous criminal record, etc.) would also be reduced by a fixed proportion. Moreover, these degrees of proof were correlated with degrees of rational belief produced in the mind of the judge. The fourteenth-century jurist Lean de Legnano set forth a psychological sequence of proof: initially, the judge knows nothing, and is plunged into doubt by the conflicting claims of the two parties. Proof is then produced, and a suspicion (*suspicatio*) grows in the mind of the judge, although doubt lingers. If the proofs are confined to mere indices, the judge remains at this wavering stage. However, if more compelling proof is submitted, the judge's suspicions may crystallize into an *opinio*, which corresponds to a semiproof equivalent to that produced by a strong presumption or a single trustworthy eyewitness. Ultimately, proofs may satisfy the requirements for a complete proof, impelling the judge to firm conviction, erasing all doubt.[107] In this system of legal proofs, belief was tightly reined in by evidence, to which the judge responded like "a piano, which reacts inevitably when certain keys are hit."[108]

Jurists drew distinctions between the grades of proof assigned various sorts of evidence. Some consequences could be deduced almost necessarily from the evidence, as in the case of violent presumptions, although even full proofs in law did not pretend to the status of the "fixed and immutable" truths of demonstrative science. Inferences based on what happened "naturally and in the common course of things" might, by the same token, also be accepted without caveat: the presumption of a father's love for his children, for example, required no further proof. However, other conjectures were less reliable, "depending on causes whose effects are uncertain." In general, legal rules of evidence stipulated that the certainty of a conjecture varied as "the link which might exist between the known facts, and those that must be proved"—that is, between evidence and presumption, between sign and judgment, between effects and their causes.[109]

The examples and vocabulary in the first three chapters of Part IV of the *Ars conjectandi* are drawn almost exclusively from the legal literature on proofs. Pallor under questioning, blood-stained swords, witness reports, and confessions all make their appearances. As the general rules of chapter

[107] Jean Philippe Lévy, *La hiérarchie de preuves dans le droit savant du Moyen Age*, in *Annales de l'Université de Lyon*, 3ᵉ séries: Droit, fascicule 5 (Paris: Société Anonyme, 1939), pp. 28–29.

[108] A. Esmein, *Histoire de la procédure criminelle en France* (Paris, 1882), p. 260.

[109] Domat, *Loix civiles*, p. 284.

2 of Part IV make clear, Bernoulli was interested in a general theory of rational decision under uncertainty, not just a mathematization of legal practice. However, it was that legal practice that supplied him with the problems and concepts (along with Latinate cast of characters) that serve as the model for other kinds of conjecture in civil life. Did Titius kill his enemy Maevius? Did Sempronius take the stolen object? Was Gracchus the black-cloaked murderer in the fray? What circumstances should incline the judge toward one verdict or another, and by how much?

Of course, Bernoulli's "how much" was not the "how much" of the jurist. The juridical fractions of certainty were assigned by convention rather than measurement; Bernoulli's were in principle genuine proportions of cases, with a nod in the direction of the Port Royal *Logique*.[110] In fact, Bernoulli, like the Port Royal authors, was obliged to fall back on guesswork to supply his numbers: how else could one estimate the number of cases in which, say, pallor was a sign of guilt, as opposed to the number of cases where it was caused by something else? Yet Bernoulli looked forward to a day when encyclopedias of such statistics would be available, and the remainder of Part IV explained just how such statistics are mathematically related to the probabilities. No jurist schooled in the Roman-canon theory of legal proofs would have conceived of such a program. But Bernoulli was nonetheless at one with the jurists in thinking about certainty as having degrees, in representing these degrees by fractions, and in inventing a formal set of rules to connect evidence with rational belief. The legal signature is writ large over the approach as well as over the problems of the early chapters of Part IV.

The precise source of Bernoulli's legal turn is not clear. Leibniz seems a likely conduit for such ideas, but we have no direct evidence that he was Bernoulli's advisor here, although he corresponded with him about other matters in Part IV of the *Ars conjectandi*.[111] Still, Leibniz's ideas are of some independent interest, for they suggest the extent to which degrees of certainty were a theme in late seventeenth-century philosophy. Although

---

[110] Jakob Bernoulli, *Ars conjectandi*, p. 249; Arnauld and Nicole, *Logique*, Part IV, chapters 12, 16.

[111] See Ivo Schneider, "Leibniz on the probable," in *Mathematical Perspectives*, Joseph Dauben, ed. (New York: Academic Press, 1981), pp. 201–219; also his "Why do we find the origin of a calculus of probabilities in the seventeenth century?" in *Pisa Conference Proceedings*, J. Hintikka, D. Gruender, and E. Agazzi, eds. (Dordrecht and Boston: Reidel, 1980), vol. 2, pp. 3–24.

comments on probability are scattered throughout Leibniz's writings and correspondence, the most interesting observations occur in his rebuttal to Locke's *Essay Concerning Human Understanding* (1689), the posthumous *Nouveaux essais sur l'entendement humain* (composed 1703–1705). Locke's own discussion "Of Probability" and the related "Of the Degrees of Assent" lay squarely within the premathematical tradition of probability and abounds with legal terms and analogies.[112] Between "certainty and demonstration, quite down to improbability and unlikeness, even to the confines of impossibility" Locke interpolated "degrees" of probability, associated with "degrees of assent from full assurance and confidence, quite down to conjecture, doubt, and distrust."[113] A proposition accrues probability from our own observation and experience, or from the testimony of others. Like the authors of the Port Royal *Logique* (which Locke had read and may have helped to translate into English), Locke devoted considerable space to the evaluation of testimony, invoking legal criteria: number of witnesses, their skill and integrity, and private interest.

Philalethe, the character representing Locke in the *Nouveaux essais*, repeated much of this discussion verbatim. Leibniz's mouthpiece Theophile seconded the claim that many judgments must be made in the "twilight of probability," but construed this constraint more optimistically. After all, Leibniz argued, jurists habitually make fine distinctions between presumptions, conjectures, and indices in order to arrive at a "provisional Truth." Leibniz combatted Locke's nominalism here and elsewhere by defending the reliability of probabilistic reasoning. Again and again, he supported this claim with legal examples of sound, albeit nondemonstrative, reasoning. In response to Philalethe's advice to "proportion our assent to degrees of probability," Theophile pointed out that jurists, with their elaborate hierarchy of full and half-proofs, presumptions, conjectures, and indices, had already made great strides in this direction: "And the entire form of these procedures in justice is in fact nothing else but a type of logic, applied to questions of law."[114] In the next breath, Theophile mentioned the contributions of Pascal, Huygens, and De Witt to a mathemat-

---

[112] John Locke, *An Essay Concerning Human Understanding* (1689; all editions dated 1690), collated and annotated by Alexander Campbell Fraser (New York: Dover Publications, 1959), vol. 2, Book IV, chapters 15–16.

[113] Locke, *Essay*, vol. 2, pp. 364–365.

[114] Leibniz had once planned to write a synopsis of Menochius (see n. 105, above), and was presumably well acquainted with the subject of legal indices. Allard, *Histoire*, p. 440.

ical version of this alternative logic. Jurists, aided by mathematicians, should serve as models for a new branch of logic based on probabilities.[115]

Although Leibniz's project for a logic of probabilities, like his plans for a Universal Characteristic and a perpetual peace, was never realized in the form he envisioned, the work of the early mathematical probabilists, particularly on the Continent where the Roman-canon tradition flourished, bore the stamp of legal influences. The most obvious example is Nicholas Bernoulli's *De usu artis conjectandi in jure* (1709), which applied mathematical probability to numerous legal questions. Like Liebniz, Nicholas Bernoulli was trained both as a mathematician and a jurist (*De usu* was his dissertation for the degree of doctor of jurisprudence at Basel), and presumed a community of interest and assumptions between the two disciplines.[116] Although they met with only mixed success even by contemporary standards, the attempts of John Craig and others[117] to apply the calculus of probabilities to witness credibility excited the interest of mathematicians through Poisson. Probabilists turned their new mathematical tools to older problems within what was perceived as a single coherent tradition.

The legal approach to probability, through only crudely quantitative, did furnish mathematical probabilists with a ready-made interpretation of their calculus, and also suggested applications to jurisprudence which remained a feature of classical probability theory until the mid-nineteenth century. *Pace* later critics, late seventeenth- and eighteenth-century probabilists did not condemn themselves to complete subjectivism by adopting the epistemic view of the jurists, which made probability a measure of belief. Following the jurists a step further, the probabilists assumed that rational belief sprang from evidence; stronger evidence, both from things

---

[115] Gottfried Wilhelm Leibniz, *Nouveaux essais sur l'entendement humain* (composed 1703–1705; publ: 1765), in *Sämtliche Schriften und Briefe*, Akademie der Wissenschaften zu Berlin, Sechste Reihe: Philosophische Schriften (Berlin: Akademie Verlag, 1962), vol. 6, pp. 460–465.

[116] Nicholas Bernoulli, *De usu*, chapter 9.

[117] John Craig, *Theologiae christianae principia mathematica* (London, 1699). For a sympathetic analysis of Craig's views, see Stephen M. Stigler, "John Craig and the probability of history: From the death of Christ to the birth of Laplace," Technical Report No. 165, Department of Statistics, University of Chicago (September 1984; revised February 1985). See also the anonymous "A calculation of the credibility of human testimony," *Philosophical Transactions of the Royal Society of London* 21 (1699): 359–365 (probably by George Hooper), which is noteworthy for its application of probabilistic expectation to the problem.

and from testimony, carried firmer conviction. Associationist psychology tightened the bond between evidence and belief by proposing a quasi-frequentist mechanism for belief.[118] From the standpont of the late seventeenth-century thinker, a systematic treatment of probability already existed within jurisprudence. The calculus of probabilities was just that: a calculus applied to extant legal approaches to partial certainty.

1.5 Conclusion

In this chapter I have argued from the problems and concepts contained in the early works on mathematical probability for the formative (and lingering) influence of two legal doctrines, aleatory contracts and partial proofs. By themselves, these doctrines hardly constitute an explanation of why probability theory emerged just when and how it did. It is true that both aleatory contracts and partial proofs attracted considerable legal attention in the sixteenth and seventeenth centuries—the one because of ever more widespread commercial activities like charging interest, and the other because of the criticisms of the natural-law jurists and because a decrease in torture (and the confessions thereby extracted) made the circumstantial evidence of indices more important.[119] But the fundamental arguments had been rehearsed for several centuries beforehand. Moreover, these legal doctrines say nothing about elements like combinatorics and statistics that also very soon came to play a leading role in classical probability theory. Not surprisingly, probability baffles attempts at a single-stranded explanation.

Instead, I have concentrated on the narrower problem of how some notions of probability came to be quantified more readily than others. Here, both expectation and evidence in the legal sense had the signal advantage that they were already expressed in numbers, expectation in money and evidence in fractions. These quasi-quantifications would hardly have been sufficient for mathematical treatment, but in the climate of near quanti-

---

[118] See Section 4.2 below.

[119] Noonan, *Usury*, chapter 10; Paul Foirers, "La conception de la preuve dans l'école de droit naturel," in *La preuve. Deuxième partie: Moyen Age et temps modernes. Recueils de la Société Jean Bodin pour l'histoire comparative des institutions* 17 (1965): 169–192; John H. Langbein, *Torture and the Law of Proof* (Chicago and London: University of Chicago Press, 1976), chapters 1–3.

phrenia that reigned in the latter half of the seventeenth century (recall Leibniz's Universal Characteristic or Borelli's iatro-mechanism) they were undoubtedly suggestive. Equally venerable and current probabilistic notions like verisimilitude, still prominent in the Port Royal *Logique*, had never been so formulated, and the mathematicians steered clear of them.[120] Of course, what the mathematicians touched, they changed drastically and irrevocably. By the time Nicholas Bernoulli took up the problems of expectation and evidence in his 1709 dissertation, both had assumed a new mathematical form never dreamed of by (and probably unacceptable to) the jurists. Expectation now clearly emerged as the product of probability and outcome value, and the evidentiary force of testimony was the proportion of times the truth was told to total times borne witness. But the fact remains that throughout the eighteenth century, mathematical probability was less a theory than a set of applications, and these applications often betrayed their legal pedigree in both problem and approach.

With the publication of the *Ars conjectandi*, these applications came to center quite explicitly on the problems of rational decision under uncertainty of all kinds. Although computing expectations for aleatory contracts and weighing trial evidence are both examples of such decisions, they hardly exhaust the field. This is a fortiori true of the one kind of problem the early probabilists were in fact equipped to solve, expectations for comparatively simple games of chance. In the next chapter we examine how reasoning by expectation, especially of the wagering sort, became paradigmatic for a new kind of rationality or "reasonableness" in the latter half of the seventeenth century. As a consequence, the potential domain of applications for mathematical probability widened considerably, but this very breadth placed the central concept of expectation under intolerable strain, for "reasonableness" turned out to be a highly ambiguous notion.

[120] See L. Jonathan Cohen, "Some historical remarks on the Baconian conception of probability," *Journal of the History of Ideas* 41 (1980): 219–231, for another seventeenth-century variety of nonmathematical probability.

CHAPTER TWO

Expectation and
the Reasonable Man

2.1 Introduction

Jakob Bernoulli entitled his treatise the "Art of Conjecture," and the rules he offers to guide deliberation in Part IV, chapter 2, often appeal more to judgment than to reckoning: take into account all the relevant factors; in doubtful but urgent matters choose the course of action that seems "safer, wiser, or more probable"; don't judge the value of human actions by their outcomes; and so forth. Conjecture remains more an art than a science. Yet like the Port Royal "Art of Thinking" that inspired Bernoulli, it was an art governed by rules. With more than a faint echo of René Descartes' *Rules for the Direction of the Mind*, the Port Royal authors sought to make logic more useful by "giving the rules for good and bad reasoning."[1] Bernoulli went further and tried to reduce at least some of these rules to calculation. In most cases he was in no better position than the Port Royal authors to put a number to the variables in his formulas, but he provided the beginnings of a mathematical rationale for deriving those numbers from statistics, and in the interim made do with estimates: the probability that *this* document was antedated, given the shady character of the notary, is 49/50; the chances of dying from a lightning bolt are less than one in two million; the probability that a man of Titius's constitution will die within ten years is 2/3. Thus encouraged, Bernoulli's eighteenth-century followers set about transforming his "art of conjecture" into a "calculus of probabilities," while retaining the program of applications suggested by his title. Conjecture in Bernoulli's sense meant making rational choices

---

[1] Antoine Arnauld and Pierre Nicole, *La logique, ou l'Art de penser* (1662), Pierre Clair and François Girbal, eds. (Paris: Presses Universitaires de France, 1965), p. 21.

under uncertainty—choices about which witness (or creed) to believe, which venture to invest in, which candidate to elect, what insurance premium to charge, which scientific theory to endorse. Since risk and uncertainty are our constant companions, this was effectively a rationality of daily life, or "reasonableness," as it was called. The classical theory of probability was intended by its practitioners as a mathematical model for this brand of mundane rationality, or in Pierre Simon Laplace's famous phrase, "good sense reduced to a calculus."[2]

A calculus was a considerably more ambitious project than the "rules" and "methods" that proliferated in late seventeenth-century philosophical texts and in the practical handbooks and manuals of the early eighteenth century. The rules guided judgment in every field from natural philosophy to horticulture to child rearing, but inevitably left much to the discretion of the reader. No general maxim could anticipate all the extenuating circumstances of the individual case, especially in the risk-laden situations that became the domain of mathematical probability. Estienne Cleirac's handbook on sea-going practices recommends adjusting insurance premiums according to the cargo, season, distance, latest reports of pirates, reputation of the captain, and other specifics;[3] Charles Du Moulin offers some rules of thumb for pricing annuities with the caveat that the "too great and continual uncertainty and risk" renders exactitude impossible, and at other points suggested recourse to a prudent judge instead;[4] the authors of the Port Royal *Logique* gloss Descartes' last two rules of method with an "autant qu'il peut," admitting that application may be difficult in specific cases;[5] Jakob Bernoulli warns that general proofs hold only for general events and that individual events demand more concrete details.[6]

The rule makers presumed a general but not unexceptionable regularity

[2] Pierre Simon Laplace, *Essai philosophique sur les probabilités* (1814), in *Oeuvres complètes*, Académie des Sciences (Paris, 1878–1912), vol. 7, p. cliii; English translation from Frederick W. Truscott and Frederick L. Emory, trans., *A Philosophical Essay on Probabilities* (New York: Dover, 1951), p. 196.

[3] Estienne Cleirac, *Les us, et coutumes de la mer* (Rouen, 1671), pp. 271–272.

[4] Charles Du Moulin, *Sommaire du livre analytique des contractz usures, rentes constituées, interestz & monnoyes* (Paris, 1554), f. 187r; Nicholas Bernoulli, *De usu artis conjectandi in iure* (1709), in *Die Werke von Jakob Bernoulli*, Basel Naturforschende Gesellschaft (Basel: Birkhäuser, 1975), vol. 3, pp. 304–305.

[5] Arnauld and Nicole, *Logique*, p. 334.

[6] Jakob Bernoulli, *Ars conjectandi* (1713), in *Die Werke*, vol. 3, p. 242.

in events: as Francis Bacon so vividly put it, nature trod her wonted paths day in and day out, but on occasion was known to stray into strange and secret byways.[7] For the calculus makers, however, nature's habits had become laws, and deviations were no longer permitted. Late seventeenth-century determinism rid mathematical probability of Gerolamo Cardano's daimon, and it also raised hopes that rules could be rid of exceptions. Rules, like Aristotle's idea of the probable, hold only "most of the time" and admit equivocal interpretations; calculations hold always and admit a unique solution.

Of course, this staunch determinism was a metaphysical conviction, which accounts for its resilience and fruitfulness. No mere empirical regularity could have been so persuasive or have justified the almost unbounded optimism of Enlightenment scientists, mathematicians, and philosophers seemingly bent on translating the whole of experience into the necessary grammar of mathematics. On the contrary: their deterministic faith often required that they close their eyes to or explain away the patent variability of every kind of data, even astronomical. But there is reason to believe that certain key areas of experience did in fact become more regular and predictable during this period. I have in mind here not only emblematic achievements like Newton's celestial mechanics and the prediction of the return of Halley's comet (although these were certainly of great propaganda value for the determinists), but also more prosaic matters like travel and mortality that touched everyone's life.[8] Even human conduct became less erratic, at least in the economic realm. Joyce Appleby has argued that "with the widening of the market, uniform and known prices replaced the face-to-face bargaining of the local market," and that these changes and the baffling economic success of the Dutch prompted seventeenth-century English writers to ascribe "to human nature a constancy that permitted them to treat it as a dependable factor in analysis."[9] Albert Hirschman traces the separation of the traditional canon of vices into two distinct sorts, passions and interests, in the sixteenth and seventeenth centuries: originally equally reprehensible, the interests came to be seen as

[7] Francis Bacon, *The Advancement of Learning* (1605), in *The Works of Francis Bacon*, Basil Montagu, ed. (London, 1831), vol. 2, p. 102.

[8] See Section 3.5 below.

[9] Joyce Oldham Appleby, *Economic Thought and Ideology in Seventeenth-Century England* (Princeton: Princeton University Press, 1978), pp. 79, 93.

ever more innocuous by European moralists, who preferred the predictable calculations of self-interest to the rash outbursts of passion.[10] Economic factors and moral teachings conspired to make human actions more calculating, and therefore calculable.

Some such belief in the regularity of human conduct was a prerequisite for the probabilist seeking to reduce good sense to calculus. Quantification presumed regularity and unambiguous answers, at least for eighteenth-century mathematicians. Nor could the probabilists be content with regularities of a purely prescriptive kind, that is, what the rational agent *should* do under the given circumstances. Although the line between prescriptive and descriptive objectives was never sharply drawn in classical probability theory, its practitioners were by and large committed to a descriptive account of rational choice among uncertain alternatives. Not everyone was rational, of course, and the probabilists studied only that small minority certified as reasonable. Once they had made explicit and mathematical the intuitions and judgments that guided this elite, the probabilists hoped to present their calculus as a prescriptive aid for the muddle-headed majority. Before the French Revolution, very few probabilists doubted that the rational sheep could be separated at a glance from the irrational goats, that the rational minority reasoned by implicit and immutable calculations, and that these calculations were the same for all rational people. Their deterministic faith made it possible for them to further assume that experience is regular and repetitive enough to make such inflexible calculation indeed a rational strategy. Mathematical probability was to be a formal description of that strategy, just as mathematical hydrodynamics was a formal description of fluids in motion.

This mandate to describe must be emphasized, for it shaped the development of classical probability theory in crucial ways. Like the "moral" or human sciences it was intended to supplement, the mathematical art of conjecture aimed at both a theoretical understanding of human conduct *and* a practical guide to action. Two sets of tensions strained this alliance between classical probability theory and the moral sciences. First, the descriptive and prescriptive objectives that directed the calculus of probabilities to both reflect and correct the psychology of decision making raised uncomfortable questions when the two conflicted. If the conduct of reasonable

---

[10] Albert Hirschman, *The Passions and the Interests* (Princeton: Princeton University Press, 1977), part 1.

men contradicted the mathematical results, which was at fault? Second, even if the conduct of reasonable men were acknowledged as the ultimate standard, what were the criteria for reasonableness? Both a learned magistrate and a canny merchant apparently belonged to that select company, but the dictates of equity and prudence did not always coincide.

Almost without exception, the eighteenth-century probabilists resolved the first dilemma by siding with the practice of reasonable men. If the mathematical results did not agree with practice, they rearranged or modified the mathematics to reconcile the two, as in the case of the St. Petersburg problem. Daniel Bernoulli was exceptional in championing mathematics over common sense: he argued on probabilistic grounds that, despite public apprehension, smallpox inoculation was a rational preventive measure. Jean d'Alembert rebuked Bernoulli not because he opposed smallpox inoculation, but because Bernoulli's life expectancy calculations conflicted with the actual psychology of risk taking. The second issue splintered mathematicians into opposing factions, each advocating somewhat different mathematical versions of reasonableness. Some emphasized equity, others prudence, still others the psychology of risk or the intuitive appreciation of physical possibility. Each slant yielded a different mathematical formulation of expectation, and the resulting proliferation of definitions prompted d'Alembert to challenge the validity of the theory of probability itself. However, none of these mathematicians, even critics like d'Alembert, ever relinquished the fundamental assumption that mathematical probability, in particular probabilistic expectation, should describe the way in which reasonable men conducted their affairs.

The tenacity with which eighteenth-century probabilists clung to this assumption, even when it threatened to undermine their theory at its foundations, stemmed from related assumptions regarding the goals of mathematics in general and probability theory in particular. Most eighteenth-century mathematicians devoted the bulk of their professional efforts to what was termed "mixed mathematics." Although the modern designation of "applied" mathematics comes closest, it does not precisely coincide with the eighteenth-century category. First, eighteenth-century mathematicians inverted the twentieth-century hierarchy that sets pure mathematics at the pinnacle of the discipline. Research in pure mathematics per se was rare and almost always pursued with an eye toward application, or, more accurately, realistic interpretations. The wave equation and first partial differential equations, for example, arose from d'Alembert's physical investigations: "D'Alembert's chief concern was in making this [mathematical] language not merely descriptive of the world, but congruent to it."[11] In the classification of human knowledge that prefaced the *Encyclopédie*, mixed mathematics took pride of place. Pure mathematics, comprising geometry and arithmetic, was dwarfed by mixed mathematics, which embraced all of mechanics, optics, acoustics, pneumatics, astronomy, and the "art of conjecture." Mixed mathematics dominated the discipline, both in prestige and proportion of published research.

Second, mixed mathematics included not only applications but whole disciplines that would now be classified as integral parts of physics or astronomy, such as celestial mechanics and hydrodynamics. Mathematics defined by the *Encyclopédie* consisted of three parts: pure mathematics, which studied quantity independent of both "real" individuals (e.g., the moon, giraffes) and "abstract" ones (e.g., color, heat, motion); mixed mathematics, which studied quantity as manifested in such individuals; and physico-mathematics, which studied quantity as it bore on the real and hypothetical causes of observed events.[12] All of mathematics, including pure mathematics, studied some*thing*: geometry treated extension; arithmetic (broadly defined to include algebra and the differential and integral calculus) was the science of the numerable. The many branches of mixed mathematics corresponded to all the varieties of phenomena that could be investigated with respect to quantity. Mixed mathematics did not simply translate the subject matter of physics or astronomy into the abstract, neutral language of pure mathematics: it was a genuine science in its own right, with subject matter as well as a powerful method.

Research in mixed mathematics was therefore predicated on a more intimate relation between the formal mathematical component and the class of real phenomena under investigation than that to which the modern discipline of applied mathematics would lay claim. In fact, eighteenth-century mathematicians would have found the distinction between the formal apparatus of mathematics and the subject matter it treated to be an alien one. For them, mathematics, even pure mathematics, did not exist without a real interpretation. Mathematics differed from the other sciences of nature only in the simplicity of its subject matter, and the consequent

[11] J. Morton Briggs, "D'Alembert," in *Dictionary of Scientific Biography*, C. C. Gillispie, editor-in-chief (New York: Scribner's, 1970), vol. 1, p. 116.

[12] Jean d'Alembert and Denis Diderot, eds., "Prospectus," in *Encyclopédie, ou Dictionnaire raisonné des sciences, des arts et des métiers* (Paris, 1751), p. 9.

certainty of its results. In d'Alembert's Lockean theory of knowledge, mathematics represented the farthest extreme of abstraction from immediate sensation, but it was nonetheless the end of a continuum. The mind operated on the rich, detailed particularity of sensation to subtract every feature until only the skeleton of number and extension remained. Reversing its course, the mind added back the abstracted features one by one, creating the various sciences at corresponding levels of abstraction, until the understanding reconstructed the original perception in its full complexity. For example, impenetrability and motion added to the magnitude and extension of mathematics resulted in the science of mechanics. Thus, even the most rarefied of the sciences, mathematics, was ultimately anchored in the concrete reality of sensation.[13] If other eighteenth-century mathematicians did not share d'Alembert's Lockean views, they concurred in assuming mathematics to be inseparable from its interpretation. Mixed mathematics brought the science of extended and numerable magnitude to bear on more specific aspects of experience.

As a branch of mixed mathematics, the "art of conjecture" concerned "*quantity* considered in the possibility of events,"[14] which d'Alembert elaborated into three divisions: chance, "la vie commune" (i.e., annuities, maritime insurance, etc.), and the "science de monde" (the art of conducting human affairs so as to maximize commercial advantage without violating moral obligations).[15] D'Alembert here followed Jakob Bernoulli in extending the "art of conjecture" beyond games of chance to civil society. Thus its domain overlapped with that of the moral sciences, as did its objective. The Enlightenment moral sciences were practical in orientation, aiming to derive guidelines for conduct and social reform. Even theorists committed to investigate the natural laws governing society, such as the physiocrats, sought these laws in order that social organization might be harmonized with the order of nature.[16] Consequently, the moral sciences could limit their investigations to exemplary, rather than typical, conduct in civil society. The precepts and practice of a select group of individuals

[13] D'Alembert, "Discours préliminaire," in *Encyclopédie*, vol. 1, pp. v–vii.

[14] D'Alembert and Diderot, "Prospectus," p. 9.

[15] D'Alembert, *Essai sur les élémens de philosophie*, in *Oeuvres* (Paris, 1821), vol. 1, pp. 156–158.

[16] François Quesnay, "Despotisme de la Chine" (1767); "Le droit naturel" (1765), in Auguste Oncken, ed., *Oeuvres économiques et philosophiques de Quesnay* (Paris, 1888), pp. 645–646, 362–365.

distinguished by their prudence and perspicacity sufficed to lay the foundations of a moral science conceived as a code of action. Similarly, the calculus of probabilities would quantify the rules of conjecturing "want to dictate to any man of sound mind, and which are also continually observed by more judicious men in the experience of civil life."[17]

Once the calculus of probabilities adopted the conduct of reasonable men as its subject matter, the descriptive orientation of mixed mathematics required that probabilists accept that conduct as the final test of its results. If a mathematical theory of lunar motion failed to predict the observed orbit, mathematicians challenged the theory. Similarly, when the conventional solution to the St. Petersburg problem ran counter to the judgment of reasonable men, probabilists reexamined their definition of expectation. Mathematical expectation expressed a theory about the way in which reasonable men comported themselves in the face of uncertainty, just as the mathematical formulation of the law of gravitation expressed a theory about ponderable matter. As mixed mathematics, neither had any independent claim to validity beyond describing its designated class of phenomena mathematically. In the minds of eighteenth-century mathematicians, there did not exist any theory of probabilities disembodied of subject matter.

There was, of course, no necessary connection between the calculus of probabilities and the judgments of reasonable men. As Kenneth Arrow has noted, the association was historical rather than logical,[18] although still critical for both social theory and mathematical probability of the period. A new approach to certainty, contemporary with the emergence of mathematical probability in the late seventeenth century, consolidated the alliance between classical probability theory and reasonable conduct.

In the latter half of the seventeenth century, contemporary with the advent of mathematical probability, another strain of probabilistic reasoning emerged in the writings of proponents of rational theology and the new natural philosophy which later became closely associated with the classical theory of probability. These writers, including Robert Boyle, John Wilkins, Joseph Glanvill, Marin Mersenne, Pierre Gassendi, and Hugo Grotius, simultaneously insisted upon the incorrigible uncertainty

[17] Jakob Bernoulli, *Ars conjectandi*, p. 241.

[18] Kenneth J. Arrow, "Formal theories of social welfare," in *Dictionary of the History of Ideas*, Philip Wiener, editor-in-chief (New York: Scribner's, 1973), vol. 4, pp. 276–284, on p. 277.

of almost all human knowledge and on our ability to nonetheless attain to inferior degrees of "physical" and "moral" certainty. They hoped to bridge the chasm between the absolute doubt of the skeptics and the dogmatic certainty of the scholastics. At the heart of their arguments lay a new definition of rationality, one based on the practical aspects of everyday life. A proof for a hypothesis in natural philosophy or the precepts of Christianity need not achieve mathematical rigor, but only that threshold of certainty sufficient to persuade a reasonable man to act in daily life.

Although these apologies for qualified belief were for the most part non-mathematical, some of their most frequently deployed arguments, such as Pascal's wager concerning the existence of God, appealed to the idea of expectation. The most popular illustrations, many of them drawn from the list of aleatory contracts, not only pitted one risk against another, but also compared outcome values. Maximizing expectation, often described in the crassest economic terms, was central to the new reasonableness of these worldly apologists for other-worldly values. A mathematical treatment of expectation was therefore bound to find its way into the discussions of practical rationality.

Like the mathematical probabilists, the philosophical moderates often took legal reasoning as their model of rational judgment based on incomplete proof, and made expectation the measure of prudent action in situations complicated by uncertainty. However, they sometimes parted company with the early probabilists and the jurists in substituting prudent self-interest for equity as the ultimate criterion of choice in the face of uncertain alternatives. The reasonable man, whose seasoned practical judgment guided belief in religion and science, was ambiguously identified with both the equitable judge and the shrewd merchant. Although this equivocation eventually split mathematicians into two camps over the issue of mathematical versus moral expectation, classical probabilists through Poisson hewed to the belief that their calculus was the reasonable man mathematized. Thus, the late seventeenth-century proponents of reasonableness not only campaigned for probabilistic reasoning on a greatly expanded scale while at the same time underscoring the centrality of expectation; they also bound the calculus of probabilities to a new, pragmatic notion of rationality that influenced the development of classical probability theory.

This chapter is divided into two parts: Section 2.2 traces the origins of the identification of classical probability theory with the good sense of reasonable men; Section 2.3 follows the fortunes of the concept of probabilistic expectation in the work of classical probabilists through Siméon-Denis Poisson. [19] True to the maxims of mixed mathematics, the classical probabilists accepted their mandate to tailor their mathematical results to fit the actual conduct of reasonable men. Expectation was to be the mathematical expression of that mundane rationality, the core of a general, quantitative, and prescriptive science of civil society. However, the ambiguous notion of reasonableness embroiled mathematicians in a dispute over rival definitions of expectation, mathematical and moral. The controversy widened as conflicting strains within the moral sciences subtly altered the accepted understanding of "reasonableness," and with it the mathematical formulation of expectation. Not until the French Revolution shattered the tacit consensus that the conduct of an elite of reasonable men should comprise the subject matter of both the calculus of probabilities and the moral sciences did the protracted debate over expectation gradually subside among mathematicians.

2.2 Expectation as Reasonableness

The advent of the mathematical theory of probability during the latter half of the seventeenth century coincided with the rise of a new philosophical defense of belief that conceded the inevitably imperfect state of human knowledge and reason to the skeptics, but nonetheless affirmed the existence of rational (as opposed to fideist) grounds for belief. [20] These philosophical moderates acknowledged that in science, religion, and other less lofty matters absolute certainty lay beyond the reach of human understanding. However, where reason faltered, prudence prevailed. The world of daily affairs was also fraught with uncertainty, but no reasonable man could afford to suspend judgment and action until all risk disappeared. Instead, judgment and past experience combined to form expectations that guided practice and justified belief. If the certainty of the rationalists was beyond human ken, degrees of certainty were still within grasp. The de-

[19] Some of this material has appeared previously in Lorraine J. Daston, "Probabilistic expectation and rationality in classical probability theory," *Historia Mathematica* 7 (1980): 234–260.

[20] Richard Popkin, *The History of Scepticism from Erasmus to Descartes*, rev. ed. (Assen, The Netherlands: Van Gorcum, 1964), chapters 1–2, 7.

gree of certainty sufficient to impel a "reasonable man" to action on his own behalf was, the moderates argued, also a sufficient warrant for belief and action in the contested areas of religion, science, and law. Reason per se demanded demonstration, but reasonableness was content with probabilities, or rather, with expectations.

The new doctrine of reasonableness became the in-house philosophy of an influential circle of natural philosophers and theologians. Marin Mersenne, Pierre Gassendi, Hugo Grotius, Antoine Arnauld, Robert Boyle, John Tillotson, Joseph Glanvill, and John Wilkins were of their number. Barbara Shapiro has collected numerous examples of this doctrine from the seventeenth-century English literature on history, law, religion, and natural philosophy.[21] Among the English theologians of this stripe, latitudinarians figure prominently. The latitudinarians viewed economic self-interest as the mainspring of human action, and deliberately pitched both tone and content of their sermons to, in Isaac Barrow's words, "men of business and dispatch." Patterning their arguments for Christianity on the reasoning of the law court and the marketplace, the latitudinarians shored up their points with examples drawn from daily affairs. They set new standards for rational belief in science and law as well as in religion: the empiricist modesty of Boyle and other spokesmen for the new philosophy; the doctrine of "reasonable doubt" applied to judicial decisions; the natural theology of the Boyle lectures.[22]

This worldly philosophy eschewed certainty outside of mathematics and metaphysics, but nonetheless sought to persuade. Where demonstration was not practicable, the salutary example of "reasonable men," industrious and successful, carried conviction. A new religious apologetic, brisk in tone and curt in precept, replaced the "Questions and Speculations that engender Strife" and the oratorical flourishes that embellished them. Proponents of rational theology and the new natural philosophy—and these

[21] Barbara J. Shapiro, *Probability and Certainty in Seventeenth-Century England* (Princeton: Princeton University Press, 1983); also her *John Wilkins, 1614–1672* (Berkeley and Los Angeles: University of California Press, 1969), pp. 224–248.

[22] See James Jacob, *Robert Boyle and the English Revolution* (New York: Burt Franklin, 1977), pp. 156–159; Margaret C. Jacob, *The Newtonians and the English Revolution, 1689–1720* (Ithaca: Cornell University Press, 1976), chapter 1; Barbara Shapiro, "Law and science in seventeenth-century England," *Stanford Law Review* 21 (1969): 727–766; Theodore Waldman, "Origins of the legal doctrine of reasonable doubt," *Journal of the History of Ideas* 20 (1959): 299–316; Keith M. Baker, *Condorcet: From Natural Philosophy to Social Mathematics* (Chicago: University of Chicago Press, 1975), part 3.

were often the same people[23]—emphasized that though complete certainty in matters of faith, science, and civil society was unattainable, reasonable assurance—"moral" and "physical" certainty—could be had. This polemical strategy had been elaborated by Hugo Grotius in his *De veritate religionis christianae* (1624) and eagerly exploited by admirers like William Chillingworth in his *Religion of the Protestants* (1638).[24]

Moreover, it was emphasized that it was the feebler assurance of moral certainty, not the ironclad guarantees of demonstration, which made the "world go 'round." The Royal Society theologians accommodated their chosen audience of practical men by redefining rationality in their auditors' own terms, as the pursuit of personal advantage. Historians such as Margaret and James Jacob have remarked upon the repeated appeals to self-interest in latitudinarian rhetoric, but have not analyzed the care Wilkins, Tillotson, Boyle, and other apologists took to define this tough-minded rationality of practical men. Self-interest meant not so much the headlong quest for worldly gain, but rather the deliberate weighing of probabilistic expectations. The archetypal reasonable man appraised risks as well as the profits or losses which turned thereupon.

Blaise Pascal advanced the most famous version of this expectation argument, commonly known as Pascal's wager, in his *Pensées* (1669) in order to persuade worldly doubters that religious observance was in their own best interests. Given Pascal's premises, Christianity promised maximum expectation. Although recent analyses have vindicated Pascal's reasoning (though not of course his premises) from the standpoint of modern decision theory,[25] and duly noted the links with his mathematical work on probability, none of these has explored the close connection between rational decision making and early probability theory. Arguments of the Pascal type, which likened the decision to believe (be it in God, the theory of gravitation, or the guilt of the accused) to decisions to invest or act prudently in the world of affairs, hinged on three assumptions: first, that in almost all cases human knowledge is in principle imperfect; second, that

[23] Shapiro, *Wilkins*, p. 229; also Shapiro, *Probability*, chapters 2–3.

[24] See Henry Van Leeuwen, *The Problem of Certainty in English Thought, 1630–1690* (The Hague: Martinus Nijhoff, 1963), pp. 15–32.

[25] Ian Hacking, "The logic of Pascal's wager," *American Philosophical Quarterly* 9 (1972): 186–192.

[26] Blaise Pascal, *Pensées*, Louis Lafuma, ed. (Paris: Editions du Seuil, 1962), no. 418 ("Infini-Rien"), pp. 187–190.

whatever one's intellectual scruples about belief without demonstrative certainty, the exigencies of daily life obliged one to act upon such tentative beliefs; and third, that rational self-interest was synonymous with maximizing expectation in such uncertain situations. With no choice but to play the game ("il faut parier"), Pascal argued that one might bet a certain sum against an uncertain gain "without sinning against reason," because the certainty of the wager and the uncertainty of the outcome were not infinitely separated but rather proportioned to the "chances of gain and loss."[26] To be uncertain was therefore not to be necessarily irrational: reasoning by expectation constituted a new brand of rationality.

Pascal's was not the first published version of the wager. The *Pensées* were first published posthumously in 1669, and went through numerous French editions. The first English translation by Joseph Walker, with a dedicatory epistle to Robert Boyle, appeared in 1688,[27] but French editions circulated among members of the Royal Society during the mid-1670s.[28] The Port Royal *Logique* (1662) included a similar argument, without the distinctive gambling analogy, in its final section on expectation.[29] John Tillotson had set forth a nearly identical argument in a 1664 sermon "On the Wisdom of Being Religious," and may well have been repeating a still earlier one. Tillotson took atheists to task for, among other things, demanding "more evidence for things than they are capable of," and cited with approval Aristotle's dictum that different proofs and evidence should be matched to different subject matters. Tillotson's successors would ring numerous changes on his fourfold distinction among the different types of proof appropriate to mathematics, natural philosophy, moral subjects, and matters of fact, but all insisted upon Tillotson's conclusions that although the last three do not admit mathematical demonstration, "yet we have an undoubted assurance of them."[30] Atheism, Tillotson continued, was imprudent as well as unreasonable, for the atheist risked "his eternal interest," while the believer "ventures only the loss of his lusts." So far, despite the gambling vernacular, the argument re-

[27] Albert Maire, *Pascal philosophe*, vol. 4, *Les Pensées-Editions originales, ré-impressions successives* (Paris: Giraud-Badin, 1926), pp. 195–196.

[28] John Barker, *Strange Contrarieties: Pascal in England During the Age of Reason* (Montreal and London: McGill-Queens University Press, 1975), pp. 48–57.

[29] Arnauld and Nicole, *Logique*, p. 355.

[30] John Tillotson, "On the wisdom of being religious" (1664), in *Works* (Edinburgh and Glasgow, 1748), vol. 1, p. 32.

traced the traditional lines of Christian apologetics, which contrasted infinite happiness to infinite misery (or, at best, annihilation)—in other words, a simple comparison of outcome values. However, Tillotson restated his case in unmistakable expectation terms, combining both probabilities and outcome values, in an example precisely equivalent to Pascal's cosmic game of Croix-ou-Pile:

> So that, if the arguments for and against a God were equal, and it were an even question, Whether there were one or not? yet the hazard and danger is so infinitely unequal, that in point of prudence and interest every man were obliged to incline to the affirmative; and whatever doubts he might have about it, to chuse [sic] the safest side of the question, and to make that the principle to live by. For he that acts wisely, and is a thoroughly prudent man, will be provided against all events, and will take care to secure the main chance whatever happens.[31]

Drawing his examples from everyday life, Tillotson pointed out that since men daily risked their lives and fortunes "only upon moral assurance" in such ventures as trading expeditions, consistency and self-interest must "persuade a reasonable man to venture his greatest interest in this world upon the security he hath of another," even if that security were only a probability. No longer were worldly practices and rationales distinct from—and subordinate to—spiritual appeals to reason or to faith. In Tillotson's scheme, religion took its cue from temporal self-interest, rather than from rational dogmatism or mystical enthusiasm.

Boyle and Wilkins reproduced this argument in print as early as 1675. Whether or not it was adapted from Tillotson, Pascal, or some source, the argument in any case spread swiftly to the works of other Royal Society theologians. Whatever its origins, the argument from expectation became a standard feature of religious apologetics among natural theologians, especially but not exclusively in Britain.

British apologists from Tillotson to Bishop Joseph Butler regarded the problem primarily in terms of sufficient evidence—in Boyle's words, "what degree of evidence may reasonably be thought sufficient, to make the Christian religion thought fit to be embraced"[32]—and often substi-

[31] Tillotson, "Wisdom," pp. 40–41.

[32] Robert Boyle, *Some Considerations about the Reconcileableness of Reason and Religion* (1675), in *Works of the Honourable Robert Boyle* (London, 1772), vol. 4, p. 182.

tuted legal analogies for Pascal's game of chance. All of these treatises began by ordering the types of demonstration in a hierarchy of descending certainty. Boyle distinguished among metaphysical, physical, and moral proofs; Glanvill contrasted "infallible" to "indubitable" certainty; Wilkins identified metaphysical, physical, and moral degrees of certainty. Although the distinctions were not identical, they overlapped to a large extent. Moreover, they served the same ends: to interpolate intermediate stages between the two extremes of certainty and doubt. In contrast to Pascal's wager, these attempts to span the "infinite distance" between the two poles were based on the types of evidence corresponding to each level of certainty, rather than upon the mathematical proportion of hazards. They frequently set the expectation argument within its original legal context; at other times, they appealed to a commercial context.

Boyle's *Some Considerations about the Reconcileableness of Reason and Religion*, originally published in 1675 as *Possibility of the Resurrection*,[33] is typical of the evidentiary argument from expectation, and later renditions were probably patterned upon it. According to Boyle, metaphysical certainty derived from axioms that "can never be other than true"; physical demonstration was "deduced from physical principles" which God might choose to alter or suspend; and moral proofs entered "where conclusion is built, either upon some one such proof cogent in its kind, or some concurrence of probabilities, that it cannot be but allowed, supposing the truth of the most received rules of prudence and principles of practical philosophy." By "concurrence of probabilities," Boyle did not mean mathematical probabilities, but rather the additional conviction carried by convergent evidence. He took the conjunction of legal testimony as his paradigm case:

> For though the testimony of a single witness shall not suffice to prove the accused party guilty of murder; yet the testimony of two witnesses, though but of equal credit . . . shall ordinarily suffice to prove a man guilty; because it is thought reasonable to suppose, that, though each testimony single be but probable, yet a concurrence of such probabilities (which ought in truth to be attributed to the truth of what they jointly tend to prove) may well amount to a moral certainty, i.e. such a certainty, as may warrant the judge to proceed to the sentence of death against the indicated party.[34]

[33] John F. Fulton, *Bibliography of the Honourable Robert Boyle* (Oxford, 1961), pp. 86–87.
[34] Boyle, *Considerations*, p. 182.

This passage simultaneously captures the fluid sense of the word "probability" during this period and epitomizes Boyle's principal argument. On the one hand, "probability" harkens back to the older, slightly disparaging sense of "opinion," as opposed to certainty: "though each testimony single be but probable." On the other hand, Boyle believed, along with the jurists, that such probabilities can be added together until they "amount to a moral certainty," lying somewhere between perfect certainty and total ignorance. Moral certainty, for Boyle and the Royal Society theologians as for the Port Royal authors and Jakob Bernoulli, governed civil society as "the surest guide, which the actions of men, though not their contemplations, have regularly allowed them to follow." If moral certainty was sufficient to persuade a judge to pass sentence, a ship's captain to throw cargo overboard during a storm, or the victim of gangrene to sacrifice a limb in order to save his life, then, Boyle contended, the same moral certainty should impel men to embrace Christianity, even at the expense of forsaking worldly pleasures.

Reminding his readers that Christianity not only promises a heaven to the faithful, but also threatens a hell for infidels, Boyle warned that the same prudence that informed decision in human affairs plagued by "divers hazards, or other inconveniences" must dictate religious belief. The key question was not, Boyle concluded, whether Christianity was true or false, ". . . but whether it be more likely to be true, than not to be true, or rather, whether it be not more advisable to perform the conditions it requires upon a probable expectation of obtaining the blessings it promises, than by refusing to run a probable hazard of incurring such great and endless miseries, as it preemptorily threatens."[35] The criterion for belief was not reason but what Boyle called "reasonableness"; not metaphysical or even physical, but moral certainty. Moral certainty sprang from action and prudence, as opposed to contemplation and demonstration. Boyle maintained that Christianity "requires not of us actions more imprudent, than divers others, that are generally looked upon as complying with the dictates of prudence." Prudence was to be measured against the standard set by men universally respected for their practical sagacity, those who systematically pursued the maximum "probable expectation." John Craig made this equation or rationality and expectation an axiom of his *Theologiae christianae principia mathematica* (1699):

[35] Boyle, *Considerations*, p. 185.

The endeavors of wise men are in direct proportion to the true value of their expectations. Any man who follows this calculation of endeavors is most wise; and he who follows it less exactly is rated less wise.[36]

Boyle and his fellow apologists for rational or natural theology framed their arguments for the reasonable man, one "who hath but an ordinary *capacity*, and an *honest 'mind*; which are no other qualifications than what are required to the institution of men, in all kinds of Arts and Sciences whatsoever."[37] If at least the same degree of expectation obtained in theological matters as in practical ones, then no reasonable man could refrain from faith. Like the authors of the Port Royal *Logique*, the "constructive sceptics"[38] equated prudent conduct with the weighing of alternative expectations. Expectations served as the yardstick against which almost all human beliefs were to be gauged. The decisions of the judge or merchant or natural philosopher were no longer viewed as pale copies of absolute certainty, but as ideals that carried their own imprimatur of certainty suited to their subject matter. As Willem 'sGravesande declared in his *Mathematical Elements of Natural Philosophy*, "We must look upon as true, whatever being deny'd would destroy Civil Society, and deprive us of the means of living."[39]

Expectation made it possible to compare the prosaic certainty of daily life with that attainable in questions of religion and natural philosophy. All might agree upon what constituted prudent conduct in the practical sphere, and that such conduct sufficed to guide all decisions involving uncertainty. Since the consensus among seventeenth-century philosophers, even optimists such as Bacon, had been that the human intellect was hemmed in with limitations and debilities of all kinds, this meant virtually all decisions, ranging from the mundane to the spiritual. However, problems arose in translating this pragmatic criterion to cases remote from daily life. The concept of expectation reduced prudence and sound

---

[36] John Craig, "Craig's rules of historical evidence," trans. from *Theologiae christianae principia mathematica* (London, 1699), in *History and Theory*, Beiheft 4 (The Hague: Mouton, 1964), p. 3.

[37] John Wilkins, *Of the Principles and Duties of Natural Religion*, 4th edition (London, 1699), p. 2.

[38] Popkin, *Scepticism*, chapter 7.

[39] Willem 'sGravesande, *Mathematical Elements of Natural Philosophy*, J. T. Desaguliers, trans., 5th edition (London, 1737), vol. 1, p. xv.

judgment to a common measure which could then be applied to more esoteric concerns. All reasonable men would invest in a mercantile venture with sufficiently high expectation; if Christianity achieved or exceeded that expectation it commanded assent as well. Expectation was nothing more than reasonableness rendered in mathematical—hence general—terms. The calculus of probabilities adopted the prequantitative sense of expectation used by Tillotson, Boyle, and their circle as a measure of good sense. In the form of probabilistic expectation, reasonableness could be applied to areas where disagreement reigned for want of a clear-cut criterion of belief. The calculus of expectations became a calculus of consensus.

A quantitative measure of expectation would make both comparisons and consensus possible. John Wilkins buttressed the "indubitable" certainty of moral matters with a system of *"Postulates, Definitions, and Axioms"* which laid down the rules for expectational reasoning: a greater good was to be preferred to a lesser one, and both certain and probable events might "fall under computation, and be estimated as to their several degrees."[40] Wilkins pointed out that this type of reasoning underpinned all manner of commercial and legal dealings, and proposed an example of how values could be estimated for "the fourth or fifth Expectant for an Inheritance" though in such Cases there be the odds of Three or Four to one: yet the price that is set upon this may be proportioned, as either to reduce it to an equality, or make it a very advantageous bargain."[41] In this way the "Rules of Prudence" might be made precise enough for general use in cases that required a rank ordering of expectations, rather than the simple equality of a fair contract. Moreover, numerical arguments enforced agreement in ways that rhetoric could not. Gottfried Wilhelm Leibniz's enthusiasm for his Universal Characteristic stemmed in part from a similar faith in the power of mathematics to command assent. Amidst the endless strife of academic, religious, and national factions, mathematicians alone agreed among themselves. Leibniz envisioned a day when all disputes would be settled by computation: "Let us calculate, Sir!" would be the challenge thrown down to opponents.[42] The proponents of political arithmetic held

---

[40] Compare Jacob Bernoulli, *Ars conjectandi*, part IV, chapter 2.

[41] Wilkins, *Principles*, p. 14.

[42] Leibniz, "Preface to the General Science" (1677), and "Towards a Universal Characteristic" (1677), in Philip Wiener, ed., *Leibniz Selections* (New York: Scribner's, 1951), pp. 15, 23.

similar opinions on the coercive power of number.[43] Probabilists also hoped to compel reasonable assent once reasonableness had been reduced to number.

It was expectation, rather than either probability or outcome value alone, which interested these apologists. Prudent decisions and aleatory equity balanced both factors to deal judiciously with uncertain situations. The wager argument would have been meaningless if it had considered only the probability of God's existence, and a cliché if it had only compared outcome values. In most cases, quantitative estimates for risk were impossible to come by except in the simplest case of games of chance, and even then only under idealized conditions. Mathematicians were quick to realize that statistics might provide a source of such estimates, but their eagerness far outran the available data. Apologists such as Wilkins hoped for a genuine calculus of expectations, but because they were chiefly concerned with decisions among a few alternatives, they could make do with an ordinal system instead. Because it figured so prominently in the "Rules of Prudence," and because those men "whom we esteem the most wise and the most honest"[44] were presumed capable of assessing its value intuitively, expectation remained a focus of probabilistic thinking even after the traces of legal equity had faded from the mathematical theory.

The intertwined themes of equity, prudence, and reasonableness shaped the career of classical probability theory. Taking the reasonable man as their model, the probabilists sought to translate the notions first of equity and later of prudence into mathematical terms. From these mathematical elements they hoped to construct a calculus of rational assent founded on the practical necessity of acting in the face of uncertainty. M.J.A.N. Condorcet's "social mathematics," for example, commanded agreement by freeing less sagacious men from the psychological tyranny of "vague and automatic" impressions, and so belongs to the mainstream of classical probability theory. So long as probabilists defined their task as a mathematical description of good sense, and so long as reasoning by expectation was judged synonymous with good sense, expectation continued to play an important role in the calculus of probabilities.

---

[43] See Peter Buck, "Seventeenth-century political arithmetic: Civil strife and vital statistics," *Isis* 68 (1977): 67–84.

[44] Wilkins, *Principles*, p. 15.

2.3 The Debate over Expectation

Expectation, and with it the calculus of probabilities, thus became bound up with common, albeit equivocal, reason. Laplace concluded his *Essai philosophique sur les probabilités* (1814) by underscoring this connection:

> It is seen in this essay that the theory of probabilities is at bottom only common sense reduced to calculus; it makes us appreciate with exactitude that which exact minds feel by a sort of instinct without being able ofttimes to give a reason for it.[45]

When mathematical results clashed with good sense, eighteenth-century probabilists anxiously reexamined their premises and demonstrations for inconsistencies. However, the questions of "What kind of good sense?" and "Whose?" were seldom confronted explicitly by probabilists between Jakob Bernoulli and Laplace, although the problems that exercised them ultimately reduced to these questions. With one or two notable exceptions, they assumed that good sense was monolithic, and that it was self-evident who had it and who didn't, at least to members of the same select club. In fact, the mathematical descriptions of good sense they advanced vacillated between legal and economic models, between fairness and fiscal prudence, with substantial borrowings from jurisprudence in the one case and political economy in the other. These descriptions inevitably contradicted one another, and the contradictions in turn stimulated analyses of equiprobability assumptions based on indifference, and of the psychology of risk taking. Although the meanings of both probabilistic expectation and good sense shifted in the course of these long debates, mathematicians persisted in identifying the two until well into the nineteenth century.

This is why the St. Petersburg paradox, trivial in itself, triggered an animated debate over the foundations of probability theory.[46] Unlike most mathematical paradoxes, the contradiction lay not between discrepant mathematical results reached by methods of apparently equal validity, but rather between the unambiguous mathematical solution and good sense.

---

[45] Laplace, *Essai*, in *Oeuvres*, vol. 7, p. cliii; trans. from Truscott and Emory, *Essay*, p. 196.

[46] For a comprehensive treatment of the history of the St. Petersburg problem, see Gérard Jorland, "The St. Petersburg Paradox (1713–1937)," in *The Probabilistic Revolution*, vol. I, *Ideas in History*, Lorenz Krüger, Lorraine J. Daston, and Michael Heidelberger, eds. (Cambridge, Mass.: MIT Press, 1987), pp. 157–190.

The problem, first proposed by Nicholas Bernoulli in a letter to Pierre Montmort and published in the second edition of the latter's *Essai d'analyse sur les jeux de hazard* (1713), belonged to the staple category of expectation problems. Two players, A and B, play a coin-toss game. If the coin turns up heads on the first toss, B gives A two ducats; if heads does not turn up until the second toss, B pays A four ducats, and so on, such that if heads does not occur until the nth toss, A wins 2^{n-1} ducats. Figured according to the standard definition of expectation, A's expectation is infinite, for there is a finite, though vanishingly small probability that even a fair coin will produce an unbroken string of tails, and the payoff always grows apace. Therefore, A must pay B an infinite amount (E = expectation and therefore the stake) to play the game:

$$E = 1/2(1) + 1/4(2) + 1/8(4) + \ldots 1/2^n(2^{n-1}) + \ldots$$

However, as Nicholas Bernoulli and subsequent commentators were quick to point out, no reasonable man would pay more than a very small amount, much less a very large or infinite sum, for the privilege of playing such a game. The results of the standard mathematical analysis clearly affronted common sense; hence the "paradox." The divergent solutions proposed by eighteenth-century probabilists to this dilemma reflected the tension between the equitable and prudential connotations of expectation.

By the beginning of the eighteenth century, expectation-based treatments of probability had begun to be superseded by more explicitly probabilistic ones. By 1709, Nicholas Bernoulli could reproach his fellow jurists for their imprecise understanding of expectation, holding up the mathematical "art of conjecture" as a surer guide to equity in aleatory contracts, particularly annuities.[47] In this new scheme, probabilistic expectation defined the terms for an equitable contract rather than the reverse. However, Nicholas Bernoulli still upheld the older identification of equal expectation with equitable contracts and exchanges; he simply proposed to substitute computation for qualitative reasoning and probabilities per se for expectations. Expectation retained its legal overtones of equity.

For Nicholas Bernoulli, expectation remained "the foundation of this Art [of Conjecture] as a whole," comparable to the Rule of Alligation in arithmetic or the determination of the center of gravity in statics. The "universal rule" of expectation was the mean proportion, representing the

[47] Nicholas Bernoulli, *De usu*, pp. 303–314.

maxim which enjoined one "to follow the mean in doubtful and unclear matters."[48] In the spirit of the golden mean, expectation moderated between the best to be hoped for and the worst to be feared. In contracts involving the purchase of a "venture"—for example, some future cast of the fisherman's net—or annuity, Nicholas emphasized the importance of equalizing the price and the expectation of the buyer. He argued against those who had claimed that the uncertainty of the outcome was in itself a legal justification of the agreement. In the case of annuities, he concluded, "With regard to the justice of this contract, it is required that the condition holding between buyer and seller be equal, and both be placed in equal danger of losing."[49] Trained jurist that he was, Nicholas upheld the equitable interpretation of expectation against his cousin Daniel's introduction of the prudential "moral" expectation.

In 1738, Daniel Bernoulli[50] published a resolution of the paradox in the annals of the Academy of St. Petersburg. Historians of economic theory regard this memoir as the earliest expression of the concept of economic utility, but Daniel Bernoulli and his colleagues considered the memoir to be an important contribution to the mathematical theory of probability.[51] Bernoulli's analysis did more than rechristen the problem in honor of the Academy; it also set the acceptable terms of solution for his successors. Although other mathematicians challenged the specifics of Bernoulli's approach, all agreed that the paradox was a real one that threatened to undermine the calculus of probabilities at its foundations; that the definition of expectation was the nub of the problem; and that a satisfactory solution must realign the mathematical theory with common sense.

Bernoulli's strategy was to distinguish two senses of expectation, one "mathematical" and the other "moral." Mathematical expectation corresponded to the classical definition of expectation, as the product of the probability and value of each possible gain or loss. His discussion of this

---

[48] Nicholas Bernoulli, *De usu*, p. 291.

[49] Nicholas Bernoulli, *De usu*, p. 305.

[50] For a full account of Daniel Bernoulli's work in probability theory, see O. B. Sheynin, "Daniel Bernoulli's work on probability," *Rete* 1 (1972): 273–300.

[51] Daniel Bernoulli, "Specimen theoriae novae de mensura sortis," *Commentarii academiae scientiarum imperialis Petropolitanae* 5 (1730–31; publ. 1738): 175–192; English trans. by L. Sommer, "Exposition of a new theory on the measurement of risk," *Econometrica* 22 (1954): 23–36 (all further references are to the Sommer translation). For Bernoulli's role in economic theory, see Emil Kauder, *A History of Marginal Utility Theory* (Princeton: Princeton University Press, 1965).

type of expectation was thoroughly legal in tone. He observed that the standard definition ignored the individual characteristics of the risk takers in question, and that it was premised on the equitable assumption that everyone encountering identical risks deserved equal prospect of having his "desires more closely fulfilled."[52] "Mathematical" expectation was thus identified with the legal context of aleatory contracts, balancing uncertain expectations against an immediate and certain amount: it quantified the jurist's intuitive "Sort of Equality" obtaining among parties to such contracts, retaining the vocabulary and aims of legal equity.

Bernoulli proposed to shift the perspective in his treatment of the St. Petersburg problem from that of a "judgment" of equity pronounced "by the highest judge established by public authority" to one of "deliberation" by an individual contemplating a risk according to his "specific financial circumstances." Once the concept of expectation was transplanted from a legal to an economic framework, the earlier definition lost its relevance. Whereas mathematical expectation had been purposefully defined to exclude personal circumstances that might prejudice the judicial assumption of the equal rights of contracting parties, fiscal prudence required some consideration of just such specifics.

For example, the plight of a poor man holding a lottery ticket with 1/2 probability of winning 20,000 ducats, Bernoulli argued, was in no way symmetric to that of a rich man in the same position: the poor man would be foolish not to sell his ticket for 9,000 ducats, although his mathematical expectation was 10,000; the rich man would be ill-advised not to buy it for the same amount. Bernoulli maintained that a new sort of "moral" expectation must be applied to such cases in order to bring the notion of value (i.e., the measure of possible gain or loss) into line with prudent practice. Mathematical expectation quite rightly equated outcome value with price, a method well suited to civil adjudication because it is intrinsic to the object and uniform for everyone. Moral expectation, in contrast, based value on the "utility"—or "the power of a thing to procure us felicity," in the words of the eighteenth-century economic theorist Ferdinand Galiani—yielded by each outcome, which may vary from person to person. Moral expectation was the "mean utility" (*emolumentum medium*), or product of the utility of each possible outcome and its probability.[53]

[52] Daniel Bernoulli, "New theory," p. 24.

[53] See Kenneth Arrow, "Alternative approaches to the theory of choice in risk-taking situations," *Econometrica* 19 (1951): 404–437.

In order to estimate utility, Bernoulli supposed as the most general hypothesis that infinitesimal increments of utility were directly proportional to infinitesimal increments in wealth and inversely proportional to the amount of the original fortune. Thus, the function relating utility to actual wealth would be logarithmic. In other words, the richer you are, the more money it takes to make you happy:

$$dy = b(dx/x), \text{ where } dy = \text{increment in utility}$$
$$x = \text{actual or ``physical'' wealth}$$
$$dx = \text{increment in actual wealth}$$
$$b = \text{constant, } b > 0;$$

or

$$y = b \log x/a, \text{ where } a = \text{the initial fortune.}$$

In order to apply this equation to real situations, Bernoulli was obliged to incorporate a theory of value into his analysis, defining wealth as "anything that can contribute to the adequate satisfaction of any sort of want," including luxuries. Here he appears to have been indebted to the then ongoing controversy over luxury in political economy sparked by Bernard Mandeville's *Fable of the Bees* (1727), and fanned by Jean-François Melon's *Essai politique sur le commerce* (1734; enlarged 2nd edition 1736) and Voltaire's polemical poem *Le mondain* (1736). Although the critique of luxury from a moral standpoint dated back to classical sources,[54] only in the first half of the eighteenth century did economic and moral philosophers mount a systematic defense of luxury. Mandeville's maxim "Private vices; public virtues" neatly epitomized the chief argument of these writers: whatever moral toll a sybaritic life took on the individual level, the manufacture and consumption of luxuries was a necessary stimulus to the economic growth and prosperity of the nation. By defining luxuries as any commodities not essential to subsistence, Mandeville could assert their ubiquity. Other apologists for luxury countered protests that material indulgence spawned sloth with the argument that luxury industries provided work for the surplus population of laborers and goaded better-heeled citizens to work harder in order to acquire more. These points were elaborated and propagated by Melon and Voltaire, generating a great deal of heated debate in reviews such as the *Journal des Savants*,[55] which had a wide Continental

[54] George Boas, "Primitivism," in *Dictionary of the History of Ideas*, vol. 3, pp. 577–598.
[55] André Morize, *L'Apologie du luxe au XVIIIe siècle et "Le Mondain" de Voltaire* (Geneva: Slatkine Reprints, 1970, reprint of 1909 edition), pp. 78–80.

circulation. Numerous eighteenth-century thinkers, including Montesquieu, Hume, Diderot, and Rousseau, as well as Voltaire, examined the theme of luxury from both a moral and an economic standpoint.[56]

Mandeville, Melon, and other political economists accepted the tenet that consumption was proportional to needs, but further posited that needs might expand, creating new necessities. The defenders of luxury pointed to the relativity of distinctions between luxuries and necessities: what was superfluous for one person might be essential for another of different cultural background or higher social standing. As Voltaire observed in his article "Luxe" in the *Dictionnaire philosophique*, in a land where all went barefoot, the first to wear shoes would be accused of opulence.[57] Melon made the same point when he queried whether white bread ranked as a necessity or as a luxury.[58] By accepting luxuries as at least psychologically equivalent to basic needs such as food and clothing, Mandeville and his followers opened the door to unbounded consumption, in the name of national prosperity.

Luxuries also assured the continual upward climb of Daniel Bernoulli's logarithmic utility curve. Although the amount of utility (*emolumentum*; roughly meaning "satisfaction" or "benefit") to be purchased by a given unit of money leveled off as the baseline fortune grew, Bernoulli's equation relating utility to actual wealth required that each increment of wealth buy some further increment of utility, however minute. Only if psychological needs expanded to covet ever more refined luxuries would this utility function be credible. Bernoulli's logarithmic curve implied that satiety would eventually curb but not quench acquisitive appetites: as wealth increased, utility achieved a plateau. In terms of the utility function, the curve was bounded by an asymptote. However, the logarithmic formulation required that the curve continuously approach the asymptote, a condition fulfilled by the ceaselessly renewed and enlarged desire for luxuries posited by the political economists: "Any increase in wealth, no matter how insignificant, will always result in an increase in utility which is inversely proportional to the quantity of goods already possessed." Bernoul-

---

[56] See Morize, *Apologie*, pp. 177–189, for a partial bibliography of post-1736 works on luxury; also Voltaire, "Luxe," in *Dictionnaire philosophique, Oeuvres complètes* (Paris, 1785–89), vol. 41, pp. 501–506; Diderot, "Luxe," in *Encyclopédie*, vol. 9, pp. 763–771.

[57] Voltaire, "Luxe," pp. 518–519.

[58] Jean-François Melon, *Essai politique sur le commerce* (1734; revised edition 1736), in Eugene Daire, ed., *Economistes-financiers du XVIIIe siècle* (Paris, 1843), p. 744.

li's mathematical analysis of utility and moral expectation thus hinged on prevailing theoretical assumptions regarding value, luxuries, and consumption.

Political economy entered into Bernoulli's mathematical treatment through other channels as well. In order for the utility function to be everywhere defined, the actual fortune a must always be greater than zero, since

$$y = b \log a.$$

In order that a will always be positive, Bernoulli declared that "There is then nobody who can be said to possess nothing at all in this sense unless he starves to death. For the great majority the most valuable of their possessions so defined will consist in their productive capacity."[59] Wealth consisted therefore not only of money and goods, but also of labor, the potential to create such values. Even a beggar, he contended, earns enough money to keep body and soul together.

Bernoulli here made use of another theory that had gained wide currency in political economy during the latter half of the seventeenth century. C. B. MacPherson has traced the origins of this theory of "possessive individualism" to the works of Hobbes and Locke, who viewed society as an aggregate of individuals, each defined chiefly as the "proprietor of his person and capacities," and related to one another through a network of exchanges. In contrast to a "customary or status" society, in which work and rewards are distributed by a central authority according to rank, laborers being considered inseparable from their task or the land, the possessive market society presumes the existence of a market for labor as well as for goods: "If a single criterion of the possessive market society is wanted it is that man's labor is a commodity, i.e. that a man's energy and skill are his own, yet are regarded not as integral parts of his personality, but as possessions, the use and disposal of which he is free to hand over to others for a price."[60] Bernoulli's analysis of the mathematical relation obtaining between utility and wealth rested on the assumption that every individual possessed at very least himself, and that this "productive capacity" was fully interchangeable with other sorts of goods or with money.

Bernoulli defended moral expectation with examples designed to show that mathematical results derived from this alternative definition con-

[59] Daniel Bernoulli, "New theory," p. 25.

[60] Colin MacPherson, *The Political Theory of Possessive Individualism: Hobbes to Locke* (Oxford: Oxford University Press, 1972), p. 48.

curred with sensible conduct and beliefs. Gambling, even in fair games, is frowned upon by the prudent. Because of the concavity of the utility curve, Bernoulli concluded that games of chance judged fair by mathematical expectation—that is, those in which the stake equaled expectation—in reality entail negative moral expectation for *both* players: "Indeed this is Nature's admonition to avoid the dice altogether." Moral expectation also sanctioned business decisions which were "universally accepted in practice." For example, it is sound practice according to computations based on moral expectation to divide cargo subject to a uniform risk among several ships. Moreover, any man of affairs would endorse the mathematical result derived from moral expectation that the advisability of investing in a risky venture depended on one's financial resources.

In Bernoulli's eyes, the strongest confirmation of moral expectation came from its resolution of the St. Petersburg problem. Since A's original fortune set an upward limit to the moral expectation, A can risk only a finite amount. The utility of a gain of 2^{n-1} ducats relative to A's original fortune a was

$$b \log \frac{(a + 2^{n-1})}{2},$$

and the moral expectations would be the product of this utility and its probability. For the St. Petersburg game, the value of A's opportunity to play would be

$$(a + 1)^{1/2}(a + 2)^{1/4}(a + 3)^{1/8} \ldots (a + 2^{n-1})^{1/2n} \ldots - a.$$

Clearly, the finite value of a severely restricted A's moral expectation. As Bernoulli noted, a fortune of 100 ducats would make the gamble worth only 4 ducats; a fortune of 1,000 would raise the price to 6. Hence, moral expectation affirmed the reasonable man's reluctance to play the St. Petersburg game except for trifling stakes.

Daniel Bernoulli was fully aware that this redefinition of expectation had hinged upon assumptions about the nature of value and labor, and that these assumptions were at odds with the precepts of equity embodied in the standard definition of expectation. As modern economists have recognized in paying homage to Bernoulli as a predecessor in utility theory,[61]

[61] See Nicholas Georgescu-Roegen, "Utility and value in economic thought," in *Dictionary of the History of Ideas*, vol. 4, pp. 450–458.

his theory of moral expectation was first and foremost a redefinition of value: "The determination of the *value* of an item must not be based on its *price*, but rather on the *utility* it yields."[62] It was therefore natural that Bernoulli should draw upon contemporary themes in political economy. Nonetheless, Bernoulli presented his memoir as a contribution to the calculus of probabilities, as the solution to a mathematical paradox. Mathematics turned to political economy in search of the reasonableness that probabilistic expectation was supposed to express. Proponents of the earlier definition of mathematical expectation such as Nicholas Bernoulli, who was professor of both Roman and canon law at the University of Basel, objected that moral expectation failed "to evaluate the prospects of every participant in a game in accord with equity and justice." His cousin Daniel replied that because his "propositions harmonize perfectly with experience, it would be wrong to neglect them as abstractions resting on precarious hypotheses."[63] The locus of reasonableness had shifted from jurisprudence to economic deliberation, and mathematics reflected that shift.

Daniel Bernoulli's solution to the St. Petersburg problem was by no means universally accepted by other classical probabilists. Even those who, like Buffon, agreed that the trouble lay in the relationship between money and utility sometimes arrived at a different mathematical expression of that relationship. If anything, Bernoulli's memoir widened rather than resolved the controversy, which continued among mathematicians until the nineteenth century and thereafter raged among economists. In what follows, I shall focus on those eighteenth- and early nineteenth-century reactions that shed the most light on the problem of capturing "reasonableness" in mathematical form, namely those of Jean d'Alembert, George Leclerc Buffon, M.J.A.N. Condorcet, Pierre Simon Laplace, and Siméon-Denis Poisson. As we shall see, the issues raised by the St. Petersburg problem grew into a more general examination of rational risk taking.

The French mathematician Jean de la Rond d'Alembert agreed with Daniel Bernoulli that probability theory must explain prudent action in uncertain situations, but challenged Bernoulli's solution to the St. Petersburg problem on the grounds that moral expectation, while more accurate than classical expectation, still oversimplified the actual experience of risk taking. D'Alembert accepted Bernoulli's premise that "moral considera-

---

[62] Daniel Bernoulli, "New theory," p. 24.
[63] Daniel Bernoulli, "New theory," pp. 33, 31.

tions, relative to either the fortune of the players, their circumstances, their situation, or even their strength" refined the results of probability theory, but despaired of quantifying all of these factors, or even of ordering the relative importance of the diverse variables. Moreover, Bernoulli had failed to specify the rules for determining whether mathematical or moral expectation applied to a given case. Since ordinary computations of expectation appeared to give satisfactory results without any consideration of financial status, Bernoulli's introduction of moral expectation for the St. Petersburg problem seemed suspiciously ad hoc to d'Alembert.[64]

For d'Alembert, the St. Petersburg paradox arose from a betrayal of good sense on a different front. Whereas Bernoulli believed that the value term of classical expectation overestimated the gains, d'Alembert countered that it was the probability term which absurdly inflated the expectation. Although it might be "mathematically" or "metaphysically" possible for a fair coin to turn up tails 100, 1,000, or n times in a row, experience dismissed such outcomes as "physically" impossible. Should such a run of consecutive tails actually occur, d'Alembert claimed that observers would rightly posit some underlying, uniform cause, such as an asymmetric coin. Mathematical probability, which based the postulate of equiprobable outcomes on our ignorance of any cause which might tip the balance in favor of heads or tails, would therefore no longer apply to the situation. D'Alembert rejected the naturalist Buffon's suggestion that all probabilities less than .0001 be excluded as "morally impossible"[65] as mathematically naive because of the discontinuous method of estimating probabilities. However, d'Alembert himself attempted to construct a series where the probabilities would converge and continuously decrease at a rate derived from a more physically plausible model of coin tossing.[66]

He often returned to the notion of expectation in his critical discussions of mathematical probability. Although dissatisfied with Bernoulli's alternative to classical expectation, d'Alembert acknowledged the failure of

[64] D'Alembert, "Croix ou pile," in *Encyclopédie*, vol. 4, p. 513.

[65] George Leclerc Buffon, "Essais d'arithmétique morale," in *l'Histoire naturelle. Supplément* (Paris, 1774), vol. 4, pp. 52–60.

[66] D'Alembert, *Opuscules mathématiques*, vol. 7 (Paris, 1780), pp. 49–58. For a detailed discussion of d'Alembert's views on probability theory, see L. J. Daston, "D'Alembert's critique of probability theory," *Historia Mathematica* 6 (1979): 259–279; also Zeno G. Swijtink, "D'Alembert and the maturity of chances," *Studies in History and Philosophy of Science* 17 (1986): 327–349.

mathematical expectation to capture the salient features of reasonable conduct, even in simple gambling situations. For example, a lottery with an enormous prize, say a million francs, but with only a tiny chance of winning, say .0001, offered an attractive expectation of 100 francs for each ticket, according to the classical formula. Yet d'Alembert felt that prudence would counsel against such an investment, and concluded that this discrepancy between mathematical and psychological expectation pointed to serious flaws in the foundations of probability theory.

D'Alembert's critique of probability theory merits a closer examination for at least two reasons. First, his criticisms, particularly those leveled at probabilistic expectation, dramatized the tensions between the competing definitions of reasonableness incorporated into probability theory. Second, his analyses, scattered throughout his mathematical writings over a thirty-year period, provided Laplace and Condorcet with a departure point for their own more orthodox work in probability. Although neither mathematician ultimately concurred with d'Alembert's pessimistic conclusions concerning mathematical probability, his reservations and challenges stimulated their initial interest in the subject and influenced their choice of problems and their philosophical orientation. Committed to a mathematics which was experiential both in its origins and objectives, d'Alembert sought to strike the balance between simplicity and verisimilitude which he believed to characterize mixed mathematics in general. Insofar as probability theory was within the compass of mixed mathematics, it too must set the proper ratio between abstraction, which guaranteed the certainty of its results, and experience, which guaranteed their relevance. D'Alembert faulted conventional probability theory for its neglect or distortion of the latter element. Through his alternative formulations of mathematical probability, he hoped to achieve a closer match between mathematics and experience. Although he shared this goal with most other eighteenth-century probabilists, they parted company over the precise meaning of the experience they sought to codify.

D'Alembert's critique of probability theory as applied to games of chance and other physical problems first appeared in the relevant articles he wrote for the *Encyclopédie*, beginning with the article on the coin game "Croix ou pile." This article contained the most explicit arguments against conventional probability theory of his *Encyclopédie* pieces, and foreshadowed many of his later criticisms. In this and other writings on the subject, d'Alembert showed himself to be thoroughly conversant with the conventional theory. Indeed, his earlier articles on the subject, such as "Absent" and "Bassette," had accepted its results without demure. However, in the "Croix ou pile" article, he was willing to challenge even the most elementary results of the theory in order to reveal the crucial experiential aspects overlooked by the conventional theory. In this article, the St. Petersburg problem highlighted the flaws of conventional probability theory as it was customarily applied to the mathematical analysis of games of chance, and to more important problems in physics and astronomy. D'Alembert often returned to the St. Petersburg problem—and to the questions it raised about probabilistic expectation—in his subsequent discussions of probability theory.

The St. Petersburg problem served d'Alembert as a tool with which to sharpen the contradictions he found within conventional probability theory, and with which to specify the relations of mathematical probability to experience more accurately. He often reiterated his contention that probabilists had confused mathematical and physical possibility by failing to take adequate notice of physical experience, in which, he claimed, a long sequence of identical events never occurred purely by chance. The price for this confusion, d'Alembert warned, was paid in absurd results, such as the infinite expectation computed for the St. Petersburg problem.

D'Alembert characteristically defended his exclusion of the metaphysical possibility of n tails in a row by recourse to a model of physical reality. His discussion of this model and its implications for probability theory provided a striking example of the mixed mathematician's art of compromise between the competing demands of abstraction and verisimilitude. D'Alembert supposed that there exist n different physical ways for heads (H) and tails (T) each to occur given a fair coin, thus conceding initial equiprobability to the side of simplicity. Imagine that the outcome of the first toss is T: what will be the respective probabilities for H and T on the second toss? D'Alembert argued that since nature is overwhelmingly complex, and because there are "causes continually acting to change their state at each instant," once one such physical route to a T outcome has been exhausted, there remain only $n - 1$ ways to obtain T, but H is still the endpoint of n possible routes.[67] Therefore, concluded d'Alembert, H is marginally more probable: $P(H)/P(T) = n/n - 1$ on the second toss. Even if n is infinite, implying that for any finite number m trails, $n - m$ still is

---

[67] D'Alembert, *Opuscules*, vol. 7, p. 40.

infinite, so that H and T remain equiprobable, d'Alembert contended that as m increases, it becomes ever more likely (*vraisemblable*) that the next toss will be found among the sequences yet to be executed. In short, probability theory must attend to our experientially based conviction that no event in nature is ever duplicated exactly.

To emphasize the relevance of this model of nature to problems of the St. Petersburg genre, d'Alembert imagined a massive experiment in which 2^{100} people each toss a fair coin one hundred times, and compare results. D'Alembert maintained that the all-H and all-T sequences will never occur, but that some of the mixed sequences will be repeated two or more times, thus experimentally contradicting the prediction of conventional probability theory of equiprobability ($P = 1/2^{100}$) for any and all sequences. By including the merely metaphysical possibility of a uniform sequence on an equal footing with more physically plausible ones, d'Alembert claimed that the St. Petersburg problem was based upon manifestly false premises (i.e., as tested by d'Alembert's thought experiments).

D'Alembert's series estimated the value of each successive term separately: for the first toss, $P(H) = P(T) = 1/2$; for the second toss, however, $P(H) = 1 + a/2$ and $P(T) = 1 - a/2$, assuming that the outcome of the first toss was T; for the third, $P(TTH) = (1/2)(1 - a/2)(1 + a + b/2)$, where the limit of the sum $a + b + c \ldots$ is 1 as the number of terms approaches infinity. Hence, for the St. Petersburg problem, the revised expectation would be

$$(1/2)[1 + 1 + a + (1 - a)(1 + a + b) + (1 - a)(1 - a - b)(1 + a + b + c) + \ldots].$$

Since the term $(1 - a - b - c \ldots)$ ultimately approaches zero, the last factor estimating the value for an infintely long sequence of H's or T's is also zero. D'Alembert explored other properties of the sum as well, such as the relative rates of convergence.[68]

Note that d'Alembert here not only rejected the assumption of equiprobability but also that of the independence of consecutive coin tosses. D'Alembert admitted that his solution was not only cumbersome but also begged the question as to exactly how the small increments $a, b, c, \ldots$ were to be evaluated. But he nonetheless maintained that this revised version of probability theory was superior to the conventional formulation

[68] D'Alembert, *Opuscules*, vol. 7, p. 58.

because of its greater verisimilitude, leaving it to other mathematicians to develop some reliable method for assigning such probabilities.[69]

D'Alembert's criticisms against the conventional definition of expectation were of two types: first, although it was a plausible method for comparing certain stakes with uncertain gain, there existed other equally plausible methods; second, it failed to capture the true precepts of prudent risk taking. D'Alembert proposed several alternative expressions for both expectation and stake (which he contended need not always be set equal to one another) in a game of chance, differing according to whether the players' financial circumstances were taken into account or not. He did not defend any of his alternative formulations as superior to the conventional definition, observing only that each could be supported "with plausible reasons."[70] According to d'Alembert, the conventional definition of expectation erred by weighting both probability and outcome value equally, whereas the latter should actually be somehow "subordinated to the degree of probability." For example, given a choice between a .99 probability of 1,000 écus and a .01 probability of 99,000 écus, he queried ". . . where is the man so foolish as to prefer the offer of 99,000 écus? The *expectation* in the two cases is not *really* the same; even though it is the same according to the rules of probability."[71]

As in the case of the St. Petersburg problem, d'Alembert concluded that if the probability were small enough, no outcome value was large enough to persuade a prudent man to take the risk or small enough to dissuade him. He cited with approval Pascal's remark that only a fool would hesitate to play a game with three fair dice in which a toss of three sixes, repeated twenty times in a row, condemned the player to death, and all other outcomes crowned him emperor. Yet if the losing outcome were "physically possible," d'Alembert argued, the reasonable decision would not be so obvious. Trusting the intuitive conclusions of common sense, he scrutinized mathematical expectation for covert assumptions, such as the confusion of mathematical and physical possibility, which brought its results into conflict with prudence.

In his later writings, d'Alembert questioned whether any sort of risk

[69] D'Alembert, *Opuscules*, vol. 7, p. 60.
[70] D'Alembert, *Opuscules*, vol. 4 (1768), p. 82.
[71] D'Alembert, *Opuscules*, vol. 4, p. 83.

taking truly qualified as prudent. In the unpublished ninth volume of his *Opuscules mathématiques*, d'Alembert underscored his earlier reservations and voiced new doubts concerning the bases of probabilistic expectation. Was it possible, d'Alembert wondered, "to know if the *probability* 1/100,000 of winning 100,000 écus is or could be deemed equal, as is commonly supposed in the analysis of hazards, to the *certainty* of gaining one écu, or even the *probability* 1/2 of winning two écus?"[72] Here, d'Alembert went beyond his objections to the distorted weightings of very small probabilities to question any equation between certain and merely probable values, or even between unequally probable values. By "equality" between the two states, d'Alembert did not mean mathematical equality, easily verified by computation, but rather a psychological equality sufficient to prompt a prudent man to make such an exchange. In asking whether probable and certain gains could be "perfectly the same," d'Alembert implied that the prudent man might well subscribe to the proverb that "A bird in the hand is worth two in the bush."

D'Alembert expanded these reflections on the psychology of risk taking in the context of an all too real risk facing eighteenth-century Europeans, the threat of smallpox.[73] Here he departed not only from the rather rarefied realm of lottery and gambling examples, but also from the assumptions that had guided both his own and fellow mathematicians' attempts to codify good sense. For the first time a rift opened up between what even d'Alembert agreed was the rational course of action, namely inoculation against smallpox, and what he nonetheless insisted must remain on the subject of mathematical probability, the actual psychology of risk taking. Previously (and indeed for some time subsequently), probabilists had assumed that rational action and the action chosen by citizens certified as reasonable were synonymous. In the case of the St. Petersburg problem, mathematicians, including d'Alembert, used the latter to *define* the former, and thereby to challenge the conventional mathematical solution. It is no doubt significant that unlike the St. Petersburg problem, the issue of inoculation was of urgent practical interest at a time when over 10 percent of the Parisian and London populations were estimated to have died of the disease. It was hotly debated throughout Europe during the period

[72] D'Alembert, Bibliothèque de l'Institut, MS 1793 (unpub. vol. 9 of *Opuscules mathématiques*), f. 374.

[73] See John McManners, *Death and the Enlightenment* (Oxford and New York: Oxford University Press, 1985), pp. 46–47.

1750–70, especially in France, where it became a pet crusade of the *philosophes* following Voltaire's favorable observations on English experiments with inoculation.[74] Inoculation, which meant impregnating the skin with live smallpox pustules, was by no means a risk-free procedure during this period; some died of it. Roughly, the choice facing the average Parisian circa 1750 was between a one in seven chance of dying of smallpox spread over the long term, and a one in two hundred chance of dying of the inoculation over the short term (one or two months). Like the St. Petersburg problem, the inoculation dilemma served to focus d'Alembert's objections on an example upon which specific aspects of experience could be brought to bear, but here the experience was immediate enough to discourage idealization.

Once again, d'Alembert's point of departure was a memoir by Daniel Bernoulli, this one advancing a probabilistic analysis of the relative advantages and disadvantages of smallpox inoculation.[75] In his memoir, Bernoulli argued that while the data on smallpox mortality was woefully deficient, one could still apply the theory of probability to the problem of inoculation on the assumption that "the simplest laws of nature are always the most plausible."[76] Using the incomplete mortality tables and assuming that a smallpox victim's chance of dying from the disease was a constant regardless of age, Bernoulli derived an expression for the number of people likely to succumb to smallpox in a given time period and computed the average gain in life expectancy from inoculation for any given age. In computing life expectancies, Bernoulli employed the formula analogous to that of mathematical expectation for a lottery: the area under the curve of mortality (equivalent to the product of the number of gamblers with equal chances and their stakes) divided by the number of living persons (the number of gamblers).[77] Although Bernoulli refrained from endorsing inoculation explicitly, the favorable thrust of his analysis was unmistakable,

[74] Voltaire, *Lettres philosophiques* (1734), Raymond Naves, ed. (Paris: Éditions Garnier Frères, 1964), 11th letter, pp. 48–53; also Robert Favre, *La mort dans la littérature et la pensée française au siècle des lumières* (Lyons: Presses Universitaires de Lyons, 1978), pp. 259–264.

[75] Daniel Bernoulli, "Essai d'une nouvelle analyse de la mortalité causée par la petite vérole et des avantages de l'inoculation pour la prévenir," *Histoire et Mémoires de l'Académie des Sciences* (1760; pub. 1766), part 2, pp. 1–79.

[76] Daniel Bernoulli, "Essai," p. 6.

[77] See Section 3.4.2 for alternative methods of computing life expectancy and their history.

as were the implications of his request that La Condamine, the foremost French proponent of inoculation, consider Bernoulli's arguments.[78]

In an introduction written five years after the original memoir was submitted to the Paris Académie des Sciences in 1760 and appended to the published paper in 1766, Bernoulli defended his simplifying assumptions as consistent with, if not proven by, current medical knowledge of the disease and the admittedly incomplete tables of mortality. After all, Bernoulli argued, such agreement with phenomena as far as known was the only basis for belief in the universal law of gravitation. As for those critics who missed the point of this lofty comparison, Bernoulli exhorted them to "take the trouble to apply themselves to the facts of the matters which they propose before making criticisms."[79]

There can be little doubt that Bernoulli had d'Alembert in mind. D'Alembert had read a paper on the application of probability theory to the inoculation problem before a public session of the Académie des Sciences on 12 November 1760, which was a long and detailed critique of Bernoulli's memoir on the subject. The unidentified critic addressed in Bernoulli's introduction is described as a person of "merit and great reputation." Bernoulli and d'Alembert were lifelong rivals in mathematics,[80] and Bernoulli was understandably piqued to see his memoir so harshly criticized by his old opponent in a public (versus academic) address before his own paper had been published. On his side, d'Alembert was stung by Bernoulli's patronizing suggestion that he apply himself to the "facts of the matter," since d'Alembert perceived attention to the facts of experience to be at the heart of his own alternative approach to probability theory. D'Alembert retorted that he was not surprised that those ignorant of analysis had prematurely attempted to compute the advantages of inoculation, but was indeed dismayed that they should count "a man such as M. Daniel Bernoulli" as one of their number.[81]

Bernoulli went so far as to formally protest ill treatment to the Académie des Sciences. Although Bernoulli's letter has been lost, its content and angry tone can be reconstructed from d'Alembert's reply, read before the 7 December 1762 session of the Académie.[82] D'Alembert pleaded the

---

[78] Daniel Bernoulli, "Essai," pp. 52–53.   [79] Daniel Bernoulli, "Essai," pp. 3–7.

[80] Thomas L. Hankins, *Jean d'Alembert: Science and the Enlightenment* (Oxford: Clarendon Press, 1970), pp. 44–50.

[81] D'Alembert, *Opuscules*, vol. 4, p. ix.

[82] Archives de l'Académie des Sciences, Dossier 7, December 1762.

"prodigiously delayed" publication schedule of the *Mémoires de l'Académie des Sciences* in response to Bernoulli's complaint that d'Alembert had unfairly attacked his memoir before it had appeared in print. Matters of priority and protocol aside, d'Alembert reprimanded Bernoulli for prematurely and misleadingly applying the calculus of probabilities to so important a question as inoculation while still lacking crucial data: "He reproaches me for being locked within an abstract analysis; that is, I did not believe it necessary, as he did, to build grand calculations upon vague hypotheses, in a matter concerning human life." According to d'Alembert, mathematicians still lacked both sufficient information on smallpox inoculation and mortality rates, and also a reliable method with which to apply the theory of probabilities to such problems.

Personal animosity between the two mathematicians may have intensified the dispute over the inoculation problem, but it is difficult to ascribe d'Alembert's interest in the problem wholly to pugnacity. D'Alembert's criticism of Bernoulli's solution to the inoculation problem was entirely consistent with his criticisms of conventional probability theory's treatment of physical problems, and ultimately rested on the same belief in the primacy of (in this case, psychological) experience.

Moreover, d'Alembert risked his standing as a leading *philosophe* by opposing Bernoulli on the inoculation issue, suggesting that his convictions on the subject went deeper than professional rivalry. Voltaire had advocated inoculation in a chapter of his famous *Letters philosophiques*, and by 1755 inoculation had become a central part of the *philosophes'* campaign for rational social reform against the reactionary forces of superstition. At the height of the controversy, inoculation became a favorite theme in popular French literature and almost a symbol of social and intellectual liberalism.[83] Although d'Alembert took great pains to endorse inoculation and to discount the theological arguments mustered against it, his critique of Bernoulli's memoir must have been viewed by many of his fellow *philosophes* as an attempt to undermine the strongest arguments in favor of the procedure.

D'Alembert's criticisms of Bernoulli's memoir have a familiar ring: his major objections take the conventional treatment of the problem to task for insufficient fidelity to relevant experience. The conventional method of

[83] See Arnold Rowbotham, "The *Philosophes* and the propaganda for inoculation of smallpox in eighteenth-century France," *University of California Publications in Modern Philology* 18 (1935): 265–290.

computing life expectancies, contended d'Alembert, ignored crucial distinctions. Imagine two mortality curves AOCD and AQCD, which have identical integrals (Figure 1). According to Bernoulli's method, the average life expectancy is the same for both curves. However, d'Alembert asserted the destinies of persons on the two curves differ significantly: AOCD is preferable since the number of deaths at an early age are fewer.[84] The conventional formula made no provisions for factors that are critical from the standpoint of the individual.

Moreover, mean duration was not the only plausible measure of life expectancy. D'Alembert pointed out that the table of mortality in Buffon's *Histoire naturelle* used a different, although apparently reasonable, method. Instead of dividing the area under the mortality curve by the total number of people of the same age, Buffon had computed the number of years after which exactly one-half of a given population, all the same age, would be dead. The two methods yielded widely disparate results: the better-known

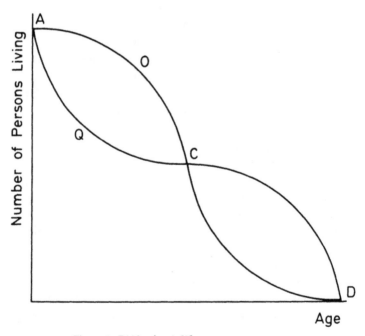

Figure 1. D'Alembert's life-expectancy curves.

[84] D'Alembert, *Opuscules*, vol. 4, p. 93.

mean duration method used by Halley, Nicholas Bernoulli, and Daniel Bernoulli gave a twenty-six-year average lifespan for a newborn; Buffon, using the same data, estimated the "half-life" for a population of newborns at only eight years.[85] D'Alembert argued that the two competing methods betrayed the lack of any "sure method" for estimating life expectancies; furthermore, both methods were intrinsically flawed. Bernoulli's mean duration method failed to distinguish between curves with the same integral but vastly different (from the standpoint of the individual) desirability. Buffon's method suffered from analogous defects.[86] Neither method paid sufficient attention to the dips and slope of the mortality curve, although the information was of critical interest to the individual facing inoculation.

Similarly, d'Alembert claimed that the conventional probabilistic treatment neglected elements of psychological experience which were essential for evaluating the situation mathematically. Even if Bernoulli's assumption that the risk of dying from smallpox does not vary with age were accepted (d'Alembert was highly skeptical of this postulate and called for more detailed mortality tables to settle the question), his solution did not, d'Alembert held, fully characterize the quandary of the person facing inoculation. The missing elements might be described as quality of life considerations: is the risk for a thirty-year-old, who might expect to live thirty more years naturally and thirty-four more years with inoculation, comparable to that of an older person who stands to gain the same increment in life expectancy? According to d'Alembert, the 1/200 risk of dying from the inoculation itself in the prime of life must be balanced against the advantages of four more years added to the nether extreme of life, when one is less capable of enjoying them.[87]

For d'Alembert, the chief flaw of conventionally derived expectations was their failure to coincide with psychological expectation, that is, "moral experience." In this case, the gap between mathematics and reality stemmed largely from the difficulty of balancing clear and present (although small) dangers such as inoculation against a greater risk of the disease itself spreading out over a longer period. The integral of the mortality curve contains no information on where the curve dips and swells, but it is just such information which is of paramount importance to the individual. D'Alembert summarized his objections to Bernoulli's method

---

[85] D'Alembert, *Opuscules*, vol. 2 (1761), pp. 74–75.

[86] D'Alembert, *Opuscules*, vol. 4, pp. 92–98.

[87] D'Alembert, *Opuscules*, vol. 2, p. 33.

with an analogy drawn from more familiar applications of probability theory. He described a lottery in which all citizens were required to participate. After the drawing, half of the participants would immediately be put to death, while the fortunate remainder would be guaranteed a lifespan of a hundred years. By conventional estimation, the average life expectancy of the entire population of newborns would be fifty years. Yet d'Alembert doubted whether anyone would voluntarily participate in such a lottery, even if the average life expectancy might ordinarily be less than fifty years.

It is important to note that d'Alembert did not condone this form of reasoning, which prefers large, long-term risks to smaller, more immediate ones, and which he called the "common logic . . . half good, half bad."[88] However, he did despair of changing it. As in the case of the St. Petersburg problem, mixed mathematics must follow, not lead, common experience—in the inoculation problem, the actual psychology of risk taking. Mathematics is descriptive, not prescriptive. On a metaphysical plane, the conventional theory of probability could not be faulted for its treatment of either the inoculation or St. Petersburg problems. Yet both treatments, d'Alembert maintained, were irrelevant to our actual experience of the physical and moral worlds.

D'Alembert deepened his critique of Bernoulli's solution of the inoculation problem by including still other aspects of moral experience. The advantages of inoculation must be judged with respect to three different types of life duration, according to d'Alembert: physical life or ordinary duration (the only type recognized in the accepted theory); "real life" or that portion of physical life, which is lived fully, without suffering; and "civil life" or that portion of physical life in which one is useful to the state.[89] Depending on the vantage point chosen, the relative benefit to be derived from inoculation would vary widely, as d'Alembert demonstrated by constructing mortality curves for real and civil lifetimes. The slope of real life curves varied as a function of enjoyment, leveling off both in early childhood and old age; the curve of civil life dipped below the abscissa to represent those periods during which the individual is a ward of the state and therefore of negative usefulness. While d'Alembert was sensitive to the difficulties of such estimations, which were certain to fluctuate with

[88] D'Alembert, *Opuscules*, vol. 2, p. 35.

[89] D'Alembert, *Opuscules*, vol. 2, pp. 82–88.

individual circumstance, he nonetheless insisted that a sound probabilistic treatment of moral problems like inoculation could not afford to ignore them. D'Alembert's list of requirements for a valid mathematical theory of inoculation was an appeal to experience at many levels: accurate mortality tables; a more sensitive method for computing life expectancies; a mathematical way of weighing small short-term risks against large long-term ones; a theory for comparing physical, real, and civil lifetimes.

D'Alembert attempted to rectify the shortcomings he saw in the analyses of Daniel Bernoulli and other probabilists until the very end of his mathematical career. He suggested both methodological and empirical refinements, but discovered that his exacting distinctions enmired him in ever more complicated calculations, which he felt compelled to amend and refine.[90] Although his persistent efforts suggest that he did not altogether despair of a probabilistic approach to such problems, his growing frustration made him pessimistic about the prospects of quantifying "moral matters in a precise and rigorous manner."[91] Because the relative appreciation of short- to long-term risks varied from individual to individual, and even for the same individual under different circumstances, the obstacles to a general mathematical treatment seemed insuperable:[92] "How are we to compare the present risk to an unknown, remote advantage? About this the analysis of hazards can teach us nothing."[93]

D'Alembert no longer unambiguously identified common sense with rationality. While he heartily recommended smallpox inoculation to his fellow citizens as a sound preventive measure, he recognized the force which the uneven appreciation of short- and long-term risks carried "for the common run of men, however ill-founded it may be."[94] For d'Alembert, the calculus of probabilities was primarily a description of the *common* psychology of risk, rather than a corrective to this sometimes faulty reasoning. Hence, his criticisms opened another crack in the façade of the calculus of probabilities. His colleagues had disputed the meaning of rea-

---

[90] Mémoire 35, "Réflexions sur la théorie de l'inoculation," of the unpublished ninth volume of d'Alembert's *Opuscules*, for example, reformulates his earlier corrections to the standard calculation of life expectancies. Bibliothèque de l'Institut, MS. 1793, ff. 460–485.

[91] D'Alembert, *Opuscules*, vol. 4, p. 91.

[92] D'Alembert, *Opuscules*, vol. 2, p. 30.

[93] D'Alembert, *Opuscules*, vol. 2, pp. 33–34.

[94] D'Alembert, *Opuscules*, vol. 2, p. 40.

sonableness, but had agreed in restricting this virtue to an elite. Their mathematics was to reflect the reasoning and practice of these *hommes éclairés*, and to mathematically compel the less enlightened to follow the example of their betters. Ideally, reasonableness would thus become the property of all. D'Alembert asserted that probability theory should broaden its base to express the "common logic," however dubious its validity. Previously, only the definition of right reason had been at issue; now other sorts of reason, even more heterogeneous, threatened to fragment mathematical probability still further.

D'Alembert evidently realized that his criticisms had opened a Pandora's box for mathematical probability, hence his contention that although the advantages of inoculation were real enough, they "were not of right nature to be appreciated mathematically." He chided "the mathematicians who made too great haste to reduce this matter to equations and formulas."[95] However, his warnings went for the most part unheeded, although his protegés Condorcet and Laplace continued to labor over the puzzles he had posed. As J. Lalande commented apropos of d'Alembert's objections to the theory of probability: "I admit that if this claim is well founded, it would be necessary to strike the theory of probability from the list of mathematical theories, and I could have dispensed with so lengthy a treatment . . . but it appears that in spite of the plausibility of d'Alembert's arguments, they have not shaken, in the minds of mathematicians in general, the generally accepted theory of probabilities."[96]

Can perceptions of risk be measured and compared? The objective size of various risks could be appreciated statistically, but the classical probabilists were equally interested in the possibility of subjective units and weightings. D'Alembert concluded from the inoculation case that individual differences in risk perception made the project impossible, but other eighteenth-century probabilists, notably Buffon and Condorcet, were more optimistic. Their optimism depended on assumptions d'Alembert would not have accepted: that there existed a single minimum psychological unit of risk for at least all reasonable men; that this minimum unit could be discovered and quantified with reference to the statistical likelihood of certain touchstone events; and that comparable statistical risks would evoke comparable

[95] D'Alembert, *Opuscules*, vol. 2, p. 45.

[96] J. F. Montucla (completed and edited by J. Lalande), *Histoire des mathématiques* (Paris, An X/1802), vol. 3, pp. 405–406.

perceptions of risk. As usual, the descriptive project shaded off imperceptibly into a prescriptive one. If reasonable men did not in fact fulfill the conditions of the last assumption, the probabilists hoped that they would change their ways for consistency's sake, once acquainted with the relevant statistics. The trick was to find a fundamental unit of perceived risk that could be both agreed upon by the reasonable elite and quantified, to be used as a standard in gauging more controversial risks.

Given that what scanty statistics then existed concerned mostly human death rates, it is not surprising that Buffon found his unit in the mortality tables. George Leclerc Buffon was better known as a naturalist than as a mathematician, but he contributed analyses and applications of probability theory in several sections of his monumental *Histoire naturelle* and its supplements, particularly in the "Essai d'arithmétique morale" (1777).[97] Buffon based his attempts to quantify moral certainty/impossibility (as opposed to the more rigorous mathematical and physical sorts) on the assumption that "all fear or hope, whose probability equals that which produces the fear of death, in the moral realm may be taken as unity against which all other fears are to be measured."[98] Because no healthy man in the prime of life fears dying in the next twenty-four hours, Buffon took the probability of such sudden deaths, reckoned from the mortality tables to be about .0001, as the zero point ("moral impossibility") on the scale of moral probability. Buffon contended that since no reasonable person gave more than a passing thought to the small but discernible probability that he will die tomorrow, he must be equally indifferent to the expectation produced by a .0001 chance of winning the lottery. Indeed, Buffon observed, his expectation, while numerically equivalent to his risk of imminent death, hardly balances the psychological intensity of the latter, "since the intensity of the fear of death is a good deal greater than the intensity of any other fear or hope."[99] Although the mortality tables showed that one out of ten thousand people actually does die in the prime of life, the relevant measure of probability was the fear or hope such risks evoke in the reasonable man under normal conditions. Extremes of indifference or concern calibrated the scale against which straightforward probabilistic computations, such as chances in a fair lottery, were to be assessed.

[97] See L. E. Maistrov, *Probability Theory: A Historical Sketch*, Samuel Kotz, trans. (New York and London: Academic Press, 1974), pp. 118–123.
[98] Buffon, "Essai," p. 56.

Buffon was not the first or the last to attempt a quantitative, probabilistic estimate of a minimum psychological unit. While Buffon sought the threshold probability which kindled fear or hope, the English Newtonian John Craig had gauged the minimal unit of credibility produced by a single eyewitness in his probabilistic treatment of the credibility of the New Testament. The sum of these psychological units amassed belief proportionate to the probability to testimony:

> For great probability is composed of many testimonies of primary witnesses just as a great number is composed of many unities. It can indeed happen that such a degree of probability may be so small that our mind can scarcely perceive its force, just as in the movement of bodies the degree of velocity is sometimes so little that we are unable to discern it with our eyes. But (even as the latter degree of velocity, so too) this degree of probability is of a determined magnitude and when repeated many times, produces a perceptible probability.[100]

Condorcet, borrowing a leaf from Buffon, also hoped to create a common measure of moral magnitudes from mathematical probabilities, although he did not share Buffon's confidence that these minimal units of probability could be assessed at "a fixed value" independent of particular circumstances. In the case of judicial decisions, Condorcet maintained that it was unjust to submit anyone to a judgment whose probability of error exceeded that which the accused himself "being supposed to enjoy his reason and composure, possessing knowledge and intelligence, would expose himself to an equal danger for a slight interest, for amusement, without need of bravery.[101] Condorcet suggested the risk of sailing on the Dover/Calais packetboat as a plausible measure of this minimal psychological tremor.

Just as his projects for a universal language or a universal system of weights and measures promised to rationalize communication and commerce by reducing them to a homogeneous medium which permitted comparisons of otherwise incommensurable entities,[102] so Condorcet and his fellow probabilists envisioned a universal dictionary of probabilities that would translate all moral experience into a common tongue. The per-

---

[99] Buffon, "Essai," p. 58. [100] Craig, "Rules," p. 5.

[101] M.-J.-A.-N. Caritat de Condorcet, *Essai sur l'application de l'analyse à la probabilité des décisions rendues à la pluralité des voix* (Paris, 1785), p. lxxiv.

[102] Baker, *Condorcet*, p. 65.

sonal assessments of reasonable men, suitably quantified with the help of Buffon's mortality tables or Condorcet's Dover/Calais shuttle, would become the lingua franca of the moral realm, the yardstick against which more doubt-ridden or controversial risks (like that of a false conviction) might be measured.

The measurement of such psychological units was key to the probabilist program in the moral sciences. Following Descartes, eighteenth-century French thinkers ascribed certainty to the comparison of clear and distinct ideas. For Condillac, for example, all mental operations ultimately consisted in the decomposition, recombination, and comparison of complex ideas: a mathematical proof was nothing more than a string of identities, each validated as such by a mental comparison of the two clear and distinct ideas represented.[103] The incommensurability of other ideas hindered the comparison of moral elements such as sentiments and judgments, which seemed irreducible to common terms. Probabilists conceived their calculus as a uniform measure of such elements, just as money constituted the uniform measure of value,[104] and therefore as the beginnings of a rigorous approach to the moral sciences.

Buffon viewed the calculus of probabilities as a mathematical rendering of subliminal mental operations. Although he cautioned that probabilities, rather than vague feelings, should be the "true measure" of both hopes and fears, he nonetheless affirmed the reliability of sentiment as a guide to action. Hence his redefinition of expectation confirmed the psychological fact that one is always, according to Buffon, more sensitive to loss than to gain, since even the smallest wager risked a relatively greater loss than gain (by his computations): ". . . sentiment is in general nothing but a less clear, implicit form of reasoning which is nonetheless more refined, and always more sure than the direct product of reason."[105] Daniel Bernoulli applauded Buffon's method of estimating the values of moral probabilities by considering "the nature of man by his actions . . . this is without doubt to reason as a Mathematician-Philosopher," but Buffon was no behaviorist.[106] His treatment of moral expectation probed beneath the

---

[103] E. Condillac, *Essai sur l'origine des connaissances humaines* (1746), in *Oeuvres* (Paris, An VI/1798), vol. 1, pp. 96–101.

[104] Condorcet, "Mémoire sur les monnaies," in *Oeuvres*, F. Arago and Condorcet-O'Connor, eds. (Paris, 1847–49), vol. 11, p. 587.

[105] Buffon, "Essai," p. 71.

[106] Buffon, "Essai," p. 57.

level of explicit reasoning to the nuances of sentiment and implicit judgment which governed action.

Buffon advanced his own version of moral expectation in the service of the same antigambling sentiments expressed by Daniel Bernoulli. As every sober man of affairs knew, all games of chance, even so-called fair ones, reduced to "a misconceived pact, a contract disadvantageous to both parties." Through mathematical arguments Buffon hoped to succeed where a "vague moral discourse" might fail, in curing those rash unfortunates who had succumbed to the "epidemic" of gaming. Although Buffon accepted Bernoulli's distinction between the utility of equal amounts of money for a pauper and a rich man, he posited a linear rather than logarithmic relation between utility and original fortune:

$$y = \frac{x}{a + x}, \text{ where } y = \text{utility}$$
$$x = \text{gain}$$
$$a = \text{original fortune.}$$

Hence, Buffon computed expectations as

$$p(x/a) - q\left(\frac{x}{x + a}\right), \text{ where } p = \text{probability of losing}$$
$$q = \text{probability of winning,}$$

as opposed to the classical expectation of

$$p(a - x) - q(a + x).$$

Since x/a is always greater than $\frac{x}{x + a}$, even if $p = q = 1/2$, the expectation of loss preponderates.

Like Bernoulli, Buffon relied on political economy at several points. Built into Buffon's analysis was a theory of threshold economic necessity, defined as that level of affluence (or lack thereof) to which one was accustomed. Buffon echoed the arguments of Mandeville, Melon, and Voltaire on the relativity of luxury. Necessity was nothing more than a habitual amenity: "But what is necessary; what is excess? I understand by necessary *the expense which one is obliged to make in order to live as one has always lived*, with this necessity one may have comforts and even pleasures."[107] Only new pleasures counted as luxuries, and, however high one's standard of living, the loss of habitual "necessities" would be "infinitely" regretted.

---

[107] Buffon, "Essai," p. 73.

Therefore, to risk any large portion of the income which purchased these necessities of habit was to risk an infinite amount, for which no expectation could be adequate compensation.

Buffon's notion of utility presumed a stratified society, hence the need for proportional values. Only excess wealth could be expended on risky ventures, and even then, loss loomed larger than gain in thought and therefore in computation. Buffon's redefinition of expectation aimed to express these economic ideas and psychological shadings mathematically, for "these delicate sentiments, depend on exquisite and refined ideas, and are in reality nothing else but the results of several combinations often too fine to be clearly perceived and almost always too complicated to be reduced to the reasoning which could demonstrate them" without the aid of calculation.[108]

Buffon's solution to the St. Petersburg problem (which he claimed to have discovered in 1730, eight years before Bernoulli's memoir, and sent to the Swiss mathematician Cramer) faulted conventional expectation for not neglecting probabilities less than the .0001 value for moral impossibility, and for tacitly assuming that amount of money was directly proportional to the advantages it brought. By correcting these two errors, he claimed to have discovered a definition of expectation which "did not fly in the face of good sense, and at the same time conformed to experience."[109]

Buffon's treatment synthesized Bernoulli's distinction between quantity and utility of money (an insight Buffon claimed to have achieved independently of Bernoulli's work) and d'Alembert's call for a lighter weighting of very small probabilities.[110] For Buffon, both of these refinements mirrored the true psychology of risk taking. The satisfaction that money can buy depended on one's threshold of satisfaction; moral probabilities must be gauged psychologically, according to the fear or the hope they arouse in reasonable men. Even Buffon's notion of value stemmed from the psychology of habituation: we need what we are accustomed to having; the more opulent the standard of living, the greater the increment needed to make a perceptible improvement.

---

[108] Buffon, "Essai," pp. 73–74.   [109] Buffon, "Essai," p. 81.

[110] Alone among the classical probabilists, Buffon actually made an experiment, however inconclusive, on the proper expectation in the St. Petersburg problem. He employed a child to play the game 2,084 times, and computed an "actual" expectation of about 5 écus; see Buffon, "Essai," pp. 87–88.

Condorcet's attempt to salvage the classical definition of expectation, which he reinterpreted as an average valid only over the long run, partook of all three types of "moral" considerations—legal, economic, and psychological. Condorcet's interest in probability theory was apparently first kindled by d'Alembert's critique of the theory. A protégé of d'Alembert, Condorcet undertook a never-published defense of his mentor's views against defenders of the conventional theory of mathematical probability. Many of his later writings on probability theory, although they accepted the major tenets of the theory, reveal the influences of d'Alembert's criticisms on Condorcet's philosophical interpretation of probability in terms of a "motive for belief."

Two unpublished manuscripts by Condorcet[111] shed light on the development of his views on the theory of probability in general and on the concept of expectation in particular. The ideas introduced in these manuscripts were reworked and developed in his memoirs to the Académie des Sciences on probability and in his *Essai d'analyse sur la probabilité des décisions* (1785). Although neither manuscript is dated, the earlier of the two appears to have been composed in the early 1770s as the above-mentioned rebuttal on d'Alembert's behalf to a certain Massé de la Rudelière's rejoinder to d'Alembert's criticisms of conventional probability theory and of Bernoulli's smallpox computations. Massé de la Rudelière, described by Condorcet as "absolutely unknown" (presumably to mathematicians), had accused d'Alembert of "a kind of felony" in undermining the principles of a branch of mathematics by intellectual pyrrhonism.[112] Keith Baker judges this manuscript to be the piece mentioned by Condorcet to Ann-Robert-Jacques Turgot in a letter of 3 September 1772 as a "small book" on the subject of probabilities.[113] The second manuscript could well have been a draft for Condorcet's 1781 memoir, "Réflexions sur la règle générale qui prescrit de prendre pour valeur d'un événement incertain . . . ," published in the *Mémoires de l'Académie des Sciences*. Both the topic (expectation) and the approach are strikingly similar to those aired in the published memoir.

Condorcet's defense of d'Alembert in the first manuscript was not un-

---

[111] Condorcet, Bibliothèque de l'Institut MS. 875, ff. 110–112; MS. 883, ff. 216–221.

[112] Massé de la Rudelière, *Défense de la doctrine des combinaisons* (Paris, 1763), pp. 148, 53.

[113] Charles Henry, ed., *Correspondance inédite de Condorcet et de Turgot, 1770–1779* (Paris, 1883), pp. 97–98; Baker, *Condorcet*, pp. 176–177.

qualified. He disagreed, for example, with d'Alembert's suggestion in the *Encyclopédie* "Croix ou pile" article that the probability of getting two heads in a row in the game proposed was really 1/3 rather than 1/4.[114] After alluding to the debate over expectation sparked by the St. Petersburg problem, Condorcet observed that whichever side of the debate he defended, "I am sure to have for me some very great mathematicians. However, one may disagree with their views with impunity," since mathematicians were "the only men who have adversaries without enemies."[115] If Condorcet was not d'Alembert's unconditional advocate, he did nonetheless take d'Alembert's arguments seriously. The manuscript appears to have been an introduction to an analysis of the concept of probabilistic expectation, which d'Alembert was the first, according to Condorcet, "to call into question." Condorcet briefly summarized the elementary definition of probability and the Huygenian rule of expectation, introducing the latter in the context of "distinguishing an equitable game from one that is not." The text ends before Condorcet made good his promise to analyze the rule and its ostensible proofs, a task to which he returned in other writings. We are left with only a hint of Condorcet's approach: his claim that all applications of mathematics presumed a "common measure." Geometry calculated the relations of lines, surfaces, and volumes once each of these had been reduced to a common unit; the calculus of probabilities must likewise be reduced "in order to calculate relations and to first of all find a common measure."[116] Presumably, Condorcet sought in expectation a common measure for the disparate states of individuals facing unequal risks.

The importance of a natural, universal measure was a recurring theme in Condorcet's thought. Not only mathematical magnitudes, but currency, weights and measures, and laws demanded homogeneous expression. In his commentary on Montesquieu's *L'Esprit des lois* (1748), Condorcet united the reform of weights and measures and of civil and criminal codes under the heading "Ideas of Uniformity," which, Condorcet contended, "please all minds, especially precise minds."[117] Condorcet praised Turgot's projects for a system of uniform weights and measures, a uniform code of conduct deduced from "the general principles of natural law," a

[114] Bibliothèque de l'Institut MS. 833, f. 219v.

[115] MS. 883, ff. 217v–218r.   [116] MS. 883, f. 216r.

[117] Condorcet, "Observations sur le vingt-neuvième livre de *l'Esprit des lois*," in *Oeuvres*, vol. 1, pp. 376–381.

uniform system of orthography, and so on. According to Condorcet, all of these standardizations would serve to equalize the capacities of citizens and to thus prove that "the establishment of public education worthy of the name is not a chimera."[118] Intellectual force meant the ability to analyze, compare, and recombine ideas, operations facilitated by a homogeneous mental measure. Citizens not graced by nature with strong faculties might make up the difference with the help of artificial systems which simplified and standardized all aspects of life.

The second manuscript explored the problems of assuming homogeneous expectations in greater detail. Here, Condorcet abandoned all hope of finding a mathematical expression which would render the states of two players in a game of chance with unequal probabilities of winning "exactly the same." Instead, he attempted to minimize the inequality between the net gains and losses of both players over the long run. Condorcet defended the conventional definition of expectation on four points, which he later elaborated in his Académie des Sciences memoir on the subject: the classical definition insured that the single most probable outcome would be that which produced "the least inequality possible" between players; the longer the game continued, the more probable it became, supposing the conventional definition, that the ratio of the actual loss to the greatest possible loss for each player would be very small; for a large enough number of rounds, this probability could be made larger than any given probability that the ratio was smaller than any given quantity; and finally, that even if the conventional hypothesis was not the only one to satisfy these stipulations, it was the only one which could be expressed "by a simple formula, suitable for ordinary usage."[119]

Condorcet's approach was clearly modeled on Jakob Bernoulli's limit theorem, presented in Part IV of the *Ars conjectandi*. Like Bernoulli, Condorcet envisioned the problem in terms of numerous trials and limiting values, rather than individual outcomes. In Bernoulli's theorem, the probability that observed frequencies ever more closely approximated a priori probabilities grew with the number of trials; in Condorcet's account of expectation, the probability that the actual disparity between players would approach zero also increased without bound. Rigorously speaking, expectation did not apply to single cases any more than Bernoulli's theo-

[118] Condorcet, *Vie de Turgot*, in *Oeuvres*, vol. 5, pp. 204–206; Baker, *Condorcet*, p. 65.
[119] Bibliothèque de l'Institut MS. 875, ff. 110v–114.

rem could be applied to the results of a single trial. (This observation alone was grounds for rejecting the logic of Pascal's wager, according to Condorcet.) Nor did it apply to cases that required an infinite number of rounds, as in the St. Petersburg problem. In both cases, average expectation became meaningless.

Evidently, Condorcet had originally intended to recast the definition of expectation so as to embrace both sorts of exceptions: at the end of the manuscript fragment, he posed both problems. Judging from his published memoir, however, his attempts to generalize the average expectation rule to anomalous cases in the end simply strengthened his conviction that average expectation provided the only sure guide to comparing the heterogeneous states of players with equal probabilities of winning. Cases where the rule could not be applied were dismissed as not "real." Condorcet insisted that expectation must be regarded as an average, summed over many trials. Bernoulli's theorem showed that the probability that the actual ratio differed from the a priori probability by less than any given amount approached certainty as the number of trials approached infinity. Therefore, mathematical expectation was a valid guarantee of fairness in the long run.

In a six-part series of memoirs addressed to the Académie des Sciences and published in the *Mémoires de l'Académie des Sciences* (1781, 1783, 1784), Condorcet examined the foundations and applications of probability theory with special attention "to those results too far removed from those given by common reason."[120] Condorcet conceded that mathematical expectation sometimes gave absurd results when applied to individual cases, and concluded that the conventional definition held good only for average values. In order to substitute these "average" values for the "real" values of individual expectation, Condorcet appealed both to jurisprudence and political economy. According to Condorcet, there existed two possible ways of replacing real values for average ones: "voluntary" (a willing exchange of possible for certain gain); and "involuntary" (an unavoidable risk compensated by a certain amount). Condorcet admitted that the two cases, which paralleled the legal categories of voluntary and involuntary contracts,[121] did not differ mathematically. In both an uncertain gain is traded

[120] Condorcet, "Suite de mémoire sur le calcul des probabilités. Article VI," *Mémoires de l'Académie des Sciences* (1784; pub. 1787), p. 456.

[121] See Jean Domat, *The Civil Law in Its Natural Order*, William Strahan, trans. (London, 1737), 2nd edition, p. xli.

for a certain one, or two unequal and unequally probable sums are exchanged. However, the two cases were treated separately because they involved different conditions for equity, although both reduced to the same mathematical conditions.

Voluntary substitutions provided Condorcet with the meat of his analysis. Involuntary substitutions adhered to "the laws of equity," since one need only follow "the sum total of similar conventions, and seek to arrange matters so that the least possible inequality results." In voluntary substitutions, "if one wishes to act with prudence, if the object is important," legal convention guided action only insofar as the two parties agreed to a weaker form of equity ("une égalité suffisante").[122] Like Daniel Bernoulli, Condorcet distinguished between the claims of justice and prudence, between the legal and the economic spheres. Voluntary substitutions belonged to the latter domain, and Condorcet's analysis of these relied heavily on the theory of value advanced by his friend and mentor Turgot in the latter's *Réflexions sur la formation et la distribution des richesses* (1766).[123]

Although not an orthodox disciple of the economic doctrines of François Quesnay, Turgot did accept the physiocratic maxim that in exchange, equal value is always traded for equal value. By defining exchange value as price, the physiocrats turned this precept into a tautology.[124] Turgot, however, understood the equality between parties to an exchange in psychological terms, as a balance of the needs and desires on both sides. This balance was tipped in favor of a trade by mutual personal interest which raised the psychological valuation of the other party's goods above that of one's own goods or money. In specific transactions, the value of the goods exchanged had no other measure "than the need or desire of the contracting parties balanced one against the other and is fixed only by their voluntary agreement." However, in the aggregate, the individual motives which prompted individual exchanges for similar goods tended toward an

---

[122] Condorcet, "Réflexions sur la règle générale qui préscrit de prendre pour la valeur d'un événement incertain, la probabilité de cet événement, multipliée par la valeur de l'événement en lui-même," *Mémoires de l'Académie des Sciences* (1784; pub. 1787), pp. 711–712.

[123] For Condorcet's relations with Turgot, see Baker, *Condorcet*, pp. 55–65; also Condorcet's *Vie de Turgot*, especially pp. 42–45, regarding Turgot's economic theories.

[124] Hannah Robie Sewall, "The theory of value before Adam Smith," *Publications of the American Economic Association*, series 3, vol. 2, no. 3 (August 1901), p. 89.

average price (*prix mitoyen*) that equalized the advantages of buyers and sellers over the long run.[125]

Condorcet's discussion of voluntary substitution of expectations turned upon a comparison between this type of transaction and the exchange of goods in "all other markets." In each exchange of expectations, in which like commodities have different intrinsic values and therefore could not be set rigorously equal, there must be personal or subjective *motif de préférence* on both sides which impels a trade. Just as the "relation of reciprocal needs" established equality among the free agents of the marketplace, so in probabilistic expectation, "neither he who exchanges a certain value for an uncertain one, or reciprocally; nor he who accepts the exchange, find in this change any advantage independent of the particular motive of convenience which determined the preference." Averaged over many such exchanges, the expectations, like the common price, would tend toward "the greatest equality possible" between parties.[126]

In order to satisfy Turgot's rules of marketplace exchange, the definition of expectation must meet the following conditions: over the long run, the most probable outcome should be a net loss or gain of zero for both sides; and as the number of cases becomes very large, the probability of gain or loss on both sides should approach 1/2. Condorcet argued that the conventional definition of expectation, conceived as an average value, uniquely satisfied these criteria and offered "the greatest possible equality between two essentially different conditions." Expectations of individual cases no longer made sense in Condorcet's analysis.[127]

Hence Condorcet dismissed the St. Petersburg problem, as originally presented, as an "unreal" case, since the probability of zero net gain or loss for both sides does not equal 1/2 unless the game is repeated an infinite number of times to equalize the states of the players.[128] He then turned to cases in which a "reasonable man" might refuse to pay a stake b to win with a probability p, even if $b < ap$, a being the prize; or in which an

[125] Turgot, *Les réflexions sur la formation et la distribution des richesses*, in *Oeuvres*, Gustave Schelle, ed. (Tauners: Detlev Auvermann, 1972; reprint of 1914 edition), vol. 2, pp. 552–553.

[126] Condorcet, "Réflexions," p. 710.

[127] Condorcet, "Probabilité," *Dictionnaire encyclopédique des mathématiques*, Bossut et al., eds. (Paris, 1789), vol. 2, pp. 654–655.

[128] Condorcet, "Réflexions," pp. 713–718.

equally reasonable man might give up b' to win a' with a probability p', $b' > a'p'$. In the first case, if p is very small and b is large relative to the prospective player's total fortune, the revised definition of expectation justified a refusal to play, since the stake b would deplete the player's resources before he played the game often enough for the average expectation to redress his losses. Moreover, since p is very small, he would be likely to lose b in a single trial, a significant loss which would deprive him of customary pleasures. The second case obtained when both b' and p' were very small, and the player thus risked only slight inconvenience.

Condorcet justified the slight edge, as computed by classical expectation, normally accorded the bank in games of chance as just compensation for the enormous risks taken and the small margin of long-term gains (since the probability of zero net gain or loss increases with the number of games played). Gamblers consented to this inequality as the price paid for the pleasure of gambling. Entrepreneurs, on the other hand, demanded profits greater than those to which mathematical expectation would seem to entitle them since their risks brought no intrinsic pleasure. Therefore, in order to have a *motif de risquer* they must be guaranteed a very large probability of not losing more than a given portion of their investment. All of the apparent contradictions drawn from the everyday conduct of reasonable men could be thus reconciled with expectation viewed as an average, where the probability of genuine equality approached certainty as the number of trials approached infinity, in accordance with Bernoulli's theorem.

Condorcet's elaborate and often convoluted defense of the conventional definition of expectation had drawn upon the legal doctrine of voluntary and involuntary contracts, Turgot's theory of exchange value, a version of moral expectation (the reasonableness of a gamble depends on the proportion of stake to fortune), Bernoulli's theorem, and certain psychological considerations like the intrinsic pleasure derived from gambling. It was perhaps the single most comprehensive attempt among the classical probabilists to square the problematic notion of expectation with the even more problematic notion of good sense. It was also the most neglected. Perhaps the very variety of Condorcet's sources and rationales gave his solution the mathematically unattractive appearance of a miscellany. His murky exposition cannot have helped. But it was nonetheless a typical strategy, albeit an exaggerated one, in its mixture of prescriptive and descriptive elements, its extensive borrowings from the moral sciences, its sensitivity to the psychology of risk taking, and of course in its endless appeals to "good sense" and "common reason." Condorcet added nothing truly original to the discussion, and his mathematical conclusions were conservative, but he effectively summarized all the contributions of his predecessors, thereby revealing how tangled the problem of expectation had become. Even Condorcet appears to have been relieved when the young Laplace proposed a radical simplification of the problem in a seminal paper of 1774.

Laplace attacked the St. Petersburg problem in terms closer to the spirit of d'Alembert's analysis, using a new tool, the analytic formulation of what became known as the Bayes-Laplace theorem on inverse probabilities, which he had developed in the same memoir. Laplace accepted d'Alembert's claim that the conventional method of computing expectation in the St. Petersburg problem unrealistically assumed a perfectly fair coin, "a supposition which is only mathematically admissible, because physically there must be some inequality." To the objection that because both players were equally ignorant as to which way the coin was biased their respective advantages remained equal, Laplace retorted that such specious reasoning simply showed that "the science of hazards must be used with caution, and must be modified in passing from the mathematical to the physical case." [129]

Extrapolating from his solution to the St. Petersburg problem in terms of hidden physical asymmetries, Laplace predicted that a whole genre of such problems would open up to probabilists who are alerted to the "precautions one must take in applying mathematical considerations from the calculus of probabilities to physical objects."[130] Only for single events could the Principle of Indifference be invoked to assume equiprobable outcomes, for want of any information regarding the direction of bias. For a sequence of events, however, the mathematical theory could and should be modified to take account of observed asymmetries in the relative frequencies of outcomes. Laplace was especially eager that this "aberration from the ordinary theory" be heeded in applying probability theory to "civil life." Probabilistic treatments of social phenomena, no less than of coin tosses, should follow observed frequencies of events.

D'Alembert embraced Laplace's innovation and briefly attempted to re-

[129] Laplace, "Mémoire sur la probabilité des causes par les événements," in *Oeuvres*, vol. 8, p. 54. See also *Théorie analytique des probabilités* (1812), in *Oeuvres*, vol. 7, chapter 6.
[130] Laplace, "Mémoire," p. 61.

fine the inverse probabilistic analysis.[131] Laplace's injunction to modify probabilities in light of physical experience harmonized with his own complaint that conventional probability theory conflated mathematical and physical possibility. Condorcet was equally impressed, exercising his prerogative as Perpetual Secretary of the Académie des Sciences to signal Laplace's achievement in the introductory *Histoire*, and perhaps to rush the entire memoir into print.[132] Condorcet also wrote Turgot of a bright new future for probability theory and social mathematics, mentioning Laplace's 1774 memoir.[133] Keith Baker has argued that Laplace's memoir exerted an important influence on Condorcet's conception of both probability theory and social science.[134]

Despite the enthusiasm of d'Alembert and Condorcet for this novel, inverse approach, Laplace himself eventually abandoned it in favor of moral expectation, in his magisterial *Théorie analytique des probabilités* (1812), at least as far as the St. Petersburg problem was concerned. Perhaps he realized the overwhelming difficulties of computing the discrepancy for each and every coin. In any case, in the chapter devoted to moral expectation, Laplace adhered closely to Daniel Bernoulli's analysis, citing many of the same examples, including the advisability of dividing cargo among as many vessels as possible. Laplace noted that in the limit, where the number of ships grew very large, moral expectation approached mathematical expectation. Laplace addressed not only the St. Petersburg problem, but also problems pertaining to the relative advantages of individual and joint annuities with respect to moral (as opposed to physical) fortune. He concluded this section of the *Théorie analytique* by urging governments to promote such schemes, which fostered "the most gentle tendencies of [human] nature." Furthermore they were tainted with none of the hidden pitfalls of gambling, which moral expectation exposed as a perpetual losing proposition even in fair games, and which sober moral reflection condemned.

[131] D'Alembert, *Opuscules*, vol. 7, p. 60.

[132] Baker suggests that Condorcet was so impressed with Laplace's 1774 memoir that, in his capacity as assistant to the Perpetual Secretary of the Académie des Sciences, he had it rushed into print before the logically prior 1773 (pub. 1776) memoir. Baker, *Condorcet*, pp. 168–169.

[133] Henry, *Correspondance*, p. 197. See also C. C. Gillispie, "Probability and politics: Laplace, Condorcet, and Turgot," *Proceedings of the American Philosophical Society* 16 (1972): 1–20.

[134] Baker, *Condorcet*, p. 171.

Although Laplace advanced these recommendations on the strength of results derived from moral expectation, the critiques and counterproposals of d'Alembert and others still colored his views on the subject. While he clung to Bernoulli's hypothesis that the richer one is, the less advantageous a small gain becomes, he warned of the unmanageable complexity of a more comprehensive moral expectation: "But the moral advantage that an expected sum can procure depends on an infinity of circumstances peculiar to each individual, which are impossible to evaluate."[135]

The last major work in the classical tradition, Poisson's *Recherches sur la probabilité des jugements* (1837), added little that was new, for Poisson avowed himself a disciple of Laplace. Although he questioned several of Laplace's assumptions regarding the probability of tribunal decisions, Poisson fully accepted Laplace's treatment of moral expectation, "which accords with the rules which prudence indicates on the manner in which each [individual] should conduct his speculations."[136] While Poisson's solution to the St. Petersburg problems deviated slightly from that of Bernoulli and Laplace in assuming that the critical limiting factor was the bank's rather than the player's fortune—it is worth noting that a discussion of the problems was still *de rigueur* for probability texts— this variant did not alter the definition or application of moral expectation. Like his predecessors, Poisson believed that classical expectation must be supplemented by moral expectation in cases of economic deliberation. Moral expectation embodied the worldly wisdom of the prudent, sober citizen, who understood the difference between pernicious games of chance and sound investment despite apparent similarities in the exchange of secure funds for an uncertain prospect of gain. Poisson rang the changes on Daniel Bernoulli's argument that even fair games of chance are ruinous, and added a further economic stricture against gambling: according to Poisson, gambling created no value, but commerce increased the value of goods by transporting them to market. A social boon, this margin of extra value simultaneously rewarded the merchant with profit and served the interest of the consumer.

[135] Laplace, *Théorie analytique*, vol. 7, p. 449. Laplace expressed similar views on the limitations of moral expectation in an earlier memoir, citing d'Alembert's criticisms explicitly: "Recherches sur l'intégration des équations différentielles aux différences finies et sur leur usage dans la théorie des hasards" (1773; pub. 1776), in *Oeuvres*, vol. 8, pp. 147–149.

[136] Siméon-Denis Poisson, *Recherches sur la probabilité des jugements en matière criminelle et en matière civile* (Paris, 1837), p. 72.

Poisson carried forth his work in mathematical probability under Laplace's motto that the theory "is at bottom only common sense reduced to calculus," to the point of amending some of Laplace's own solutions under the cover of this cardinal principle.[137] However, Poisson recognized that support for the probabilist program for the moral sciences, as a mathematical description of social experience and a guide to rational conduct, was flagging even among mathematicians. Louis Poinsot, for example, attacked Poisson's work on the probability of judgments as "a false application of mathematical science." Poinsot and Charles Dupin, both Poisson's colleagues in the Académie des Sciences, pointed to Laplace's own reticence in applying probability theory to situations of such formidable complexity, and warned that the notion of "a calculus applicable to those things where insufficient enlightenment, ignorance, and human passions were mingled" could constitute an invitation to abuse.[138]

Although Poisson's mathematical approach to the social sciences found new support in the work of Adolphe Quetelet, the role of mathematical probability in the study of society had changed. The eighteenth-century probabilists had viewed the theory as a description and guide to social action based on the example of the reasonable man. The dictates of "good sense" supplied the data which the calculus of probabilities was to systematize and explain. To the extent that good sense was already codified in jurisprudence and political economy, probabilists found it natural to incorporate assumptions and concepts taken from these disciplines into the mathematical theory. The traffic between mathematical probability and the moral sciences was especially heavy in cases, such as the St. Petersburg problem, where the mathematical results were at odds with good sense. With the exception of d'Alembert, eighteenth-century mathematicians did not believe that the complexity of the moral realm, as it was reflected in the psychology of decision making, posed insurmountable obstacles to the probabilist program.

However, the upheaval of the Revolutionary and Napoleonic era appears to have shaken the confidence of probabilists in a way that d'Alembert's persistent criticisms had not. The conduct of reasonable men no longer seemed an obvious standard, nor a comprehensive basis for a theory of

---

[137] Poisson, "Recherches sur la probabilité des jugements," *Comptes rendus hebdomadaires des séances de l'Académie des Sciences* 1 (1835): 477–478.

[138] *Comptes rendus* 2 (1986), on p. 380.

society. Distinguishing prudent from rash behavior in post-Revolutionary France was no easy matter, and just what constituted "good sense" was no longer self-evident. With the demise of the reasonable man, the probabilists had lost both their subject matter and criterion of validity.

Laplace and Poisson marked the transition from this eighteenth-century probabilist program to the more statistical orientation of Quetelet. Despite his use of moral expectation, Laplace was notably reluctant to extend this sort of analysis further. His treatment of probability applied to judgments, a favorite eighteenth-century topic and the subject of treatises by Condorcet and Poisson, was hedged about with warnings and caveats about the extreme delicacy required in subjecting such intricate problems to mathematical analysis. Laplace did not wholly abandon probabilist hopes for a "social mathematics," but rather directed probability toward a different measure of social experience, statistics.

Instead of making the conduct and opinions of reasonable men their subject, the new breed of probabilists focused on compilations of facts about other aspects of society, such as the annual rates of conviction in civil and criminal courts published by the French Ministry of Justice.[139] Applied to this data, probability theory would reveal the universal laws that governed social phenomena. Quetelet chose a characteristic passage from Laplace's *Essai philosophique sur les probabilités* for the frontispiece quotation to his treatise *Sur l'homme* (1835), significantly subtitled *Essai de physique sociale*: "Let us apply to the political and moral sciences the method of observation and calculation which has served us so well in the natural sciences." It was not the combination of observation and mathematics which was novel to the application of the calculus of probabilities to the sciences of society, but rather the choice of observations. Because the concept of probabilistic expectation had been shaped by legal and economic considerations, eighteenth-century probabilists took the conduct of prudent men as an index. For them, mathematical probability promised a mathematical expression of the rules underlying good sense. Their nineteenth-century successors still envisioned probability theory as a mathematical description of society, but understood society to be the aggregate of all behavior, reasonable or not, catalogued in the burgeoning store of

---

[139] These *Comptes généraux de l'administration de la justice criminelle* were first published in 1825 under the auspices of the French Ministry of Justice and greatly influenced the work of both Poisson and Quetelet. See Section 6.4, below.

statistics. Within this mass of data, probability theory would uncover macroscopic regularities about social processes like population growth and the crime rate.

Daniel Bernoulli and his contemporaries had concentrated on the psychology of the rational individual; Quetelet insisted that probability theory sanctioned the neglect of individuals, who exerted "little or no force on the mass," the level at which he expected laws to emerge.[140] For the nineteenth-century probabilists, the emphasis shifted from the measure of expectation to the study of distributions. Although both schools stressed the special fitness of probability theory as a mathematical tool for the study of society, they conceived the objectives and content of the moral sciences in very different ways.

2.4 Conclusion

Throughout the eighteenth century, probabilists regarded expectation as a mathematical rendering of pragmatic rationality. The calculus of probabilities reflected the thought and practice of a small elite of perspicacious individuals who exemplified—rather than defined—the virtue of reasonableness. Although the probabilists, with the exception of d'Alembert, did not waiver in their conviction that their theory should mirror the opinions of the elite of reasonable men, they could not agree on a suitable membership criterion. Were the elect evenhanded judges or prudent merchants? Did economic theories of value or the psychological promptings of intuitions too delicate for gross reason constitute the essence of reasonableness? Insofar as contemporary psychological theory measured the force of intellect by the capacity to "combine ideas in memory and multiply their combinations,"[141] it seemed plausible to eighteenth-century thinkers that the combinatorial calculus of probabilities could both describe and systematize good sense. Once codified in mathematical form, good sense could be disseminated beyond the narrow confines of the elite to the population at large, teaching citizens their true interests while consolidating social consensus. However, consensus eluded even the mathematicians, for whom

[140] Adolphe Quetelet, *Sur l'homme et le développement de ses facultés, ou Essai de physique sociale* (Paris, 1835), vol. 1, pp. 4–5.

[141] Condorcet, *Vie de Turgot*, p. 222. See Section 4.2 below.

the notion of reasonableness remained clouded with ambiguity. The history of the concept of expectation during the eighteenth century is the history of unsuccessful attempts to fix "reasonableness" in mathematical form.

Although mathematicians have long since abandoned this task as thankless and even antimathematical, economists and psychologists still grapple with the shadowy, multilayered meaning of practical rationality. Prescriptive and descriptive approaches still clash, and the arguments advanced by proponents of the latter have a familiar ring. The French economist M. Allais, criticizing the assumptions of the "American School" of utility theorists, accused latter-day advocates of the "Bernoulli principle" (moral expectation as modified by John von Neumann and O. Morgenstern) for tautologously defining "rationality" as conduct consistent with that formulation of utility. Allais claimed that although no one really follows the precepts laid down by economists, the American School believed that the rational individual should do so. Allais proposed an alternative, "experimental" definition highly reminiscent of the eighteenth-century probabilists: "One can only appeal to experience and observe *what men who are thought to comport themselves rationally actually do* . . . [that is] individuals considered rational by common opinion."[142] Other economists have seconded Allais' call for a fuller appreciation of the actual psychology of risk taking.[143] Exiled from probability theory, the "reasonable man," still identified by "common opinion," has taken refuge in economics, and more recently, in psychology.[144]

It would be tempting to conclude that economics and psychology have been his residence all along, and that to the extent that classical probabilists studied his behavior, they were engaging in economics and psychology rather than mathematics. However, to affirm that conclusion, convenient though it may be, would be to impose a twentieth-century classification of disciplines upon the very different intellectual landscape of the Enlightenment. Boundaries separating disciplines shift over the course of time,

[142] M. Allais, "Le comportement de l'homme rationnel devant le risque: Critique des postulats et axiomes de l'Ecole Américaine," *Econometrica* 21 (1953): 503–546, especially p. 521.

[143] See, for example, Paul Samuelson, "Probability, utility, and the independence axiom," *Econometrica* 20 (1952): 670–678.

[144] See, for example, Daniel Kahneman, Paul Slovic, and Amos Tversky, eds., *Judgment Under Uncertainty: Heuristics and Biases* (Cambridge, Eng.: Cambridge University Press, 1982).

and concerns which now seem peripheral or even alien to a given science may once have been central. Such was the case for mathematical probability and the moral sciences during this period.

It would be as misleading to identify the Enlightenment moral sciences narrowly with their twentieth-century descendants as it would be to confuse eighteenth-century probability theory with its twentieth-century counterpart. In both cases continuities exist, but there are striking contrasts as well. Enlightenment students of the moral sciences probed the psychology of the rational individual, seldom turning to large-scale social organization. Although political economists did study macroscopic phenomena such as the circulation of wealth described by the physiocrats, their primary interest remained the psychological mechanisms that governed individual behavior. The subjective theories of value advanced by Galiani, Turgot, and Condillac were of a piece with the epistemological orientation of eighteenth-century philosophy,[145] which made individual perceptions and judgments the departure point for analysis: "I conceive a thing to be rare, when we *judge* that we do not have enough for our use; it is abundant, when we *judge* that we have as much as we need, and it is superabundant when we *judge* that we have more than that."[146] Probability theory, as a branch of mixed mathematics, was not an alloy of a distinct, formal calculus and a set of applications and models, but rather a mathematical description—or an attempt at one—of these same psychological processes.

I have argued that the eighteenth-century probabilists tailored the mathematical theory to fit the prescriptions of rational conduct under uncertainty. They repeatedly rejected mathematical methods that contradicted the promptings of good sense. "Good sense," however, was by no means monolithic, admitting numerous interpretations. Legal, economic, physical, and psychological refinements were all proposed as adjustments to the mathematical theory in the name of good sense. In the case of expectation, mathematicians originally derived the classical definition from legal theories of contractual equity, but turned to economic theories of value as sources of an alternative, or moral, expectation for situations in which prudence rather than justice prevailed. The competing orientations

---

[145] Ernst Cassirer, *The Philosophy of the Enlightenment*, F.C.A. Koelln and J. P. Pettegrove, trans. (Princeton: Princeton University Press, 1951), chapter 3.

[146] Etienne Condillac, *Le Commerce et le gouvernement* (1776), in *Oeuvres*, vol. 4, pp. 17–18; emphasis in the original.

of jurisprudence and political economy created rival definitions of expectation, and psychological and physical considerations further complicated the issue. Nonetheless, probabilists continued to judge the validity of their mathematical results by their consonance with good sense.

Only d'Alembert suggested that social reality might elude altogether the sort of mathematical description sought by the probabilists, and even his criticisms shared their assumptions if not their optimism. Like his colleagues, d'Alembert believed that "social mathematics" would be a mathematical model of the way in which reasonable men made decisions and took risks. However, he doubted the capacity of probability theory to render full account of all the relevant variables. Although Laplace and Poisson retained many elements of the good-sense approach to probability, they also appealed to a statistical description of society which Quetelet made the cornerstone of his "social physics." Probabilists no longer cross-checked their results against the practices and beliefs of "men known for their experience and wisdom in the conduct of their affairs,"[147] and the once-neighboring disciplines of jurisprudence, political economy, and mathematical probability drifted ever farther apart.

[147] Silvestre-François Lacroix, *Traité élémentaire du calcul des probabilités* (Paris, 1816), p. 257.

The Theory and
Practice of Risk

3.1 Introduction

The mathematical probabilists were not the first theorists of risk. Sixteenth- and seventeenth-century jurists writing on aleatory contracts had already created a learned literature on the subject, and their analysis exerted a lingering influence on mathematical probability, as we have seen. But the mathematicians created a new approach to the subject that challenged the previous *practice* of risk, legal and otherwise. Whereas earlier writers on insurance, annuities, and other risky ventures had emphasized prudent judgment based on the particulars of the individual case, the probabilists proposed general rules to determine the fair price of the risk. The probabilists' habit of talking about their calculus as simply a mathematical version of such informed judgments sometimes hid this distinction, but it was nonetheless there. It was as if the jurists and the commercial class they wrote for lived in a world of fine-grained detail where regularities were partial at best: the sudden appearance of pirates on the route to Alexandria would raise the price of maritime insurance in Venice from one day to the next; a sickly young man of twenty years might command the same annuity price as a robust specimen of fifty; wine futures fluctuated with the myriad signs that presaged a good or bad season. It was not a world of constant surprises, but it was one where specific, up-to-the-minute, and above all personal knowledge counted, knowledge to be sifted and weighed by an old hand in the business. The mathematicians, in contrast, apparently lived in a world strictly governed by invariable laws that could be expressed as the function of a small number of variables: armed with tables of life expectancies and compound interest, the seller of annuities need not personally interview the annuitant, for knowing his age and the proper mathematical rule would allow the green-est of clerks to set the correct price. The world of the mathematicians was simple, stable, and predictable; the practical applicability of their rules depended on how closely this serene vision approximated the real world of risk taking in the eighteenth century.

What was that world of risks, real and perceived? The answer must be a carefully bounded historical one, both because one generation is oblivi-ous to what the next most fears, and because we know that key phenomena such as mortality rates are variable, at least over the long term. That is, we can contemplate circumstances that would vindicate either the jurists or the mathematicians—for example, periods of plague with a jagged mor-tality curve over time, or periods of peace and good harvests with a rela-tively smooth one. It is conceivable that broad areas of human experi-ence—life-span, travel, familial bonds—can be made more regular and stable, or the reverse, by changing political, economic, and social arrange-ments. For those fortunate enough to live in one of the more even-keeled periods, their own lived experience could provide a kind of empirical im-petus for an ambitious determinism that looked for regularities every-where. Of course I speak loosely here. No amount of experience could prove or disprove a metaphysical position; however, it could be suggestive at a psychological level. There is some evidence, for example, that in-creased urbanization and a wider web of economic interdependence in the late seventeenth and eighteenth centuries may have made daily conduct more predictable in parts of Western Europe: work schedules became more routinized;[1] middle- and upper-class manners were refined by a stricter code of etiquette;[2] casual, unprovoked violence on city streets decreased—all possible indicators of "a more orderly, more disciplined, and less per-sonally aggressive society."[3] Possibly these and other reductions in daily uncertainty, as much as the vaunted successes of Newtonian mechanics, fostered the optimistic view that other complicated forms of uncertainty might be reduced to rules, and fairly simple ones at that. Here the expe-rience of risk might have worked in subtle ways upon the theory of risk.

[1] E. P. Thompson, "Time, work-discipline, and industrial capitalism," *Past & Present*, no. 38 (December 1967): 56–97.

[2] Norbert Elias, *The Court Society*, trans. Edmund Jephcott (New York: Pantheon, 1983), chapters 5–6.

[3] Lawrence Stone, *The Family, Sex and Marriage in England 1500–1800* (London: Wei-denfeld and Nicolson, 1977), p. 94.

A full understanding of the eighteenth-century experience of risk would, however, take us far afield from mathematical probability, into historical demography and the *histoire des mentalités*. In this chapter, I shall concentrate on the institutionalized forms of risk taking that most interested the classical probabilists: life insurance and annuities; and gambling, especially lotteries. Gambling supplied the bread-and-butter problems for all the mathematical texts from Huygens on, and its practical interest, if not its respectability, was heightened by the widespread use of lotteries in the eighteenth century to raise money for governments and various charitable causes. Although problems concerning investments dependent on lives lacked the mathematical simplicity of calculating lottery chances, they were of even greater economic importance: in 1753 James Dodson gave a long and impressive list of the many kinds of property, "of much the greatest part of the real estates" in Great Britain, that "depend on the values of lives."[4] By 1750, the interested reader could find mathematical manuals on these topics in English and Dutch, somewhat later in French, German, and Latin, many deliberately pitched to the barely numerate. Yet until the turn of the nineteenth century, and in some locales still later, their influence on designing lotteries and annuity or insurance schemes was negligible. Despite the best efforts of the mathematicians, the practice of risk was almost wholly untouched by the theory.

No simple appeal to the conservatism of practice can explain this state of affairs. The early eighteenth century was a period of intense economic innovation in England and the Netherlands. These two countries led the field in practical probabilistic literature, and also in the new-fangled lottery, annuity, and insurance projects that were among the most daring of these fiscal experiments. Moreover, the competition was keen, and entrepreneurs invented endless bonuses, gimmicks, and ruses to win a slight edge. The case of Charles Povey, founder of the Sun Fire Office and a mutual life insurance company in London (1710), who threw in a free subscription to his newspaper, the *General Remark*, a claim to its profits, and an almshouse for needy policyholders to attract customers, is not atypical.[5] Novelty, particularly novelty that promised a profit, was a positive

---

[4] James Dodson, *The Mathematical Repository*, 2nd edition (London, 1775), vol. 2, pp. vii–viii.

[5] P.G.M. Dickson, *The Sun Insurance Office 1710–1960* (London: Oxford University Press, 1960), chapter 2; Charles Povey, "Proposals" (London, 1706), London Guildhall Library MS. 18,847.

attraction to these project makers. They did not neglect the probabilistic techniques tailor-made for them out of inertia.

For the practitioners of risk to accept the mathematical theory of risk required profound changes in beliefs and, at least in the case of life insurance, also in values. They had to replace individual cases with rules that held only en masse, and to replace seasoned judgment with reckoning. In effect, they had to expand their time frame to the size which smoothed out local perturbations into an overall uniformity; they had, in short, to believe in the reality and stability of averages. The very element of uncertainty that had distinguished the legal aleatory from the illegal usurious contract almost disappeared in this new long-term perspective. Partly because of these new beliefs, and partly because of changing attitudes toward family responsibility, life insurance evolved from a wager to its antithesis over the course of the eighteenth century. At the heart of all of these changes lay an altered conception of time and numbers. The founders of the early life insurance societies believed that more members enrolled over more time meant more risk; the probabilists asserted the opposite. The insurers thought in terms of cumulative risks, a growing sum over cases and time; the probabilists thought in terms of symmetric deviations from an average that would cancel one another out over the long run. The insurers equated time with uncertainty, for time brought unforeseen changes in crucial conditions; the probabilists equated time with certainty, for time brought the large numbers that revealed the regularities underlying apparent flux. From individual cases considered in the short term with a wary eye for sudden changes, to many cases considered in the long term with supreme confidence in the stability of events: this was the arduous conceptual transition that insurers and other dealers in risk had to make in order to apply the mathematical theory of risk to their practice. It is therefore not surprising that they came to accept that theory slowly and haltingly, if at all.

This chapter examines the changing relationship between the theory and practice of risk in three parts: the preprobabilistic practice of risk; the mathematical theory of risk; and the impact of theory on practice in the eighteenth century. In this last section I shall be primarily concerned with state lotteries and with life insurance rather than the whole field of gambling, annuities, and various and sundry forms of insurance in this period. My choice is based on the availability of evidence and on significance. Unlike most forms of gambling, the official lotteries were legal, and therefore left documentary traces. Moreover, the lotteries seem to have captured the imagination (and the money) of more people, high and low, than any other kind of gambling. Life insurance was the first, and for a long time the only, branch of this venerable business to make use of mathematical methods, and therefore merits special attention. It also provides a particularly dramatic example of the changes in beliefs and values needed to make mathematically based insurance attractive to both sellers and buyers. As in the case of its other applications, the classical theory of probability treated the problem of risk as one of rationality, and the treatises on annuities, insurance, and especially gambling were written in a crusading spirit. But here as elsewhere, rationality proved elusive to the mathematicians. They could flesh out their definitions of probabilistic expectation with exemplars, albeit competing ones—the equitable judge; the prudent businessman—but in the case of insurance, the accepted standard of practice was difficult to interpret as implicitly probabilistic. It was, more often than not, explicitly antiprobabilistic. This does not mean that it was, *pace* the probabilists and our own faith in actuaries, therefore irrational, but it does suggest conflicts between the rationalities of theory and practice. The resolution of these conflicts—what is sometimes misleadingly called "rationalizing" practice—involved more than just "seeing the light" or mastering new techniques; it required living in a different, more stable world.

3.2 Risk before Probability Theory

Risk taking in the form of gambling, insurance (chiefly maritime), and annuities was institutionalized in Europe long before the advent of mathematical probability. Gambling is an ancient pastime, and already was so in classical times.[6] Maritime insurance was perhaps practiced by the Babylonians, and certainly by the Romans in the form of the bottomry agreement, or *foenus nauticum*.[7] Annuities were also known to the Romans, as references in the *Digest* of Justinian and Ulpian's table show.[8] These

---

[6] F. N. David, *Games, Gods and Gambling* (London: Charles Griffin, 1962), chapter 1.

[7] C. F. Trenerry, *The Origin and Early History of Insurance* (London: P. S. King & Son, 1926), pp. 50–60.

[8] For Ulpian's rules, see Jacques Dupaquier, "Sur une table (prétendument) florentine d'espérance de vie," *Annales. Économies, Sociétés, Civilisations* (July-August 1973), pp. 1066–1070, on p. 1067.

multifarious forms of risk taking were given a common identity by legal attempts to distinguish them from usurious contracts. Bottomry agreements, for example, looked suspiciously like loans with interest to some Christian theologians: the insurer lends the merchant the cost of a voyage; if the ship is lost, the debt is canceled, but otherwise is repaid with a bonus. Pope Gregory IX's decretal *Navaganti* (1237) in fact prohibited this most popular kind of maritime insurance as usurious, thereby stimulating jurists to redefine these and other forms of investment that reaped gain without labor in safer terms.[9] As written policies and insurance spread in the fifteenth century,[10] the literature of these aleatory contracts proliferated as well.

As we have seen, the salient aspect of the aleatory contract was risk.[11] In place of labor or property, the parties to the contract exchanged present certainty for future uncertainty. Take away the essential element of risk, and the legal aleatory agreement immediately collapsed into an illegal usurious one. Since church authorities continued to regard some kinds of aleatory contracts with suspicion,[12] jurists and theologians in the sixteenth and seventeenth centuries were led to place still more emphasis on the redeeming feature of risk. Apologists for dubious commercial practices here balanced precariously between usury on one side and the equally reprehensible activity of gambling on the other, since a gamble was and remained almost synonymous with risk. Risk was the pivot around which new ideas about the fertility of money turned, and as such was accepted by some early modern canonists as a title to both profit and property, and later to interest.[13] Determining the fair price of risks, or the expectation, was a problem that exercised the jurists and later the mathematicians. However, here I am less concerned with theoretical than with practical pricing of annuities, insurance premiums, and gambling stakes.

Our main source for premium prices for maritime insurance in the preprobabilistic period are legal codes, practical manuals, and occasional no-

---

[9] *Decretales Gregorii Noni Pontificis* (Lugduni, 1558), lib. V, tit. xx, cap. xix, p. 1023.

[10] L. A. Boiteux, *La fortune de la mer* (Paris: École Pratique des Hautes Études, VIe Section, 1968), pp. 69 ff.

[11] See Section 1.2, above.

[12] See, for example, Pope Sixtus V's bull *Detestabilis avaritia* (1586), which condemned the triple contract.

[13] John T. Noonan, Jr., *The Scholastic Analysis of Usury* (Cambridge, Mass.: Harvard University Press, 1957), pp. 152, 241.

tarial records. With the spread of third-party policy insurance in the four-teenth century (the earliest known contract dates from Marseille in 1328),[14] important port cities from Barcelona to Antwerp passed detailed codes regulating all aspects of the trade. Indeed, the dates of the various ordinances roughly chart the migration of the commercial center of mari-time Europe over the course of some three centuries: Genoa (1369), Bar-celona (1435), Florence (1522), Seville (1543), Rouen (1556), Low Coun-tries (1570), Amsterdam (1598), Rotterdam (1604), Middlebourg (1660).[15] The Italian cities in general passed codes that only tardily and sketchily recognized insurance arrangements already long in practice, al-though Venice passed frequent statutes of extreme specificity.[16] The influ-ential Barcelona "Consulate of the Sea" is typically explicit about every aspect of maritime insurance except premiums: we learn, for example, that cargoes can be insured up to 7/8 value for Barcelonans and 3/4 value for foreigners; cargo loaded beyond the Straits of Gibraltar cannot be insured except under special conditions; the penalties for fraud; the time limita-tions on claim payments (two months for Catalonia and Valencia; three months for Naples and Sicily; six months when there is no information on which direction the ship sailed); but nothing about how much the insur-ance cost.[17] The Low Countries code sets an upper limit for the price of "resicq & peril" at 10 percent the value of goods insured but offers no further advice, in contrast to the painstaking instructions on all other mat-ters, including a fill-in-the-blank policy form.[18] The fact that premiums in Italian coastal cities fluctuated between 5 and 19 percent as a function of length of voyage and other factors suggests that the ordinances were uncharacteristically silent or vague on the subject because conditions were too labile to fix even approximate prices in general.[19]

The manuals confirm this impression. Estienne Cleirac's *Us, et coutumes de la mer* (1656), a paraphrase and update of the earlier Rouen *Guidon de la*

[14] J. N. Ball, *Merchants and Merchandise* (New York: St. Martin's Press, 1977), p. 180.

[15] Isidore Alauzet, *Traité générale des assurances* (Paris, 1843), vol. 1, pp. 69–90.

[16] Giuseppe Stefani, *Insurance in Venice from the Origins to the End of the Serenissima*, trans. Arturo Dawson Amoruso (Trieste: S.p.A. Poligrafici il Resto del Carlino, 1958), p. 30.

[17] Stanley Jados, *Consulate of the Sea and Related Documents* (University, Alabama: University of Alabama Press, 1975), pp. 287–302.

[18] *Ordonnance, Statut et Police Nouvellement Faicte par le Roy Nostre Sire, svr le faict des con-tractz des assevrances es Pays-Bas* (Anvers, 1571).

[19] Stefani, *Insurance*, p. 61.

mer (first known edition 1607),[20] stipulates the content of a standard insurance policy in the customary detail, including a 10 percent deductible clause to discourage negligence. Nowhere, however, in his comprehensive survey of maritime insurance does Cleirac offer any specific guidelines to pricing, except to observe that the greater the risks (which depended on the nature of the cargo, the season of the year, the route taken, the experience of the captain, etc.), the higher the premium, reckoned as some percentage of the value of the insured goods.[21] These instructions were repeated—in much the same words—in other contemporary treatises on the subject.[22]

Cleirac's failure to supply any rule or system for determining the premium from the level of risk reflects the actual practice of sixteenth- and seventeenth-century insurers, who relied on a combination of experience, intuition, and convention to set the price of the premiums. In general, the rates for certain well-traveled routes remained fairly constant, although with a sizable spread of minimum and maximum values as circumstances demanded. For example, Venetian registers from the period 1588–1605 show a fairly stable tariff for the Venice/Syria/Alexandria route at 3½–4 percent, but the spread is 2½–6 percent. Typically enough, rates for the return voyage were higher, since the insurer could not personally examine the state of the ship.[23] Tidings of warships and pirates produced the greatest fluctuations in premiums, and Cleirac and other guides instructed insurers to fix rates on an individual basis, according to the latest "good or bad news." Apparently a mystique grew up around estimating risks and setting premiums. Some insurers built reputations comparable to those of bookmakers believed to have a sixth sense for estimating odds. In the early fifteenth-century in Italy companies of insurers were formed under the leadership of a *toccatore* who assumed special responsibility for assessing premiums, companies which did not replace the older practice of merchants insuring one another. In some locales such companies never emerged, but even here a small minority of underwriters among the mer-

---

[20] Possibly there existed earlier editions in the period 1566–84; on its enduring influence, see Alauzet, *Traité*, vol. 1, p. 99.

[21] Estienne Cleirac, *Les Us, et coutumes de la mer* (Rouen, 1671), pp. 271–272.

[22] Boiteux, *Fortune*, p. 176.

[23] Alberto Tenenti, *Naufrages, corsaires et assurances maritimes à Venise 1592–1609* (Paris: S.E.v.P.E.N., 1959), p. 60; compare the Barcelona "Consulate," Jados, *Consulate*, pp. 289–290.

chants dominated the policy registers.[24] These companies were the ances-
tors of syndicates like Lloyds of London, and the *toccatore* eventually
evolved into the professional underwriter.[25]

While the expertise of a *toccatore* admittedly increased with experience,
insurers evidently did not regard information on the actual frequency of
shipwrecks to be relevant in pricing premiums. There exist no records of
any attempts to compile such statistics, although underwriters from the
late seventeenth century on kept registers of such mishaps on the ships
they insured. Judging from the injunctions of Cleirac and other handbook
authors, the sixteenth-century insurer might well have found such a statistical approach impractical, for it presumes stable conditions over a long
period. Insurers stressed the need to keep abreast of ever-changing circumstances, to make personal inspections of ship, cargo, and crew, and to
adjust each premium to the individual situation: hence the imperative to
monitor the latest reports of pirates and men-of-war along the route, and
the higher rates charged ships embarking from remote ports. Moreover, in
commercial centers populous enough to support whole markets of insurers, premium prices also reacted to levels of supply and demand, as well
as to bad news about storms and privateers. To the extent that premiums
were proportioned to risk, insurers did quantify risk. However, their
methods were not statistical—if anything, their individual bias was antistatistical.

Because premium insurance provides the only ready form of quantifying
risk, it should be sharply distinguished from other forms of risk sharing
known in medieval and Renaissance Europe, such as guild-administered
aid to members and their widows in case of illness or death, or the Florentine dowry societies. This distinction was systematically blurred in the
French and German insurance literature of the nineteenth century by writers eager to identify new social welfare schemes with older, romanticized
forms of *Gemeinschaft* solidarity.[26] However, it is an important one for our
purposes. These mutual aid arrangements certainly involved members in
some risk, for you might end up paying in out of all proportion to what

---

[24] Branislava and Alberto Tenenti, "L'Assurance en Méditerranée," *Annales. Économies,
Sociétés, Civilisations* (March-April 1976): 411–413; Ball, *Merchants*, p. 182.

[25] Boiteux, *Fortune*, p. 163.

[26] Jean Halpérin, *Les assurances en Suisse et dans le monde* (Neuchâtel: Éditions de la Bacon-
nière, 1946), p. 20.

you were paid out. But there was no attempt to proportion contributions, usually paid regularly in the form of dues to the "common box," to the risks. The conditions of equality between risk and price which were part of the legal definitions for insurance and other aleatory contracts played almost no role in the mutual-aid arrangements, and indeed they are classed with insurance only with the benefit of hindsight.[27] Although these philanthropies were attempts to share the burden of risks, they were not attempts to quantify it.

Annuities are exchanges of a lump sum for regular payments over a designated period, usually the lifetime of the annuitant or some third person. The annuitant is in effect betting that he will live long enough to recoup the original sum and more. For institutions of assured continuity, selling annuities was a convenient way of taking out a loan at interest without incurring the suspicion of usury, the uncertainty of the annuitant's life-span providing the exonerating element of risk. Hence by the fifteenth century they were popular ways for cities and religious orders to raise funds in Italy, the Low Countries, and Germany and later spread to England and France. Annuities were transmitted along with Roman law to most of Latin Europe, but although Ulpian's table correlating prices with ages was apparently known in some locales,[28] in general little formal account was taken of age in fixing prices. For example, of the several German cities offering annuities in the fourteenth century, only one (Nordhausen) took even the grossest account of age.[29] Amsterdam sold municipal annuities from 1402 on, charging flat rates of $9\frac{1}{11}$ percent for annuities on two heads and $11\frac{13}{17}$ percent for one, regardless of age.[30] William Purser's *Compound Interest and Annuities* (1634) computed the value of annuities at 8 percent simple interest for ten to twenty years' purchase without any consideration of differences in mortality: interest

[27] See Hans Schmitt-Lermann, *Der Versicherungsgedanke im deutschen Geistesleben des Barock und der Aufklärung* (Munich: J. Jehle, 1954); Gerald Schöpfer, *Sozialer Schutz im 16.–18. Jahrhundert* (Graz: Leykam-Verlag, 1976).

[28] Richard C. Trexler, "Une table florentine d'espérance de vie," *Annales. Économies, Sociétés, Civilisations* (January-February 1971): 137–139; Dupaquier, "Table."

[29] Heinrich Braun, *Geschichte der Lebensversicherung und der Lebensversicherungstechnik* (Nuremberg: Carl Koch, 1925), p. 37.

[30] Société Générale Néerlandaise d'Assurances sur la Vie et de Rentes Viagères, *Mémoires pour servir à l'histoire des assurances sur la vie et des rentes viagères au Pays-Bas* (Amsterdam, 1898), pp. 209–210.

rather than age appears to have been the important temporal variable. The tontine annuities sold by the English government in 1693 to raise money for the French war also ignored the age of the purchasers.[31]

The striking absence of a fixed schedule of prices scaled by age in most annuity schemes does not necessarily imply that buyers and sellers were blind to such considerations. As one historian on the subject put it: "No-where does one have evidence that the sale prices of annuities were in any way 'calculated'. Yet it is almost certainly to be assumed that some account of age was taken in sales."[32] In an age when sudden outbreaks of plague made the hour of death recklessly uncertain, carrying off young and old alike in appalling numbers, a personal interview with the annuitant may have been a better guide to pricing than Ulpian's table. The city of Nu-remberg, for example, charged variable annuity prices that could be ac-counted for by age and health adjustments on a case-by-case basis. Jurists recommended that annuities be priced through a combination of rules of thumb and, above all, the consideration of a sagacious judge who could weigh the circumstances of each case.[33] By the late sixteenth century, canny Dutch buyers of annuities had grasped that children between the ages of five and twelve were the best candidates for annuities, and these account for most of the annuities taken out in Holland and West Fresia after the new rates of 1588.[34] However, mindful of the high rate of infant mortality, they often prudently had a physician check the health of the particular child in question.[35] Like the insurers, the buyers and sellers of annuities were reluctant to substitute general rules for judgments about the individual case where the option existed.

Although actuarial mathematics has taught us to connect life insurance and annuities, the one being the inverse of the other and both being func-tions of human mortality, the merchants and jurists of early modern Eu-rope linked life insurance instead with gambling. Whereas annuities were primarily investments, disguised loans with interest, life insurance was a wager. In most cases, short-term bets were taken out on the life of a third person, often a celebrity like the pope, and occasionally on the outcome of

[31] Francis Leeson, *A Guide to the Records of the British State Tontines and Life Annuities of the 17th and 18th Centuries* (Shalfleet Manor, Eng.: Pinhorns, 1968), p. 1.

[32] Braun, *Geschichte*, p. 37.

[33] See, for example, Charles Du Moulin, *Summaire du livre analytique des contractz usures, rentes constituées, interestz & monnoyes* (Paris, 1554), f. 187r.

[34] Société Générale Néerlandaise, *Mémoires*, p. 211.

[35] Braun, *Geschichte*, p. 53.

some dramatic event like a battle or siege. The custom seems to have originated in fifteenth-century Spain and spread successively to Italy, Holland, France, and England.[36] Occasionally the lives of slaves or of pilgrims risking enslavement were insured, as were the lives of debtors by their creditors for the amount owed,[37] or the lives of merchants for the duration of a dangerous voyage. But the legislators understood life insurance as primarily a bet and outlawed it in the codes of the Low Countries (1570), Amsterdam (1598), Middleburg (1600), Rotterdam (1604), and France (1681). The Code of the Low Countries is typical in lumping insurance on lives together with "abuses, frauds and crimes . . . wagers on voyages and similar inventions," and banning the lot of them.[38] By the end of the seventeenth century, life insurance was illegal almost everywhere in Europe except in Naples and England, and remained so until the nineteenth century. In eighteenth-century England life insurance and betting remained intertwined until the Gambling Act of 1774. There is no evidence of any systematic interest in age and mortality in the pricing of these policies, which were almost always for periods of a year or less.

Gambling per se was at once the most disreputable and prototypical of the aleatory contracts, as the tendency of almost every other arrangement involving risk to degenerate into a wager shows. On the basis of several Old Testament passages, theologians since Augustine had condemned gambling as a "temptation of God," a profanation of God's chosen way of revealing his will by lot.[39] We have no better evidence of how endemic gambling nonetheless was than the repeated attempt to ban it.[40] Gambling was a passion that apparently knew neither national nor class boundaries, inflaming peasant and aristocrat alike until at least the end of the eighteenth century.[41] Illegal activities leave only casual, haphazard traces

[36] Schmitt-Lermann, *Versicherungsgedanke*, p. 63.

[37] Stefani, *Insurance*, p. 119.

[38] *Ordonnance*, Art. 32.

[39] This interpretation was based on several Old Testament passages: *Numbers* 33: 54; *Proverbs* 16: 33. For an account of the Catholic position on gambling from patristic writings through the Council of Trent (1607), see Abbé Coudrette, *Dissertation théologique sur les lotteries* (n.p., 1742).

[40] Coudrette, *Dissertation*, pp. 210–212; [Anonymous], *Reflexions on Gaming* (London, n.d.), pp. 16 ff.

[41] Jean Verdon, *Les loisirs en France au Moyen Age* (Paris: Librairie Jules Tallandier, 1980), pp. 199–203; Robert W. Malcomson, *Popular Recreations in English Society, 1700–1850* (Cambridge, Eng.: Cambridge University Press, 1973), pp. 41–43, 49–50. See also Section 3.4.1.

in the historical record, and precise information on how odds were set is hard to come by. We do know from Cardano and especially Galileo that gambling puzzles from the sixteenth and seventeenth centuries could reveal a refined sense of very small differences in odds.[42] However, the combination of skill and chance in many games, the irregular casting of dice and other gambling devices, belief in streaks of good and bad luck, and sharp dealing must have all conspired to obscure the idea of equiprobable outcomes. For some of the most popular betting occasions, like cockfighting, setting odds was necessarily a matter of informed opinion and hunches, in an atmosphere of near pandemonium, then as now. The sporting press, with its taste for statistics, seems to have been an invention of the late nineteenth century.[43]

A few points concerning these pre-probabilistic institutions for dealing with risk and uncertainty should be emphasized. First, largely because of the Catholic church's position on usury, risk took on a positive tinge as civil and canon lawyers and later Jesuits made it the basis for their defense of potentially shady commercial practices. Of course, they trod a thin line between risk sufficient to exonerate a merchant from charges of usury, and risk sufficiently great to look like wanton gambling. But on the whole, the concept of risk was so important a means of harmonizing precept with practice that even gambling became more innocent by association in the work of sixteenth- and seventeenth-century jurists. Second, insurance premiums, gambling stakes, and annuity rates all represented a kind of quantification of risk but not necessarily one based on probabilistic or statistical intuitions, much less calculations and data. Annuity rates and insurance premiums certainly reflected past experience, but it was a far more nuanced experience than a simple toting up of mortality and shipwreck statistics. It was an experience sensitive to myriad individual circumstances and their weighted interrelationships, not to mention market pressures and a pressing need for cash. To the sixteenth-century seller of insurance or annuities, statistics would have seemed a blunt-edged instrument indeed, incapable of yielding the informative minutiae about this or that case upon which he believed his trade depended. Given the highly volatile

[42] M. G. Kendall, "The beginnings of a probability calculus," *Biometrika* 43 (1956): 1–14, reprinted in *Studies in the History of Statistics and Probability*, E. S. Pearson and M. G. Kendall, eds. (Darien, Conn.: Hafner, 1970), pp. 19–34.

[43] Ross McKibbin, "Working-class gambling in Britain 1880–1939," *Past & Present*, no. 82 (February 1979): 147–178, on pp. 166–167.

conditions of both sea traffic and health in centuries notorious for warfare, pirates, plagues, and other unpredictable misfortunes, I am not persuaded that this was an unreasonable approach. In any case, it was the prevailing one. Finally, all of these ways of handling and exploiting risk—insurance, annuities, and gambling—evidently turned a profit most of the time.

3.3 The Mathematical Theory of Risk

The early probabilists took up the problems of aleatory contracts, in particular that of pricing risks. Although the jurists had regarded the problem as monolithic, essentially the same for annuities as for gambling, the mathematicians soon distinguished two cases: those in which the probabilities can be calculated a priori on the basis of equiprobable outcomes, as in certain games of chance; and those in which the probabilities can only be estimated a posteriori from statistics about the frequency of an event. The distinction did not emerge immediately, or at least not with full force. De Witt's assumption of equiprobable or hypothetically weighted probabilities for mortality as a function of age *and* his cross-check with Hudde's Amsterdam annuity data suggest that compromises were conceivable. In the absence of data, later mathematicians like Jakob Bernoulli were also forced to fall back on the gambling model, but by then with obvious reluctance.[44] In general, the probabilists were quick to see how statistics could extend mathematical probability to cover applications to other sorts of aleatory contracts besides simple games of chance. Given the at best frivolous and at worst immoral associations of gambling, this was a particularly welcome development for probabilists anxious for the reputation of their subject.[45]

However, the extension was not a wholly straightforward one. Mathematicians long grappled with the relationship between a priori and a posteriori probabilities, Bernoulli's and Bayes' theorems being the two most significant of the classical attempts to connect the two.[46] The almost instant alliance between mathematical probability theory and statistics did

[44] Jakob Bernoulli, *Meditationes*, nos. 77 and 80, in *Die Werke von Jakob Bernoulli*, Basel Naturforschende Gesellschaft (Basel: Birkhäuser, 1975), vol. 3, pp. 42–48, 66.

[45] See, for example, Abraham De Moivre, *The Doctrine of Chances*, 3rd edition (London, 1756), Preface and p. 254.

[46] See Sections 5.2–3, below.

more than broaden the latter's domain of applications; it also changed what probability meant. Philosophers still puzzle over how probability can mean both a degree of certainty and a number of observed instances, but Christiaan Huygens, Gottfried Wilhelm Leibniz, and other seventeenth-century probabilists identified the two without hesitation or justification. Moreover, whereas mathematicians agreed on the techniques for solving gambling problems ranging in complexity from the simplest lotteries to duration of play,[47] they argued for over a century about how to compute life expectancy, the shape of the mortality curve, and other statistical matters. Finally, reading probabilities off from statistics presupposed that the statistics existed, posing practical problems of data collection and subtler problems of choice and interpretation: what was worth counting?

The history of classical probability theory's treatment of risk is largely the history of mortality statistics and their applications. Gambling problems remained the staple of texts and the entry into the discipline, but they were no longer the cutting edge of research after 1700. Beginning with the work of Edmund Halley, Jakob Bernoulli, and Abraham De Moivre, the role of statistics in mathematical probability grew steadily, and for most of the eighteenth century the statistics of choice dealt with human mortality. Therefore, any account of the mathematical theory of risk in this period must begin with why, when, and how contemporaries kept track of death.

The parish was the administrative unit usually charged with collecting demographic data: the Council of Trent (1545–63) ordered each Catholic parish to keep a register of christenings and marriages and in 1614 Pope Paul V added a record of deaths. Some Protestant parishes seem to have kept such records even earlier.[48] Sixteenth-century governments also ordered local officials to keep such records for legal and medical reasons. For example, in 1538 Henry VIII commanded English parishes to record christenings, weddings, and burials, and Francis I passed a similar ordinance for baptisms in 1539; weddings and funerals were added in additional legislation of 1579. The motivation here seems to have been evidentiary, to provide official proof of age, identity, and status in law suits that hinged upon these issues.[49] The English bills of mortality served a quite different purpose, to keep a tally of deaths due to plague as an advance

---

[47] The St. Petersburg problem was a notable but rare exception: see Section 2.3, above.

[48] Braun, *Geschichte*, pp. 94–95.

[49] Claude de Ferriere, *Corps et compilation de tous le commentateurs anciens et modernes sur la coutume de Paris* (Paris, 1685), vol. 3, p. 385.

warning of a major outbreak, and therefore recorded only cause and place of death (and only after 1625 any cause of death other than plague), not age of the deceased. The first London bills appeared in 1562, although other locales were publishing them as early as 1538.[50] By the mid-seventeenth century the bills also included christenings, broken down by sex.[51] Official zeal for such record keeping varied widely: in 1749 Sweden created a central service to collect data regularly registered since 1686, and France was almost as conscientious after the civil ordinance of 1667; but large parts of central Europe did not register births and deaths until well into the nineteenth century.[52]

Thus demographic data existed from the early sixteenth century, albeit often incomplete and unreliable. (The London bills, for example, were not kept continuously until after 1603; ages were not recorded until 1728.) However, the motives and manner of collecting the information reveal not the slightest hint of any interest in the regularities that fascinated late seventeenth- and eighteenth-century analysts like John Graunt, William Petty, and John Arbuthnot. Until Graunt's *Natural and Political Observations on the Bills of Mortality* (1662), no one appears to have undertaken an empirical study of death as a function of age. Indeed, the regular intervals of Graunt's mortality table were as much invented as discovered. The work was more empirical in its inspiration (Graunt conceived the work as a continuation of Bacon's natural history of life and death) than in its contents. Due to the gaps in the data available from the London bills, Graunt could construct a table correlating the number of deaths with age only by making several assumptions. He posited that one-third of the deaths registered were due to diseases—*"Thrush, Convulsion, Rickets, Teeth,* and *Worms . . . Abortives, Chrysomes, Infants, Livergrown,* and *Over-laid"*—that generally struck before the age of six, and further assumed that thereafter there was an equal chance of dying in any of the seven decades between ages six and seventy-six.[53]

---

[50] Albert Rosin, *Lebensversicherung und ihre geistesgeschichtlichen Grundlagen*, in *Kölner Anglistische Arbeiten*, vol. 16 (1932), p. 12.

[51] See, for example, Company of Parish Clerks of London, *London's Dreadful Visitation: Or, A Collection of All the Bills of Mortality for this Present Year* (London, 1665), under "Christned."

[52] Jean Meuvret, "Les données démographiques et statistiques en histoire moderne et contemporaine," *Cahiers des Annales*, no. 32 (1971): 313–340, on pp. 313–316.

[53] John Graunt, *Natural and Political Observations Mentioned in a Following Index and Made Upon the Bills of Mortality* (London, 1662), pp. 29–30, 69.

Graunt was less concerned with patterns of mortality than with shedding new light on burning policy issues of the day: he used his supposed mortality table to calculate the number of able-bodied males in London who could do military duty; he concluded from the balance of males and females that polygamy was not a rational means to increase fertility; and he recommended that quarantine in times of plague was not worth the trouble it caused. He also seems to have been wholly ignorant of Huygens's Latin treatise on mathematical probability published a few years before, relying rather "upon the *Mathematiques* of my Shop-Arithmetique."[54] Although Graunt's table of deaths by age effectively assumed that the chances of dying in any decade between the ages of six and seventy-six followed a regular pattern, his reasoning was not probabilistic. After deciding on the basis of the type of disease that thirty-six out of every one hundred deaths were those of children under six years of age, Graunt simply found, as he says, the "mean proportional" for each of the next seven decades, rounding off to the nearest whole person. However, alert mathematical observers all over Europe (including Christiaan and Ludwig Huygens, Johann De Witt, Halley, Leibniz, Jakob and Nicholas Bernoulli) were quick to recognize the relevance of Graunt's mortality table as a way of applying the fledgling theory of probability to life expectancy problems. After reading Graunt's *Natural and Political Observations* in 1669, Ludwig Huygens drew his brother Christiaan's attention to the problem of determining the life expectancy of a newborn from Graunt's table.[55] Christiaan was later enlisted to evaluate De Witt's proposals on annuities to the Estates-General of Holland and West Friesland.[56] Nicholas Bernoulli used Graunt's figures;[57] Leibniz also worked on them to find the curve of mortality.[58] William Petty's unsigned review of Graunt's work in the August 1666 issue of the

[54] Graunt, *Observations*, p. 7.

[55] Société Générale Néerlandaise, *Mémoires*, pp. 58 ff.; the correspondence can also be found in Société Hollandaise des Sciences, *Oeuvres complètes de Christiaan Huygens* (The Hague, 1895), vol. 6, pp. 482 ff.

[56] Ian Hacking, *The Emergence of Probability* (Cambridge, Eng.: Cambridge University Press, 1975), pp. 49, 117.

[57] Nicholas Bernoulli, *De usu artis conjectandi in iure* (1709), in *Die Werke von Jakob Bernoulli*, vol. 3, pp. 292–300.

[58] Kurt-Reinhard Biermann, "Eine Untersuchung von G. W. Leibniz über die jährliche Sterblichkeitsrate," *Forschungen und Fortschritte* 28 (1955): 205–208; "G. W. Leibniz und die Berechnung der Sterbewahrscheinlichkeit, bei J. de Witt," ibid. 33 (1959): 168–173.

Journal des Sçavans seems to have won the *Natural and Political Observations* a select audience of Continental mathematicians.[59]

Graunt believed the numbers in his table to be "practically near enough the truth";[60] some of his readers, like Ludwig Huygens, went so far as to describe them as "observations made in London with much precision."[61] They were of course no more than shrewd guesswork, like De Witt's even more drastic simplification of mortality statistics in his 1662 brief on pricing annuities. De Witt assumed that the chances of dying between the ages of three and fifty-three were equal for any six-month period therein, and assigned probabilities to the remaining decades up until age eighty-one by simple weighting factors.[62] Edmund Halley's 1693 memoir on mortality in Breslau[63] provided the first complete table based on actual data for age at death. Despite his avowedly empirical approach, Halley was also convinced that "Irregularities in the Series of the Ages" shown in his table "would rectify themselves, were the number of years much more considerable, as 20 instead of 5"; that is, Halley believed more in the regularity of mortality than in his somewhat irregular data.[64]

Jakob Bernoulli made some of the assumptions underlying this belief explicit in a correspondence with Leibniz (October 1703–April 1704).[65] Bernoulli was interested in mortality statistics as an important example of how a posteriori statistics might be converted into a priori probabilities. His model was taken from the a priori case of an urn filled with different colored pebbles, drawn with replacement, which stand for the diseases of the human body that bring about death. Leibniz objected that the urn and mortality cases were disanalogous in significant ways: that the number of

---

[59] *Journal des Sçavans* 1 (1665–66): 585–590 (Amsterdam edition).

[60] Graunt, *Observations*, p. 69.

[61] Société Générale Néerlandaise, *Mémoires*, p. 62; Hudde was more skeptical: ibid., pp. 77–78.

[62] Johann De Witt, *Waerdye van Lyf-Renten* (1671), in *Die Werke von Jakob Bernoulli*, vol. 3, pp. 337–342.

[63] See Schmitt-Lermann, *Versicherungsgedanke*, p. 55, on how the data came to be collected.

[64] Edmund Halley, "An Estimate of the Degrees of the Mortality of Mankind, drawn from the curious Tables of the Births and the Funerals at the City of Breslau; with an Attempt to ascertain the Price of Annuities on Lives," *Philosophical Transactions of the Royal Society of London* 17 (1693): 596–610.

[65] C. I. Gerhardt, *G. W. Leibniz Mathematische Schriften* (Hildesheim: Georg Olms, 1962; reprint of 1855 edition), vol. 3, part 1, pp. 11–89.

diseases, unlike the pebbles in the urn, might be indeterminate, or infinite, or variable; and that extrapolation based on a curve drawn through points representing past instances assumed one curve out of an infinite number of possibilities. Bernoulli's reply was to insist that "nature follows the simplest paths"; she in effect prefers determinate and stable ratios, and shuns innovation. Throughout the eighteenth century, mathematicians and political arithmeticians felt free to play fast and loose with the available data and to brush aside deviations, in the conviction that human demography and especially mortality were governed by simple regularities.[66]

Where did this conviction, so different from that which informed the contemporary sale of insurance and annuities, come from? It is difficult to understand why mortality should have been assumed to be regular and other phenomena of equal practical interest, such as the incidence of fires not, in an age where both were subject to wild fluctuations: witness the plague and the Great Fire of London in 1665–66. This was a selective, not a sweeping faith in the simplicity and regularity of phenomena. It encompassed both the rate and geography of mortality: most probabilists assumed that death carried off portions of the population in arithmetic progression, and that this law held the world over. There were certainly dissenting voices on both scores, especially the latter, for eighteenth-century observers were convinced of what Halley called "the different *salubrity* of places" and therefore suspected tables that purported to hold for town and country alike.[67] The Dutch probabilist Nicholas Struyck complained in 1740 that "mortality doesn't listen to our suppositions" and that many of the tables allegedly based on observation were in fact "pure hypotheses."[68] But on the whole, the tendency was to neglect deviations in the data in favor of the expected regularities that had attracted them to the data in the first place. Johann Süssmilch's insistence upon "a constant, general, great, complete, and beautiful order" revealed by his sometimes quite irregular data was only the most influential eighteenth-century example of this unshakable conviction in demographic regularity.[69]

[66] Harald Westergaard, *Contributions to the History of Statistics* (London: P. S. King & Son, 1932), pp. 65, 72.

[67] See, for example, Thomas Simpson, *The Doctrine of Annuities and Reversions* (London, 1742), Preface; Thomas Short, *New Observations, Natural, Moral, Civil, Political, and Medical on City, Town, and Country Bills of Mortality* (London, 1750), Preface and pp. 1–60.

[68] Société Générale Néerlandaise, *Mémoires*, p. 89.

[69] Johann Süssmilch, *Die göttliche Ordnung in den Veränderungen des menschlichen Geschlechts*,

Both natural theology and a form of proto-quantification may have contributed to this selective confidence in selected statistical regularities such as mortality. The former is certainly a dominant theme in Süssmilch's work, which was after all entitled *Die göttliche Ordnung* and which used purported demographic regularities as evidence for the argument from design and the constant activity of providence in human affairs. Karl Pearson has maintained that belief in the stability of statistical ratios stemmed from the predilection of the natural theologians to see a beneficent order everywhere, from the construction of an insect's wing to the numerical balance of the species, citing William Derham's 1711–12 Boyle lectures as an early and influential example.[70] However, we must be wary of crediting Derham and followers like Süssmilch with the original idea of stable and demographic regularities, although they did much to popularize it. Just as the natural theologians appropriated Newton's work in celestial mechanics (with his wholehearted support) to make their points about design in nature, but did not themselves originate the idea of astronomical regularities, so Derham borrowed from the works of Graunt, Halley, Arbuthnot, and others to support his claim of a constant ratio of births to deaths and males to females. And Derham relied at least as heavily on Biblical citations as on articles in the *Philosophical Transactions of the Royal Society of London* to argue, for example, that longevity is inversely proportional to the size of the population—hence the much longer lives of the Biblical patriarchs at the beginning of the world and immediately after the flood, to compensate for depopulation: "For, by this means, the peopled World is kept at a convenient Stay, neither too full, nor too empty."[71] As the quotation suggests, Derham was less concerned with the regularity of the mortality curve than with the overall balance between the earth's resources and the population of the species which consumed them. He was willing to countenance both sudden bursts of fecundity or catastrophic bouts of mortality (e.g., the biblical flood, or plagues) so long as this steady state of affairs was reinstated by God's "admirable and plain Man-

---

3rd edition (Berlin, 1775), vol. 1, p. 49; Karl Pearson, *The History of Statistics in the 17th and 18th Centuries*, E. S. Pearson, ed. (London and High Wycombe: Charles Griffin, 1978), pp. 320–323.

[70] Pearson, *History*, chapter 9.

[71] William Derham, *Physico-Theology, or, A Demonstration of the Being and Attributes of God from His Works of Creation*, 4th edition (New York: Arno Press, 1977; reprint of 1716 edition), p. 173.

agement" by such further remarkable irregularities as the temporary lengthening of human life-span from seventy to nine hundred years. In a fashion that was as typical of the Royal Society as was his natural theology,[72] Derham was at least as interested in anomalies as in regularities as evidence of divine management. His footnotes for the chapter on demography are crammed with individual examples of extraordinary human longevity or fecundity, mostly drawn from contemporary sources like the *Philosophical Transactions*. Apparently Derham was torn between astonishing regularities and astonishing anomalies as the best evidence for God's hand in human affairs. Natural theology certainly fortified and exploited the belief in demographic regularities, but it did not create that belief.

Proto-quantification seems a more likely source of the selective belief in certain statistical regularities such as mortality, as opposed to other possibilities, such as fires. Neither Graunt nor Halley seems to have been animated by theological concerns (it is particularly dubious to impute such motives to Halley), although Kaspar Neumann, the Protestant pastor who collected Halley's Breslau data, wrote to Leibniz that he hoped such figures would eventually be used to help in the fight against superstition.[73] Nor did new regularities jump out at them from the extant data: as we have seen, they constructed and molded what little data they had in the image of the regularities they already believed in. Two aspects of mortality might have prepared the way for that belief: its inevitability and its connection with age. Unlike fires, death befalls us all, and since at least biblical times it has been conveniently linked with a continuous quantity, age at time of death. Perceptions about just when death was likely to strike ranged widely, from the canny Dutch who bought annuities on the heads of healthy infants, to the obtuse French, who preferred to back dotards,[74] but everyone knew that death "will have his day." This meant that a complete enumeration was in principle possible, and age was the obvious candidate for the independent variable. It is even possible that disasters like the plague may have in some ways prompted this new way of thinking about death, for it was a grim and vivid reminder that everyone was vulnerable

---

[72] Katharine Park and Lorraine J. Daston, "Unnatural conceptions: The study of monsters in sixteenth- and seventeenth-century France and England," *Past & Present*, no. 92 (August 1981): 20–54, on pp. 47–51.

[73] See n. 63, above.

[74] Société Générale Néerlandaise, *Mémoires*, p. 211; Antoine Deparcieux, *Essai sur les probabilités de la durée de la vie humaine* (Paris, 1746), pp. 74–75.

to sudden death, not just those approaching the limit of their biblical three score and ten years.[75] Thus it may have become clear that the mortality curve had some structure, even if the plague did little to suggest the regularity of that structure.

Of course neither the inevitability nor the age-linked character of death was a sufficient condition for believing in its regularity. But in an age predisposed to look for regularities everywhere, with special emphasis on those that might be quantified, these features may have been suggestive to those intellectuals who subscribed to the new determinism. Viewed from this angle, fires and shipwrecks presented far less tractable problems for the statisticians. They could certainly be counted, but not every ship or house suffered such an accident, and the obvious independent variables (season, route, captain's experience; building material, trade housed) did not easily lend themselves to quantitative expression. (Neither did certain key variables connected with mortality, such as health, and, as we shall see, these proved to be obstacles to applying statistics and mathematical theory to annuity and insurance practice.) As in the cases of equity and credibility, proto-quantification may have played a more important role than any a priori plausibility. After all, it is hardly obvious at first glance that the credibility of witnesses and historical records was a better prospect for mathematical treatment than, say, the frequency of hail storms. Yet classical probabilists attacked the problem of testimony, not the weather; of mortality, not shipwrecks, apparently on the basis of what had already been cast in semiquantitative terms.[76]

When probabilists all over Europe seized upon Graunt's table for their own purposes, they made expectation the conceptual bridge between his political arithmetic and their mathematical theory. In 1669 Ludwig Huygens proposed a problem in probabilities to his brother Christiaan based on "that table of the English book of the Bills of Mortality": find the expectation of life. Ludwig's own solution followed Christiaan's method for computing ordinary expectations to the letter, and became one standard eighteenth-century method of computing life expectancies: multiply the number of people in each of Graunt's decades by the average number of years they survive, and divide by the total number of people. This gives

---

[75] See Daniel Defoe, *A Journal of the Plague Year* (London, 1722), for a vivid, if fictional, evocation of this universal susceptibility; also Jean Delumeau, *La peur en Occident (XIVe–XVIIIe siècles)* (Paris: Fayard, 1978), pp. 105–116.

[76] See Section 1.2, above.

a life expectancy of about 18 years, 2 months. However, Christiaan was not convinced by his brother's reasoning, even though it followed his own principles in the early propositions of *De ratiociniis in ludo aleae* closely. He wrote back that Ludwig's answer in effect paid insufficient attention to the structure of the mortality table, which Christiaan had characteristically taken the trouble to graph.[77] Christiaan pointed out that a very different curve—for example, one in which 90 percent of the people died before the age of 6 and the rest survived to 152 years, 2 months—would also give a life expectancy of 18 years, 2 months by Ludwig's method. Instead, he used his curve to determine when half of the original cohort would have died, that is, the age to which a newborn had even odds of surviving, about 11 years. Christiaan's curve permitted him not only to interpolate how many people are alive at any given age, but also to compute life expectancy for all ages. For any age *A* we can locate *B* on the curve, telling us the number of people from the original cohort of newborns who have survived to that age. We then take half of the length of *AB*, and translate

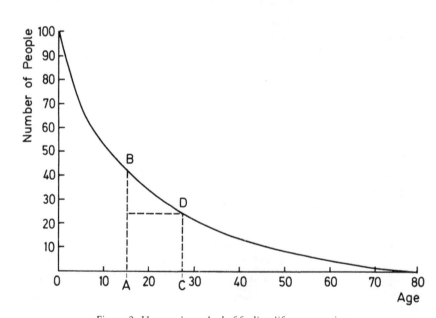

Figure 2. Huygens's method of finding life expectancies.

[77] The original graph is not reproduced in the correspondence, although Christiaan's graph of life expectancy as a function of age is. This is my reconstruction of his first graph.

along the abscissa until it intersects the curve (*CD*). Then *AC* is the life expectancy of someone aged *A*.

Christiaan saw his graph as a natural way of supplementing Graunt's table, which only gave discrete values of the number of survivors per decade, "without however encumbering myself with any calculation."[78] In fact the transition from table to curve, from discrete to continuous values, had the effect of introducing a further element of regularity into discussions of mortality. Graunt had made no assumptions about the behavior of mortality within his decades; Huygens's smooth curve implied that people died in regular increments, however small the time interval chosen. Moreover, Huygens's musings over the rival methods of computing life expectancy forced him for the first time to distinguish clearly between probability and outcome values as separate elements within expectation.[79] Although intuition tells us that probabilities are easiest to calculate for simple gambling devices like coins and dice, it was in fact mortality statistics that gave rise to the first explicit expressions of probability per se. Both the mortality curve and the new understanding of expectation of a composite are examples of how seemingly neutral techniques and applications can react back upon theory and significantly alter its conceptual content.

For Christiaan, the two methods of computing life expectancy corresponded to different practical purposes. As he explained to his brother: "They are thus two different things: the expectation or the value of a future age of a person, and the age to which there is an equal appearance that he will or will not attain. The first is for regulating annuities, and the other for wagers."[80] (As we shall see below, the remark about wagers referred to the most widespread form of life insurance in the seventeenth and eighteenth centuries.) However, classical probabilists did not find the distinction to be so clear-cut and wavered between the two methods. Nicholas Bernoulli, for example, used Ludwig's method to compute the value of annuities;[81] De Moivre favored Christiaan's for the same purposes.[82] Under the assumption of an arithmetic progression in mortality of the sort un-

---

[78] Société Générale Néerlandaise, *Mémoires*, p. 61.

[79] Société Générale Néerlandaise, *Mémoires*, p. 63.

[80] Société Générale Néerlandaise, *Mémoires*, p. 68.

[81] Nicholas Bernoulli, *De usu*, in *Die Werke von Jakob Bernoulli*, vol. 3, p. 293.

[82] Abraham De Moivre, *A Treatise of Annuities of Lives*, 3rd edition (London, 1750), p. 288.

derlying Graunt's table and reaffirmed by Halley, the two methods yield the same results for annuities on single lives, but never for joint lives. The confusion over how best to compute life expectancy divided writers on the subject throughout the eighteenth century.[83]

Their disagreements did not prevent the problem of pricing annuities, the inverse of pricing life insurance premiums, from becoming a part of the probabilist's repertoire of applications by the first decade of the eighteenth century. As both Halley's and Nicholas Bernoulli's treatment of the subject makes clear, the mathematicians still approached the problem from a legal standpoint, as subject in De Moivre's words "to the Rules of that Equity which ought to preside in Contracts."[84] Halley used his Breslau mortality table to calculate the number of able-bodied men who bear arms and the price of annuities, charging that heretofore the latter had been priced only "by an imaginary valuation," and were therefore inequitable, "for it is plain that the *Purchaser* ought to pay for only such a part of the values of the *Annuity*, as he has Chances that he is living."[85] Nicholas Bernoulli's *De usu artis conjectandi in jure* (1709) also emphasized the importance of the new mathematical methods and mortality data in guaranteeing the fairness of aleatory contracts. Using Graunt's mortality table and an expectation rule like Ludwig Huygens's, Bernoulli computed the values of annuities, as well as of reversionary payments and when a missing person might be legally declared dead.

Nicholas Bernoulli was the only mathematician to attempt to apply probability theory to maritime insurance, and it was in this context that he broached the problem of life insurance. His discussion of pricing premiums for maritime insurance was necessarily theoretical, for he lacked data on the percentage of ships that navigated a given route safely. (He also proposed applying this same statistical method to the credibility of witnesses, conceived as the ratio of the number of times the truth was told to total number of times testimony was given, so there was apparently no special presumption of regularity in the maritime case.) In default of such

---

[83] Thomas Simpson, for example, followed De Moivre, but Richard Price opted for the other method: Richard Price, *Observations on Reversionary Payments*, 3rd edition (London, 1773), pp. 170 ff. For an overview of these controversies, see Price, *Observations*, especially pp. 170 ff., 205 ff., 227 ff. For a defense of De Moivre and Simpson, see Samuel Clark, *A Letter to Richard Price, D.D. and F.R.S.* (London, 1777).

[84] De Moivre, *Annuities*, 1st edition (London, 1725), p. 2.

[85] Halley, "Estimate," p. 602.

data, Nicholas assumed equiprobability of outcome, but with the usual warning to attend to whether the ship was old, "badly equipped with sails and sailyards, run by inexperience and a lack of familiarity in a skipper, etc." He waved aside the legal objections to life insurance (here, term insurance for the life of a merchant during a voyage) as a form of buying and selling a free man on the grounds that it was the probability of human life rather than the life itself that was to be evaluated.[86] It is clear from the method that what was being insured was a business investment contingent on the merchant's survival, rather than the life itself, and this was almost the only form of life insurance sold until the last quarter of the eighteenth century. Since such policies could be envisioned as a kind of bet on the merchant's survival, they were not easily distinguished from pure "wager" policies on the short-term survival of a third party, further confusing the categories of gambling and insurance already identified under the rubric of aleatory contracts in legal treatises like Bernoulli's.

The eighteenth-century mathematical literature on annuities is large, but since De Moivre's *Treatise of Annuities on Lives* (1725) became the standard reference work on the subject, it can serve here as representative of the evolution of concepts and methods for applying mathematical probability to problems of mortality. First, De Moivre was typical in treating only annuities and other reversionary payments (and in his *Doctrine of Chances*, gambling) among the many possible kinds of aleatory contract which Jakob and Nicholas Bernoulli had considered within the mathematician's purview. No form of insurance then practiced in London (fire, life, maritime) figures in his work. Second, he followed Graunt, De Witt, and Halley in assuming that mortality statistics follow a simple pattern—in De Moivre's case an arithmetic progression, which he believed would be ever more closely approached by more data, citing Halley's Breslau table as a good first approximation.[87] For De Moivre, this confidence was buttressed by Bernoulli's theorem, interpreted in light of natural theology: "Althou' chance produces Irregularities, still the Odds will be infinitely great, that in the process of Time, those Irregularities will bear no proportion to the recurrency of that Order which naturally results from ORIGINAL DESIGN."[88] In the interest of simplifying calculations, De

---

[86] N. Bernoulli, *De usu*, in *Die Werke von Jakob Bernoulli*, vol. 3, pp. 318–319, 324–325.

[87] De Moivre, *Annuities*, 1st edition, p. v.

[88] De Moivre, *Doctrine*, p. 251.

Moivre also made assumptions that deviated more dangerously from the data, as he himself admitted.[89]

How then did the advent of mathematical probability change the *theory* of risk taking in the late seventeenth and eighteenth centuries? First, it did very little to sever the connections between gambling and other forms of aleatory contracts like annuities and insurance. The probabilists took over the notions of expectation, equity, and (at least in the case of gambling and annuities) degrees of risk, and gave them a quantitative formulation that added mathematical to legal reasons for classifying them under the same heading. As the French mathematician Antoine Deparcieux put it in 1746, an annuity was simply a form of gambling that is more than usually advantageous to the player.[90] Of course, probabilities for annuities and insurance were considerably harder to come by than those for games of pure chance, but mortality statistics proliferated in the eighteenth century, and mathematicians like Jakob Bernoulli and De Moivre legitimated turning these statistics into probabilities. Second, these statistics thus provided mathematicians with a third interpretation of probabilities that were originally conceived as either degrees of certainty or of equipossibility. Here the application changed the meaning of the mathematics. Third, probabilists assumed that simple, constant patterns underlay mortality statistics. De Moivre, Süssmilch, and Arbuthnot used the argument from design both to explain the fact and to justify the assumption of regularity, but the assumption seems to have been widely held decades before by writers like Graunt on other grounds. The new theory of risk substituted these simple, global, long-term regularities for the complicated weighting of numerous fluctuating factors as the basis for pricing uncertain ventures. The jurists and their clients had looked to experience and judgment; the mathematicians looked to tables and calculation. This was the theoretical legacy of mathematical probability to institutionalized risk taking in the eighteenth century; what was its contribution to practice?

3.4 Risk after Probability Theory

The short answer to the above question is, nil. Consider the examples of the annuities sold by the Dutch and English governments in 1671 and

---

[89] De Moivre, *Annuities*, 3rd edition, p. 326.   [90] Deparcieux, *Essai*, p. 123.

1694, respectively. Both followed directly on the heels of proposals made by mathematicians well known in government circles of each country, De Witt and Halley, for pricing annuities with respect to the age of the buyer. Indeed, De Witt's memoir on the subject had been specifically addressed to the States General of Holland as a proposal on how to organize the latest sale of annuities resolved upon on 30 July 1671. De Witt's plan was rejected, and Halley's recommendations were ignored. The Dutch instead adopted another experimental scheme already tried out by the town of Kampen in 1670, which raised 100,000 florins by a kind of tontine: four hundred annuities were sold at 250 florins apiece, and the town paid interest of 4 percent per year on the total amount (i.e., 4,000 florins) to all of the original four hundred subscribers still living.[91] The English simply sold annuities at flat prices which a 19th-century insurance writer found outrageously biased in favor of the buyer: a single life was valued at 7 years' purchase (i.e., the price was seven times the annual payment); two lives at 8½ years' purchase.[92] In neither case did the sellers take any account of the age of the buyer. Despite the burgeoning mathematical literature on such mortality-dependent reversionary payments and on gambling in the eighteenth century, much of it carefully simplified for use by the lay merchant or clerk, the practitioners of risk—buyers and sellers of annuities, insurance, lottery tickets—paid little or no attention to probability theory, even though all of these enterprises flourished as never before during this period. As the insurance writer and historian Cornelius Walford later complained after a survey of the eighteenth-century English insurance scene, "Notwithstanding the brilliant array of names we have already seen associated with the development of the *science*, we cannot escape from the fact that the *practice* had hitherto been left to blind-fold progression . . . and that the improvements which from time to time crept in were rather the result of accident than design."[93]

Why did the practitioners of risk fail to avail themselves of a mathematical technology custom-made for them? The mathematicians were at pains to make their work accessible to those who had only arithmetic, often translating all formulas into words and appending tables to relieve the burden of calculation. Moreover, in an age of fierce economic compe-

---

[91] Société Générale Néerlandaise, *Mémoires*, pp. 223–225.

[92] Cornelius Walford, *The Insurance Guide and Hand-Book*, American edition (New York, 1868), p. 36.

[93] Walford, *Guide*, p. 36.

tition and unprecedented innovation, any edge in the race for customers and profits was usually eagerly seized upon and exploited to the hilt, particularly in the insurance business. Dealers in risk had, as the mystery writers put it, the motive as well as the opportunity to take advantage of the mathematical methods purveyed by the probabilists. The London trade papers of the day are full of advertisements for astonishing schemes, most of them operating out of coffeehouses and taverns: insurance was hawked to the public for female chastity, against lying, against venereal disease, against losing at lottery.[94] London entrepreneurs were thus no strangers to novelty, and London was the undisputed center of the financial (including the insurance) world throughout the eighteenth century. Daniel Defoe called his "The Projecting Age," and ascribed its spirit of invention to necessity, since the recent war with France had ruined many a trade: "These [merchants], prompted by Necessity, rack their Wits for New Contrivances, New Inventions, New Trades, Stocks, Projects, and any thing to retrieve the desperate Credit of their Fortunes."[95] Yet even desperation was not enough to drive these ingenious projectors to probability theory, despite their penchant for "Assurances, Friendly Societies, Lotteries, and the like."

To understand their reluctance, we must return to the themes discussed in the introduction to this chapter. We must also examine their attitudes toward risk itself without the prejudices later developments created against gambling and in favor of insurance. We are the heirs of values that oppose insurance to gambling: gamblers pay to take unnecessary risks; buyers of insurance pay to avoid the consequences of necessary risk. Since roughly the beginning of the nineteenth century, gambling has come to be seen as irrational as well as immoral, and insurance, particularly life insurance, as both prudent and tantamount to a moral duty. Yet eighteenth-century practice regularly conflated gambling with insurance in general and life insurance in particular, and neither legal nor mathematical theory did much to clarify the distinction. Indeed, both theories tended to have the opposite effect by emphasizing the structural affinities between all kinds of risks—the lawyers with the general rubric of aleatory contracts, and the mathematicians with the notion of probabilistic expectation.

I shall argue that it was only when gambling was severed from insurance

---

[94] Rosin, *Lebensversicherung*, pp. 86n, 92.

[95] Daniel Defoe, *An Essay on Projects* (London, 1697), p. 2.

that the latter came to be understood by buyers and sellers alike as a suitable subject for mathematics. Divorcing gambling from insurance in turn depended on the emergence of new attitudes toward risk, toward the stability of the natural and social orders, and toward family bonds. This section is divided into two parts: the first describes seventeenth- and eighteenth-century lotteries, with a focus on the relationship between pricing and prizes and on evolving attitudes toward gambling; the second looks at life insurance during the same period, concentrating on how premiums were determined and on how the first mathematically based company came into being.

3.4.1 Lotteries

In 1697 the publicist Gregorio Leti wrote that in "all the nations of Europe, men and women, of all ages, sexes, and conditions . . . everywhere one speaks, reasons, and is concerned only with the lottery."[96] Leti had the hack writer's flair for hyperbole, but it is true that a rage for lotteries of every description swept Europe in the late seventeenth and eighteenth centuries. Lotteries of various sorts had been known for centuries: fifteenth-century Venetian merchants held lotteries to unload costly but unsalable goods; Genoans elected their five senators from a pool of ninety by lot; Louvain and Bruges held charity lotteries in 1520 and 1538, respectively, to build churches; Augsburg held a *Glückshafen* or tombola at its annual market as early as 1470; Francis I of France issued letters patent for a lottery in 1539; Elizabeth I of England tried to raise revenue with a lottery in 1566 (drawn 1569).[97] However, lottery fever struck only at the end of the seventeenth century, primarily in Holland and England, that is, precisely at the time and places that were the foci for new developments in annuity and insurance schemes. Because an aura of risk and speculation surrounded all these projects, they were closely identified by contemporaries, as we shall see.

[96] Gregorio Leti, *Critique historique, politique, morale, économique, & comique sur les lotteries* (Amsterdam, 1697), vol. 1, pp. 12–13.

[97] Eugen Roth, *Das grosse Los* (Nordwest Lotto in Nordrhein-Westfalen, 1965), pp. 14–39; René Rouault de la Vigne, *La loterie à travers les âges et plus particulièrement en France* (Paris: Éditions Maurice D'Hartoy, 1934), pp. 14–15; John Ashton, *A History of English Lotteries* (Detroit: Singing Tree Press, 1969; reprint of 1893 edition), pp. 5–24.

Leti credits the 1694 English lottery, passed by Parliament to raise £1,000,000 with sixteen-year annuities as prizes,[98] with starting the craze. Foreign investors were at first skittish about buying tickets given England's recent political instability, but their fears were allayed by the spectacle of Britons crowding the ticket offices. According to an issue of *The London Spy* of 1698, all of London was transfixed by the new state lotteries: "The *Gazette* and *Post-Papers* lay by Neglected, and nothing was Pur'd over in the *Coffee Houses*, but the *Ticket-Catalogues*; No talking of the *Jubilee*, the want of Current Trade with *France*, or the *Scotch* Settlement at *Darien*; Nothing Buz'd about by the Purblind *Trumpeters* of *Stale News*, but *Blank* and *Benefit*."[99] The Dutch cities of Amersfort and Amsterdam quickly followed suit, this time awarding prizes of lands and money. At the Amersfort drawing, the town could barely feed and house the visitors attracted by hope and curiosity; in Amsterdam, ticket buyers caused traffic jams.[100] Throughout the eighteenth century, governments (and the papacy) in England, Holland, France, Italy, and Germany had regular recourse to lotteries to replenish the treasury and to finance special projects (for example, the construction of a bridge over the Thames in 1740 or of the École Royale de Militaire in 1757).[101]

State lotteries of this period were primarily of two types, both of Italian origin: blank and Genoan. Blank was the older form; the official French and English lotteries of 1539 and 1596 were of this sort. Players wrote their name and some "device"—a proverb, a verse, an invocation to a saint—on a slip of paper, which were registered with a number at a central office. These were all put into an urn or leather sack. A second receptacle held an equal number of slips of paper, either blank or with a certain prize designated. A child (sometimes blindfolded) would draw out a name slip, an official would read aloud the device and the number (players' identities were usually not revealed), and the child would then draw a slip from the second urn, which the official would announce as either a blank or a prize. Blank lotteries were highly individualized, through both the use of personal

[98] 5 William and Mary c. 2.

[99] Quoted in John Ashton, *The History of Gambling in England* (London: Duckworth, 1898), p. 227.

[100] Leti, *Critique*, vol. 1, pp. 183–189.

[101] Charles Florange, *Curiosités financières sur les emprunts et loteries en France depuis les origines jusqu'à 1783* (Paris: Alphonse Margraff, 1928); Rouault de la Vigne, *Loterie*, pp. 42–43; Ashton, *Lotteries*, pp. 61–62.

devices and the exhaustive drawings, creating the impression of a personal encounter with fortune. The devices often took the form of a prayer, with particulars of need, like that of William Dorghtie de Westholme for Elizabeth's 1569 lottery:

> God send a good lot for my children and me,
> which have had twenty by one wife truly.[102]

Of course, drawings which granted each player his own moment of suspense and revelation could last a very long time. For example, Amersfort was saddled with its lottery tourists for a whole month, and a 1612 lottery in Hamburg took fifty-seven days and nights to draw.[103] The Genoan-style lottery, invented by Benedetto Gentile in 1610, had the advantage of greatly shortening procedures, while still allowing a modicum of individual choice. A child drew five from ninety numbered balls out of a rotating hopper known as "fortune's wheel." Players could bet any amount up to a certain limit on one (*extrait*), two (*ambe*), or three (*terne*) numbers out of the five, with a prize of some multiple of the bet. Later wrinkles included a *quaterne* and a *quine*, and the possibility of also betting on the order of the numbers drawn. In 1757 Casanova and Florentin Calzabigi introduced Parisians to the Genoan system for the École Militaire lottery; Calzabigi's brother persuaded Frederick II to establish one in Prussia in 1763.[104]

Not all of the state lotteries succeeded. Elizabeth had to extend the subscription period to fill up the 1569 lottery; the 1726 British lottery could not sell 11,093 of its 100,000 tickets; the first Prussian lotteries collapsed; a 1753 Bavarian Genoan-style lottery went bankrupt when an Augsburg tradesman improbably won 21,000 gulden with a *terne*.[105] But on the whole, the scene reported by the beleaguered ticket seller for the lottery authorized by Parliament in 1753 to buy and house the collections that became the core of the British Museum was more typical: "People broke in on me, above me, behind me, and before me, and in at the Win-

---

[102] Quoted in Ashton, *Lotteries*, p. 19; special prayers are apparently still common among contemporary lottery players: H. Roy Kaplan, *Lottery Winners* (New York: Barnes & Noble, 1978), pp. 32–39.

[103] Leti, *Critique*, vol. 1, pp. 181–184; Roth, *Los*, pp. 66–68.

[104] Jean Leonnet, *Les loteries d'état en France aux XVIIIe et XIXe siècles* (Paris: Imprimerie Nationale, 1963), pp. 119–125 (original legislation and rules of 1757 lottery); Roth, *Los*, p. 44.

[105] Ashton, *Lotteries*, pp. 24, 60; Roth, *Los*, pp. 47, 56.

dows, with Ladders; the Partitions of my Desk were broke down, while I lay exposed to be plundered of all my money, which was quite open."[106]

Lotteries were in general a steady source of revenue for the state and for charities during the eighteenth century, as the almost annual British government lotteries from 1699 on (and its repeated attempts to ban competing private lotteries) bear witness.[107] Frederick II was won over to the idea despite initial misgivings after his wars emptied the state coffers, although he never ceased to worry that a lucky player might break the bank. The French Royal Lottery, established by Louis XVI in 1776, was conservatively estimated to bring in an annual revenue of seven million livres in 1781. The Convention abolished the Royal Lottery on 25 Brumaire an II (15 November 1793) on moral grounds, but promptly revived it during the financial crisis of 1797.[108] Lotteries were the controller's dream, a method of raising more money without imposing more taxes. Although some wits lambasted lotteries as "a Taxation/Upon all the Fools in Creation,"[109] both the buyers and sellers seemed to agree with a 1757 French plan to help build a new parish church that a lottery was a "a less onerous means" of financing than other subventions.[110] Moreover, popular demand was apparently great enough to support a plethora of private lotteries as well. The Parisian church of Saint-Sulpice was funded by a lottery in 1721; James Cox held a lottery in 1773 when he could not sell his museum of automata; a London eatery in Wych Street, Temple Bar, offered customers lottery tickets to win sixty guineas; Leti reports that even the children of Amsterdam played lottery games.[111]

Although lottery problems were the stock-in-trade of the probabilists, mathematicians played a largely peripheral role in designing the lotteries. The one notable exception was Frederick II of Prussia, patron of Voltaire and other *philosophes*, who nervously consulted Euler on several occasions concerning the probabilities of winning the Genoan-style lottery he was

---

[106] Great Britain, House of Commons, *Report from the Committee, Appointed to Examine the Book, Containing an Account of the Contributors to the Lottery of 1753* (London, 1754), p. 9.

[107] Ashton, *Lotteries*, pp. 50 ff.

[108] Leonnet, *Loteries*; see Pierre Coste, *Les loteries d'état en Europe et la Loterie Nationale* (Paris: Payot, 1933), p. 28, for the heavy traffic in foreign lotteries during the interim.

[109] Henry Fielding, *The Lottery* (London, 1732), p. 1.

[110] Archives Nationales, Paris, MS. F^{12} 795.

[111] Ashton, *Lotteries*, pp. 75–76, 88; Rouault de la Vigne, *Loterie*, pp. 42–43; Leti, *Critique*, vol. 1, p. 189.

planning to install.[112] Johannes Hudde appears to have been involved in an Amsterdam lottery to benefit French refugees, but more in his capacity as mayor than as mathematician.[113] Otherwise there was little motive for governments to consult mathematicians, since the traditional lottery methods usually returned a handsome profit and since competition from foreign and private lotteries did not cut greatly into domestic state sales.[114] Profit margins usually ranged from 15 to 20 percent, although they fluctuated from year to year.[115] Even in France where the regime depended upon the expert opinion of the Académie des Sciences in most technical matters, the mathematicians seem to have been only rarely consulted about lottery matters.[116] Not surprisingly, the pay-off rates for the Genoan-style French Royal Lottery bore only a rough proportion to the improbability of the bet: an *extrait* won 15 times the stake; an *ambe*, 270 times, and a *terne*, 5,500.[117] A *quine déterminé* (i.e., all five numbers in the order drawn) won a million times the price of the ticket. Despite the infinitesimal chances of this last, it was a great favorite among many players, ruining some.

It is clear that those who ran the lotteries had little to gain from calculating exact probabilities so long as profits rolled in; but who played at such odds, and why? The records provide us only with hints, but these are suggestive. Although the tickets could be expensive, particularly for the British state lotteries, it was possible to purchase a fraction of ticket, put-

[112] L. E. Maistrov, *Probability Theory: A Historical Sketch*, trans. Samuel Kotz (New York and London: Academic Press, 1974), pp. 101–106; Leonhard Euler, "Sur la probabilité des séquences dans la loterie Génoise" (1767); "Solution d'une question très difficile dans le calcul des probabilités" (1771); "Réflexions sur une espèce singulière de loterie, nommée Loterie Génoise" (posthumous); all reprinted in *Opera Omnia* (Leipzig and Berlin: B. G. Teubner, 1923), vol. 7, pp. 113–152, 162–180, 466–494.

[113] Leti, *Critique*, vol. 2, pp. 14–15.

[114] This is not to say that lottery patrons were wholly indifferent to prices, or at least to price changes. For example, raising the price of the tickets for the Loterie Générale de Paris from 20 to 24 sols in 1754 cut severely into sales: Archives Nationales, Paris, MS. F^{12} 795.

[115] See Leonnet, *Loteries*, for lottery revenues from the Old Regime through the Restoration.

[116] For a rare exception, see the negative report by Condorcet, Borda, and Legendre on the lottery plan of Caminade de Castres, Archives de l'Académie des Sciences, MS. *Procès-verbaux*, 12 May 1787.

[117] The actual probabilities are: *extrait*: .056; *ambe*: 2.5×10^{-3}; *terne*: 8.5×10^{-5}; *quaterne*: 1.9×10^{-6}; *quine*: 2.3×10^{-8}. Of course, probabilities for bets that specified order as well were even lower.

ting them within the reach of humble artisans and even servants. For example, the winners of the £30,000 prize for the 1792 state lottery had each bought ¹⁄₁₆ of a ticket, and included a clergyman from Lincolnshire, a tradesman in Longacre, two servants to a widow at Epsom, an innkeeper from Kent, and a gentleman from Malverton.[118] The French lotteries tended to be cheaper, especially after the introduction of the Genoan lottery made it possible in effect to price one's own ticket, and participation of the poorer classes was presumably greater. Certainly French critics of the lottery like Talleyrand insisted that it preyed most upon those who could least afford it, although he also admitted that the better classes were not immune.[119] Since the prices of the earlier lotteries ranged between 5 sols and 1,000 livres, it must be assumed that their clientele cut a correspondingly broad cross section through the social pyramid, and one similarly bottom-heavy. The occasional literature on lotteries confirms this impression. Leti describes the lottery fantasies of every calling and class, from the physician who would have his magnum opus opulently produced by a vanity press and quit his practice of unpleasant, infectious patients, to the serving girl who would dress to the nines and cut her mistress dead in the town square.[120] A 1790 English ditty in a similar subjunctive vein also suggests that high and low alike bought tickets:

> The Name of a Lott'ry the Nature bewitches,
> And City and Country run Mad after Riches:
> My Lord, who, already has Thousands a Year,
> Thinks to double his Income by vent'ring it here:
> The Country Squire dips his Houses and Grounds,
> For tickets to gain him Ten Thousand Pounds:
> The rosie-jowl'd Doctor his Rectorie leaves,
> In quest of a Prize, to procure him Lawn Sleeves.
> The Tradesman, whom Duns for their Mony importune,
> Here hazards his All, for th' Advance of his Fortune:
> The Footman resolves, if he meets no Disaster,
> To mount his gilt Chariot, and vie with his Master.

---

[118] Ashton, *Lotteries*, pp. 118–119.

[119] L'Évêque d'Autun [Charles de Talleyrand-Périgord], *Des loteries* (Paris, 1789), pp. 29–31; Pierre Dupont de Nemours, *Opinion de Du Pont (de Nemours) sur les projets de loterie, & sur l'état des revenus ordinaires de la République*, Séance du 24 germinal an V [1797], pp. 10–12.

> The Cook Maid determines, by one lucky Hit,
> To free her fair Hands from the Pot-hooks and Spit:
> The Chamber-maid struts in her Ladies Cast Gown,
> And hopes to be dub'd the Top Toast of the Town . . .[121]

As might be expected, the average player of lotteries paid even less attention to the probabilities than the cities and states that ran them. Defoe wrote a vehement antilottery tract pointing out the hopelessness of the odds, and calling upon mathematicians to contribute to "a Weekly Paper, . . . highly necessary and useful to instruct the People how to lay their Money, and very instrumental to the abolishing of Gaming."[122] Some probabilists like Daniel Bernoulli and Buffon actually did try to reason gamblers out of their passion, albeit in a more learned format than Defoe had perhaps envisioned.[123] However, the only "mathematicians" most lottery players read were numerologists who peddled sure-fire systems for choosing a winning combination.[124] Lottery critics of the day catalogued a number of familiar gamblers' fallacies, including an obsession with certain combinations, played over and over again in the forlorn hope that perseverance would finally win out, or the ruinous passion for the *quaterne* and *quine*, which paid the most.[125] In France, broadsides and almanacs poured from the presses full of magic numbers, prophetic dreams, and lottery stories with happy endings;[126] ticket sellers hawked their wares in the streets of Paris from 7 A.M. to 7 P.M., crying "je vends de la chance"; drawings were held with pomp and circumstance at the Hôtel de Ville according to a ritual that survived the Old Regime, the Revolution, the Empire, and the Restoration. A blindfolded child dressed in blue, with a red sash, drew five numbers from a huge wheel of fortune before a hushed audience of officials, dignitaries, and the public. Not even the

---

[120] Leti, *Critique*, vol. 1, pp. 2–12.

[121] Quoted in Ashton, *Lotteries*, pp. 61–62.

[122] [Daniel Defoe], *The Gamester* (London, 1719), p. 12.

[123] Daniel Bernoulli, "Specimen theoriae novae de mensura sortis," *Commentarii Academiae Scientarum Imperialis Petropolitanae* 5 (1738): 175–192; George Leclerc Buffon, "Essais d'arithmétique morale," *Histoire naturelle*, Supplément, vol. 4 (Paris, 1777), pp. 46–148.

[124] J. B. Marseille, *La pierre, et la vraie pierre philosophale des loteries impériales de France* (Paris, 1807), is typical of the genre.

[125] Talleyrand, *Loteries*, pp. 16–17, 23–24.

[126] Marcel Charpaux, *Almanach de la Loterie Nationale 1539–1949* (Paris: Imprimerie E. Desfossés-Néogravure, 1949).

storming of the Bastille on 14 July 1789 could interrupt the regularly scheduled drawing of the lottery two days later.[127]

Ritual, advertisement, illusion: all of these must have joined forces with wishful thinking to obscure even approximate ideas of likelihood among the majority of the players. But at some visceral level all knew that a big win called for little less than a miracle: hence the widely deplored and equally widely practiced expedient of invoking God and various saints for luck.[128] Why then was the lottery so popular, especially among people who had very little discretionary income to risk? Twentieth-century sociologists of gambling emphasize the elements of passivity and denial of industry. As Roger Caillois put it: "*Alea* signifies and reveals the favor of destiny. The player is entirely passive. . . . Whoever despairs of his own resources is led to trust in destiny. . . . By studying and utilizing heavenly powers over chance, they get the reward they doubt can be won by their own qualities, by hard work and steady application."[129] There were seventeenth- and eighteenth-century writers who disapproved of lotteries in very similar terms, as severing the link between merit, skill, and hard work and temporal rewards, so that "the Winner many tymes is driven to wonder how hee wanne," with no advantage of "industrie or overreaching wit."[130] However, Callois and his predecessors assume that there exists such a link between patience, thrift, industry, talent, and success. Despite the ideology of the meritocracy that blossomed in the eighteenth century—the belief that "Only talent is necessary to make one's *fortune*, only the determined resolution to do so, the patience and the audacity"[131]—the reality was quite different for the mass of people. They were imprisoned within a static social and economic order that no amount of talent, resolution, patience, and audacity could unlock. These qualities might preserve one from the worst, in the case of Beaumarchais' Figaro, but they could not suffice for the man to trade places with his master. Yet social

---

[127] Leonnet, *Loteries*, p. 26; Rouault de la Vigne, *Loterie*, p. 30.

[128] Coudrette, *Dissertation*, pp. 91–92.

[129] Roger Caillois, *Man, Play, and Games*, Meyer Barash, trans. (New York: Free Press of Glencoe, 1961), pp. 17, 48.

[130] Lambert Daneau, *Trve and Christian Friendshippe* (London, 1586), chapter 9; cp. Coudrette, *Dissertation*, p. 30; Talleyrand, *Loteries*, p. 25; Erasmus Mumford, *A Letter to the Club at White's* (London, 1750), pp. 26–29.

[131] [Jean d'Alembert], "Fortune (Morale)," in *Encyclopédie, ou Dictionnaire raisonné des sciences, des arts et des métiers* (Paris, 1757), vol. 7, p. 206.

boundaries were somewhat more fluid in the eighteenth century than pre-viously, in that money might sometimes repair the unhappy accident of birth. Thus the only hope of moving up in the world was suddenly to acquire a vast sum of money, and practically the only hope of such a wind-fall in a nonmeritocratic society was to buy a lottery ticket. Where indus-try and talent counted for little, buying a lottery ticket might have been the one escape from passivity.

Recall the lottery fantasies, particularly those of the tradesmen and do-mestics who dreamed of rubbing shoulders or even upstaging their betters: "To mount his gilt Chariot, and vie with his Master." The eighteenth-century annals of gambling are full of tales of clever rakes who fleece this duke or that earl, including one Dick Bourchier who won £500 from the Earl of Mulgrave and then revealed himself to be Mulgrave's former foot-man.[132] The message of the lottery was implicitly one of social subversion, and it was not lost upon conservative contemporaries. They decried the lottery because it defrauded the poor of what little they had, but in the next breath they worried about the social consequences should one of these penniless players actually win the grand prize: "On the other hand, these people drawn from the dust . . . are going to place themselves among the nobility. Only their money, and not their merit, elevates them to charges and functions which require talent and sensibility, which one ordinarily receives only by a gentle education [*éducation honnête*]."[133] But the gam-bler's retort was that chance had distributed the goods of rank and wealth in the first place, and it was fitting that fortune should redistribute them with an equally capricious hand: "And thus by accidental Events, Poverty and Riches are transplanted, and shift their Seat; and a Blast of Wind, than which nothing is more uncertain, drives good Fortune from one Hand to another. And since Casualties dispose of Things at this arbitrary Rate, since the World is but a kind of Lottery, why should we Gamesters be grudged the drawing a Prize?"[134]

Money was the means by which the lottery player would soar from the life style of a pauper to that of a prince, and the power of money to erase

---

[132] Ashton, *Gambling*, p. 25; Theophilus Lucas, *Memoirs of the Lives, Intrigues, and Com-ical Adventures of the Most Famous Gamesters and Celebrated Sharpers* (London, 1714); Andrew Steinmentz, *The Gaming Table: Its Votaries and Victims* (London, 1870), vol. 1, p. 120.

[133] Coudrette, *Dissertation*, p. 44.

[134] Jeremy Collier, *An Essay upon Gaming*, Edmund Goldsmid, ed. (Edinburgh, 1885; reprint of 1713 edition), p. 10.

social distinctions also troubled these critics. Anyone with the price of a ticket could play, with no questions asked; children as well as their parents, wives as well as husbands, servants as well as masters, clergy as well as laymen, "in a word persons of every age, sex, profession, and condition."[135] An English defender of gambling brushed aside objections that gaming clubs debased the peerage by a promiscuous mingling of ranks at the dicing tables, declaring that anyone with "the Dress, and Purse of a Gentleman" should be welcome to play, regardless of pedigree.[136] A legal writer who insisted that all gaming presumed equality among the players recommended not playing with social inferiors (or even in their presence) lest this very equality lead to a loss of deference and respect.[137] At all levels, the lottery was perceived as a radical equalizing force. At the moment of the drawing, all were equal before fortune (represented since ancient times as blindfolded, imagery echoed by blindfolding the child who drew the lots) however the crowd in the hall might have been physically segregated by rank and quality. Aristocrat and artisan alike trembled with suspense. Moreover, both playing and winning the lottery dissolved these distinctions in the universal solvent of money. The lottery did not create this leveling potential of money, but it was its most dramatic symbol, particularly for those at the bottom of the social heap. Just as money was "indifferent to Parties," chance was the "chief Uniter of Mankind."[138]

The polemical literature concerning lotteries is a braid of many strands; religious, moral, social, and economic. Religious writers, both Protestant and Catholic, debated over whether lotteries were blasphemous "temptations of God" to intervene in profane matters or simply a part of his ordinary providence; moralists claimed that it inflamed avarice and destroyed the normal checks and balances among the vices (e.g., greed checks sloth); social observers feared that it corroded trust within families and among friends and fostered irrational beliefs like numerology; economists branded it the most regressive of taxes and predicted that it would decrease productivity by destroying the link between work and gain.[139] No simple set of Englightenment oppositions does it justice: on this issue *philosophes*

---

[135] Coudrette, *Dissertation*, p. 76.

[136] [Anonymous], *A Modest Defence of Gaming* (London, 1754), p. 26.

[137] Jean Barbeyrac, *Traité du jeu*, 2nd edition (Amsterdam, 1737), vol. 2, pp. 474–476.

[138] *Modest Defence*, p. 40.

[139] Collier, *Essay*, p. 32; see also n. 130, above.

often joined forces with the Catholic clergy; physiocrats with mercantilists; social reformers with archconservatives. My intention here is not to present a thorough analysis of this bulky and complex corpus, but rather a discussion of a few selected issues that shed some light on our main question, the prevailing attitudes toward risk. Of course, we must proceed with extreme caution, for these sources are largely the work of intellectuals, albeit a motley crew embracing anonymous pamphleteers as well as sober theologians. They cannot be taken to mirror popular attitudes in any straightforward way, and we use them as sources for such *faute de mieux*. Yet they are nonetheless valuable, because at some points they explicitly oppose popular views (and describe these at some length), and because at others they participate in them. However self-conscious the critical turn of mind, no intellectual is wholly estranged from the cultural specifics of time and place, and the lottery writers are no exception to this rule. Finally, intellectuals may not always faithfully reflect the status quo, but since the seventeenth century they have often influenced it by either creating or popularizing new values among the reading public. I shall use them as witnesses to the beliefs and character of the gambler, first to the ideas and then to the emotions that drove him or her to play in face of the ruinous odds.

Fortune (*fortuna*, τύχη) was a conceptual category inherited from classical times and dovetailed with Christian teachings about providence by generations of scholastics. Although the most penetrating thinkers such as Augustine and Aquinas realized the philosophical difficulties of distinguishing a semi-independent fortune from divine providence and interpreted so-called chance events in terms of human ignorance, it was more typical of medieval and Renaissance Christian thought to make a place for fortune somewhere between divine providence and willed human agency. Like nature, fortune could thus be saddled with responsibility for inequities (or in the case of nature, anomalies)[140] that would be impious to charge directly to God. Boethius's allegorical evocation of Fortuna in the *Consolations of Philosopy* draws on Martianus Capella's portrait of her as a fickle jade, and follows Aristotle in assigning her the task of doling out the external goods of wealth, power, beauty, health, and so forth. The wise philosopher cannot entirely escape the capricious rule of fortune, but he

---

[140] Park and Daston, "Conceptions," pp. 41–42.

strives chiefly after the internal goods of virtue, which depend on his moral will.[141] It is probable that Boethius also incorporated popular ideas about fortune; certainly his account later influenced these by inspiring almost all medieval and Renaissance pictorial representations of that unstable semideity.[142] In the iconography of Fortuna through the eighteenth century, she is almost always depicted with a blindfold,[143] symbolic of an ambiguity that seventeenth- and eighteenth-century writers on gambling read both ways: she is blind to all distinctions, to those of merit as well as rank in the distribution of her gifts. Thus she is at once impartial (recall that justice is also blindfolded) and unfair. Depending on whether the emphasis was placed on commutative or distributive justice, fortune appeared as the champion of equality, or the spoiler of just deserts.[144]

In all her incarnations, Fortuna was the enemy of prudence, rationality, and calculation. In the *Eudemian Ethics*, Aristotle pitted good luck against good reasoning, the one snatching victory from the jaws of defeat in defiance of the other.[145] Machiavelli struck a similar note in his famous discussion of fortune and *virtù* in *The Prince* (1532). Although he believed in the power of fortune and indeed cast himself as her victim, he nonetheless claimed that, like a river, she might be partially tamed by the proper precautions of dikes and fences. Yet despite this imagery of foresight and rational planning, Machiavelli declared fortune to be better mastered by audacity than calculation:

> I think it is true, that it is better to be heady than wary; because Fortune is a mistresse; and it is necessary, to keep her in obedience to ruffle and force her; and we see, that she suffers her self rather to be mastered by those, than by others that proceed coldly. And therefore,

---

[141] Boethius, *Philosophiae consolationis*, Book II, in *Boethius: The Theological Tractates*, H. F. Stewart, E. K. Rand, and S. J. Tester, eds. (Cambridge, Mass.: Harvard University Press, 1973), Loeb Classical Library.

[142] Vincenzo Cioffari, "Fortune, fate, and chance," in *Dictionary of the History of Ideas*, Philip P. Wiener, editor-in-chief (New York: Charles Scribner's Sons, 1973), vol. 2, pp. 225–236, on p. 232.

[143] For medieval and Renaissance depictions of Fortune, see A. Doren, "Fortuna im Mittelalter und in der Renaissance," *Vorträge der Bibliothek Warburg* 2 (1922): 71–144.

[144] The rationale for using children to draw lots closely parallels this combination of impartiality and incompetence, for they possess "neither skill nor malice at such a tender age"; see Claude François Menestrier, *Dissertation des lotteries* (Lyons, 1700), p. 141.

[145] Aristotle, *Eudemian Ethics*, 1247b34 ff.

as a mistresse, shee is a friend to young men, because they are lesse respective, more rough, and command her with more boldnesse.[146]

Something of Machiavelli's view survives in the adventure literature of our own day, in which the heroes court danger with boldness and panache, always triumphing against astronomical odds. As Erving Goffman observes, these favorites of fortune succeed "only by grossly breaking the laws of chance. Among young aspirants for these roles, surely the probabilistically inclined must be subtly discouraged."[147] In the seventeenth and eighteenth centuries, fortune was therefore the ally of all those who sought a chink in either the tightened determinism of the natural order or the almost equally obdurate rigidity of the social order. Probability, statistics, and the regularities they were intended to prove were as inimical to fortune as the wary, calculating prudence they were meant to model and instill. Small wonder that the probabilists joined forces with the opponents of gambling to attack fortune (variously known as "luck" or "chance") as a chimera and a superstition, that is, what Englightenment writers were pleased to call a popular error.

Like many of the so-called popular errors attacked in the late seventeenth and eighteenth centuries, beliefs in fortune and luck had a long and learned lineage.[148] Aristotle, Boethius, Albertus Magnus, and many other, lesser Latin lights had given these words philosophical, theological, and literary substance, although there were powerful dissenting voices including those of Cicero, Augustine, and Aquinas. It is doubtful that this learned tradition shaped popular views by some simple trickle-down effect; more likely, both learned and popular views cross-fertilized one another here as in medicine and natural history. This two-way flow of influence was in fact encouraged by a conception of philosophy, dating from Aristotle[149] and still alive today in the ordinary language school, that made everyday locutions the departure point for conceptual analysis. From the mid-seventeenth century on, however, intellectuals distinguished com-

[146] Niccolo Machiavelli, *The Prince* (1532) Edward Dacres, trans. (New York: AMS Press, 1967; reprint of 1640 edition), p. 352.

[147] Erving Goffman, "Where the action is," in *Interaction Ritual* (Garden City, N.Y.: Doubleday, 1967), pp. 149–270, on p. 166n.

[148] Lorraine J. Daston, "Folklore and natural history," *Harvard Advocate* 107 (Autumn 1983): 35–38.

[149] G.E.L. Owen, "Tithenai ta phainomena," in *Aristote et les problèmes de méthode*, S. Mansion, ed. Louvaine: Publications Universitaires, 1961), pp. 83–92.

mon but false notions ever more sharply from the warranted opinions of the enlightened.[150] They recognized common sense, and the common language that embodied it, for the theory about the world that it was, and in most cases they firmly and scornfully rejected the theory and its "vulgar" proponents. Indeed, they sometimes seemed to judge a theory as much by the (low) company it kept as by its contents: Hume considered it a damning argument against miracles to observe that they were believed by the ignorant and barbarous.[151] The learned roots of the "popular" errors were forgotten, or when they were mentioned, it was only to reprimand the guilty philosopher for reifying the misguided notions of the folk, that "most deceptable part of mankind."[152] So did Jean Le Clerc, author of a 1676 treatise on lotteries, upbraid Aristotle for his discussion of chance in the *Physics*: "What is more common than to see that the people have false ideas, and that the expressions formed from such ideas are very inexact?"[153]

Le Clerc was one of several critics, both Protestant and Catholic, who took up the theme of gambling in general and lotteries in particular in this period, with special attention given to popular errors concerning fortune and luck. The original issue was the theological legitimacy of lotteries and other games of chance, but this very quickly led to broader discussions of providence, determinism, and chance. On the basis of several biblical passages,[154] various theologians had outlawed games of chance and especially the drawing of lots as a "temptation of God." Since the Jews of the Old Testament occasionally had recourse to lots to ascertain God's will in perplexing cases, early Christian theologians interpreted all such deliberate chance setups as an invitation to God to intervene in the normal causal order. Recreational or commercial gambling was therefore condemned as the equivalent of a crank telephone call to the deity.[155] Both the increased popularity of lotteries and gambling and new views of prov-

[150] Natalie Zemon Davis, "Proverbial wisdom and popular errors" in *Society and Culture in Early Modern France* (London: Duckworth, 1975), pp. 227–267.

[151] David Hume, "Essay on miracles," in *An Enquiry Concerning Human Understanding* (1749), in his *The Philosophical Works*, Thomas H. Green and T. H. Grose, eds. (London, 1882), vol. 4, pp. 96–97.

[152] Thomas Browne, *Pseudodoxia Epidemica* (1646), in *The Works of Sir Thomas Browne*, Geoffrey Keynes, ed. (London: Faber & Faber, 1964), vol. 2, p. 25.

[153] Jean Le Clerc, *Réflexions sur ce que l'on appelle bonheur et malheur en matière de loteries, et sur le bon usage qu'on en peut faire* (Amsterdam, 1696), p. 43.

[154] See n. 39, above.

[155] See Coudrette, *Dissertation*, for the history of canon law on the subject.

idence prompted a reexamination of this position in the seventeenth century. The sheer frequency with which Europeans played at cards or dice made it seem ludicrous to claim that God directly controlled each and every outcome through primary causes. Familiarity with chance events bred, if not contempt, at least a certain irreverence.[156] Hence Thomas Gataker and others separated profane from sacred appeals to lots, and made the distinction hinge on the circumstances: the magnitude of the outcome; the intent of the participants; the overall scheme of God's providence. Gataker concluded that profane lotteries were therefore lawful if "used with due Caution," equating them with other aleatory contracts "wherein men buy bare hope alone then actually ought else."[157] That is, such "lusorious" lots became instances of God's ordinary rather than extraordinary providence, in contrast to the traditional view that every lot "necessarily suppose[d] the special providence & determining presence of God."[158]

Their ambiguous status with respect to ordinary and extraordinary providence thrust lotteries into the larger debate over miracles: to what extent and under what conditions did God break his own rules? Just as intellectuals, both devout and skeptical, became increasingly reluctant to countenance such spectacular suspensions of secondary causes, so they also balked at God's direct participation in every casual coin-toss game, much less that of intermediate entities like fortune. In both cases, the wide and varied range of the arguments mustered to these ends plants the suspicion that the arguments themselves were not of cardinal importance. There are too many different and sometimes incompatible arguments converging upon the same position. That *terminus ad quem* was a more strictly determined natural order that displayed God's providence through regularity and foresight rather than through showy signs and personal appearances. The determinists were for the most part sincerely pious, but they admired

---

[156] However, some of the early charitable lotteries apparently preserved an element of this appeal to divine authority. In 1572, for example, Louis Gonzaga, the Duke of Nevers, founded a lottery on behalf of sixty poor but deserving girls without dowries: lots were inscribed with either "Dieu vous a élue" or "Dieu vous console." Rouault de la Vigne, *Loterie*, p. 22.

[157] Thomas Gataker, *Of the Natvre and Vse of Lots* (London, 1619), pp. 84, 124; Barbeyrac, *Traité*, vol. 1, p. 29.

[158] James Balmford, *A Modest Reply to Certaine Answeres* (n.p., 1623), p. 9; Daneau, *Friendshippe*, chapter 9; Jean La Placette, *Divers traités sur des matières de conscience* (Amsterdam, 1697), pp. 201–215.

God more for his rules than for the exceptions. Hence, chance was demoted to a *façon de parler*, a shorthand way of referring to our own ignorance of the causes of certain events and/or inability to predict them. While this was by no means a new doctrine, it was given a fresh lease on life by Cartesian mechanism, which made gaps in the casual chain governing the motion of physical objects like lots or dice unimaginable. Belief in fortune or in good and bad luck on this construction became at best a pathetic error and at worst a pagan personification of unknown causes, as opposed to the Christian view of a God who "governs all." Theologians, Protestant and Catholic alike, joined forces with the mathematicians to battle these vestiges of "*Ignorance* and *Superstition*."[159]

Although this became the official Enlightenment position,[160] it apparently made little headway among the lottery players themselves. Talleyrand, writing in 1789, complained of the same dogged faith in luck, helped along by talismen, prayers, and numerology, that had appalled writers like Le Clerc at the turn of the century. Subscribers to French blank lotteries continued to invoke the Virgin Mary and patron saints in their devices, a practice deplored even by those who supported charitable lotteries;[161] Henry Fielding's play *The Lottery* (1732) shows ticket buyers requesting lucky numbers, and it is a fortuneteller who persuades the heroine that she will win the £10,000 prize.[162] The mathematician Samuel Clark devoted an entire tract to exploding the pretensions of one John Molesworth's scheme for selecting winning numbers. Clark patiently explained with the help of simplified calculations how probability theory accounted for short-run fluctuations in gains and losses, "wherefore Chance alone, by its Nature, constitutes the Inequalities of Play, and there is no need to have Recourse to Luck to explain them."[163] However, neither sermons nor calculations seemed to have made much of an impression on the average player, who apparently understood chance to mean equal proportions of gains and losses over the short as well as the long run.[164] Any

---

[159] Le Clerc, *Réflexions*, pp. 41 ff.; Coudrette, *Dissertation*, p. 234; Samuel Clark, *Considerations Upon Lottery Schemes in General* (London, 1775), p. i.

[160] See [Denis Diderot], "Fortuit (Gramm.)," and [Jean d'Alembert], "Fortuit (Metaphys.)," in *Encyclopédie*, vol. 7, pp. 204–205.

[161] Menestrier, *Dissertation*, p. 116.

[162] Fielding, *Lottery*, pp. 3–11.

[163] Clark, *Considerations*, p. 11.

[164] Psychologists continue to report that such beliefs in "local representativeness" are

deviation from this balance counted as luck, a personal if brief alliance of fortune with the individual player whose personality or manner, as Machiavelli had speculated, attracted such favors.

Some eighteenth-century gaming manuals did provide calculations for games, made "Easy to those who understand Vulgar Arithmetick only," but these often involved simplifying assumptions of equal probabilities of winning even for billiards and cockfighting that might well have led players astray in practice.[165] Certainly both the reforming and popular literature on gambling suggest rather that calculations were either ignored or held in outright contempt by players. Talleyrand pointed out with some exasperation that the profit margin of the royal lottery even on the *extrait* was nearly twice that in notorious games like Belle and Biribi, which had been suppressed by the police because of public outcry against the house's disproportionate take—a calculation that players who were devoted to the one but reviled the others had evidently never made. Indeed, their suicidal attachment to the most unlikely combinations like the *quine* all but insured the lottery enormous profits.[166] Pierre Dupont de Nemours admitted that calculations might expose the lottery for the subtle theft that it was, but realized that inveterate players were not the reckoning sort, trusting in their good luck and eternally hopeful against all odds.[167]

However, even some of the antigambling writers were skeptical about the applicability of calculations: "There's no computing the Issues of Play, the Casualties are too great for Calculation."[168] In an age where these "casualties" included rampant fraud and cheating, this may have been all too true.[169] The adventurer hero Ferdinand Count Fathom of Tobias Smollett's novel suspects an easy mark in the bumptious Sir Stentor, for gaming had been one of Ferdinand's "chief studies . . . he could calculate all the

---

widespread, even among the compilers of random number tables, who delete runs that seem too patterned, and therefore insufficiently "random": Lola L. Lopes, "Doing the impossible: A note on induction and the experience of randomness," *Journal of Experimental Psychology: Learning, Memory, and Cognition* 8 (1982): 626–636.

[165] Edmund Hoyle, *The Polite Gamester* (Dublin, 1761); [Anonymous], *Directions for Breeding Game Cocks* (London, 1780); T. Gard, *The Odds and Chances of Cocking, and Other Games* (London, n.d.).

[166] Talleyrand, *Loteries*, pp. 13–17.

[167] Dupont de Nemours, *Opinion*, p. 8.

[168] Collier, *Essay*, p. 30.

[169] [Charles Cotton], *The Compleat Gamester* (London, 1680), pp. 5 ff.; Ashton, *Gambling*, pp. 235–236.

chances with utmost exactness and certainty." Alas, the doctrine of chances avails Ferdinand not against the sharpster Stentor and his associate, who empty his pockets in short order, probably with the aid of loaded dice.[170] Theophilus Lucas's collection of rollicking tales of famous gamblers recounts case after case of prodigious good luck, including the notable career of Beau Hewit. Hewit at first approaches gambling scientifically, Huygens's treatise in hand, but

> . . . finding his Rules of calculating chances most false and erroneous, he damn'd that Authour for as great a Blockhead as he was a Fool, in losing his Money upon such conceited whims; therefore learning the most profitable and surest way of tricking both at Cards and Dice, in which the Adversary could make no Calculation of Chances, be became so expert in the Dexterity of Slipping Cards, or Cogging a Dye, that in 4 years he was worth 6000 Pounds.[171]

Beau Hewit's conversion from probability theory to cheating hints at one reason why the majority of gamblers did not calculate the odds. As Cardano, perhaps the only probabilist with much hands-on gambling experience, had long before warned, probability mathematics does not eliminate the uncertainty of play. The outcome of any given roll of the dice remains incalculable. Hewit's slipping and cogging greatly reduced the short-term risks, though it had its own disadvantages: he came to a bad end when challenged to a duel by one of his outraged dupes. Of course, Hewit's methods were not readily adapted to the lottery, although there were occasional attempts to bribe the children who drew the lots from the wheel.[172] But lottery players also thirsted after systems and charms that would transform a paltry chance into a sure thing. The stronger the hope, the more intense the belief in the helping hand of fortune, who either chose her favorite freely or was manipulated by the pseudotechnologies of numerology, dreams, and fortunetelling. Far from being an "impersonal neutral power, without heart or memory, a purely mechanical effect,"[173] fortune had an all-too-human face for these lottery players. She could be notoriously capricious and cruel in her wavering preferences, but even caprice might be seen as a loophole in an otherwise calcified social system.

[170] Tobias Smollett, *The Adventures of Ferdinand Count Fathom* (Dublin, 1753), vol. 1, pp. 115–117.

[171] Lucas, *Memoirs*, p. 285.   [172] Ashton, *Gambling*, pp. 232–233.

[173] Caillois, *Games*, p. 46.

Determinism had a social as well as a natural side, and fortune was unwelcome to both. Conservatives mourned the fall of great estates and ancient families in a few hours of gaming; liberals deplored the broken link between industry, talent, and gain; philosophers and mathematicians excluded chance and its personifications from nature.[174] All viewed fortune as a dangerous disruption of the order of people as well as of things, overturning the predictability of life lines in society as well as of trajectories in mechanics. Yet where there is uncertainty there is hope, and fortune's wheel, made concrete in the lottery wheel, became the emblem of hope, especially for those whose lives were crushed by their grim predictability. The incalculability of the single outcome and a systematic conflation of probability and possibility sustained this hope. The jurist Jean Barbeyrac used a revealing lottery analogy to rebut Pierre Bayle's claim that the truths of human nature, namely that most people act from bad motives, could not be reconciled with the Christian injunction of charity. Barbeyrac conceded Bayle's point about human nature but argued that just as improbability did not wholly preclude the possibility of winning at the lottery, so our past experience with corrupt human nature should not prejudice a charitable judgment in a given case: "Thus the greatest degree of possibility is in no way the rule of judgment for the fact or even itself."[175] Here the antistatistical emphasis on the particular case converged upon an unreasonable optimism in the face of the odds, and the point of convergence was the lottery.

Although hope and the antiprobabilistic illusions it fostered were no doubt strongest among those in most desperate straits, "the poorest citizens of the towns and country . . . the cook, the servant, the artisan, the herb seller,"[176] wealth alone provided no immunity to either passion for gaming nor its accompanying superstitions. Well-heeled gamblers eagerly scanned their surroundings—the name of a café, the numbers of passing coaches—for clues to fathom the ways of fortune.[177] Given the vast sums purportedly lost at court, in "ordinaries," in *académies de jeu* and clubs, and even in private homes, and the social importance attached to immediately honoring gambling debts, even many affluent players may have been re-

---

[174] Collier, *Essay*, p. 19; *Réflexions*, pp. 14, 23–24; Jean Dusaulx, *Lettre et réflexions sur la fureur du jeu* (Paris, 1775), p. 16; Talleyrand, *Loteries*, p. 3; De Moivre, *Doctrine of Chances*, 3rd edition (London, 1756), p. 253.

[175] Barbeyrac, *Traité*, vol. 3, pp. 80–84. [176] Dupont de Nemours, *Opinion*, p. 10.

[177] Dusaulx, *Lettre*, p. 32.

duced to desperation comparable to that which oppressed the poorer sort day in and day out.[178] Reams of legislation intended to protect the profligate from their own excesses by setting limits to legally enforceable gambling debts were woefully ineffective, and many scions of good families were reduced to taking out loans at usurious rates.[179] Fortune was as much their companion as that of the maidservant who scraped and saved to buy 1/16 of a lottery ticket.

But over the course of the eighteenth century, class differences began to emerge in attitudes toward gambling. Initially the rage for taking chances at gaming tables, cockfights, lotteries, and so on was almost a social universal, providing some of the increasingly rare opportunities for high and low to mingle in a period otherwise characterized by cultural withdrawal.[180] As Samuel Pepys exclaimed over the mix of people at the cockfighting pit in Shoe Lane: "Lord! to see the strange variety of people, from Parliament man, by name Wildes, that was Deputy-Governor of the Tower when Robinson was Lord Mayor, to the poorest 'prentices, bakers, brewers, butchers, draymen, and what not; and all these fellows one with another cursing and betting."[181] Lotteries attracted a similarly motley following at the outset: "One stream of *Coachmen, Footmen, Prentice Boys*, and *Servant Wenches* flowing one way, with wonderful hopes of getting an estate for three pence. *Knights, Esquires, Gentlemen* and *Traders, Marry'd Ladies, Virgin Madams, Jilts*, etc.; moving on *Foot*, in *Sedans, Chariots*, and *Coaches* another way; with a pleasing Expectancy of getting Six Hundred a Year for a Crown."[182]

Antigambling legislation drove wealthier players indoors, to sumptuous clubs like White's or more private resorts, and the free-for-all cockfights gradually gave way to horse racing and hunting, which until the

---

[178] See, for example, Ashton, *Gambling*, chapter 5; Madame de Sévigné, *Correspondance*, Roger Duchêne, ed. (Paris: Gallimard, 1974), vol. 2, pp. 351–352 (29 July 1676); Collier, *Essay*, p. 28. It is tempting to speculate on how upper-class gambling may have provided a forum for aristocratic values no longer given military scope: display, courage conceived as the deliberate courting of large risks, a rigid and often ruinous code of honor.

[179] *Réflexions*, pp. 23 ff.; Ashton, *Gambling*, chapters 3–4 and p. 229; [Thomas Erskine], *Reflections on Gaming, Annuities, and Usurious Contracts* (London, 1776), p. 14.

[180] Peter Burke, *Popular Culture in Early Modern Europe* (London: Temple Smith, 1978), pp. 270–279; Davis, "Proverbial wisdom."

[181] Samuel Pepys, *The Diary of Samuel Pepys*, Henry B. Wheatley, ed. (London: G. Bell and Sons, 1962), vol. 3, p. 260 (21 December 1663).

[182] Quoted in Ashton, *Gambling*, p. 227.

late nineteenth century remained an upper-class preserve.[183] The lottery was open to all, but to judge from late eighteenth-century reports it had become primarily the pastime of the laboring and indigent classes. The rich gambled with equal fervor, but they played at cards and dice for dizzying stakes, or placed bets on anything and everything within the confines of their exclusive clubs. Although well-dressed adventurers of dubious backgrounds might gain entry, the fashionable gambling venues were no longer the social miscellany described by Pepys. The professional middle classes gradually withdrew from the gambling scene in the second half of the century, and became its sharpest critics. In the literature on gambling the emphasis shifted from theological and philosophical issues like the existence of fortune to moral matters. These moral elements had never been wholly absent; however, the evolving constellations of middle-class attitudes toward time, risk, and familial responsibility gave them new force.

Recurring motifs were the waste of time and money; the neglect of familial and business duties; the erosion of social trust; and the severed link between hard work, talent, and gain. But at the heart of the moral critique of gambling was an emotional portrait of the gambler as one racked by uncontrollable passions. Not only the intensity but the sudden alternation of these passions weakened the will and rendered the gambler dangerously incalculable. Neither rational self-interest nor conscience could any longer be depended upon to check wild impulses or restrain impetuous desires. Tossed about by "five or six opposing winds," the gambler acted without premeditation, and without regard to consequences. The stormy succession of "*Desire, Fear, Hope, Disappointment, Joy, Scorn, Regret*, sometimes even *Rage* and *Hate*" wrenched the gambler out of time and out of the web of social interdependence.[184] All the other evils listed above followed as corollaries from this violent uprooting. For some, the very fact that gambling inflamed the passions was grounds for condemning it, as a violation of Christian discipline; for others less hostile to the passions *in se*, it was the denaturing of honest passion, "for in the alternate whirls of fortune there is no time for any sensation but uneasiness; the cup of their pains and pleasures is so mixed and dashed with each other, that

---

[183] Brian Harrison, "Religion and recreation in nineteenth-century England," *Past & Present*, no. 38 (December 1967): 98–125, on pp. 116–119.

[184] Barbeyrac, *Traité*, vol. 2, pp. 335–336.

it is one, continued, nauseous, brackish dose."[185] Although women were generally granted more emotional license on the alleged physiological grounds of softer, more impressionable brains, the exaggeration and profanation of these passions in gambling was deeply deplored.[186] The sudden swings from euphoria to despair, as unpredictable as the spin of fortune's wheel, struck moralists as almost a form of madness (*fureur*), a paralysis of reason and will that went hand-in-hand with a total absorption in the present moment.

Loss of self-control and the eclipse of past and future: these were the aspects of the gambler's character that simultaneously fascinated and horrified the moralists. Here the passions had overwhelmed the interests both in strength and numbers, and thus rendered the gambler both incapable of calculation and incalculable. The calculation of long-run probabilities and consequences—to one's family, one's business, one's reputation—presupposed a future that "exists not at all for gamblers."[187] People who refuse to calculate by the rules of rational self-interest, including long-run probabilities, become incalculable. Unruly passions, a kind of civil war of the soul, undermined the constancy of human conduct upon which a secure society was believed to depend.[188] As Erving Goffman has observed, a steady, predictable life harmonizes well with a dense network of social interdependence;[189] eighteenth-century moralists perceived the gambler as a tear in that network. Lottery players were particularly threatening, for they might cultivate their vice in secret, deceiving friends and family until it was too late. Since almost anyone might nurse a covert lottery passion, the moral uncertainty spread like an infection: suspicion eroded trust within families, among friends, between employer and employee.[190] What fortune was to the natural order—a gap in the causal chain—the gambler was to the social order: the one was unthinkable; the other intolerable.

These sentiments eventually led to the abolition of state lotteries in Britain and France in 1826 and 1836, respectively. There is some reason to believe that the moral opposition might have succeeded earlier had lot-

[185] La Placette, *Traités*, p. 236; Erskine, *Reflections*, p. 6.

[186] [Richard Steele], *The Guardian* (London, 1714), vol. 2, p. 191 (no. 120, 29 July 1713).

[187] Dusaulx, *Lettre*, p. 50n.

[188] Albert O. Hirschman, *The Passions and the Interests* (Princeton: Princeton University Press, 1977), pp. 52–53.

[189] Goffman, "Action," p. 174.   [190] Talleyrand, *Loteries*, pp. 30–31.

teries not been essential to national finances before these dates.[191] Moreover, once the lottery, like cockfighting, was abandoned by first its middle-class and then its upper-class patrons for other pursuits, it became a far more vulnerable target for the moralists.[192] Whether the poor actually monopolized lottery play is an open question,[193] but lottery critics from the last quarter of the eighteenth century onward firmly believed so. Perhaps precarious financial circumstances bred a foreshortened attitude toward time that reminded the moralists of the gambler's: both inhabited "a world without a past and without a future . . . an environment where time was encapsulated, causation muddled, and the future looked after itself."[194] Foresight, prudence, and the other timely virtues had no place in lives made uncertain either by necessity or choice. It was the short run or the individual event that mattered, not the outcomes computed over the long run on the assumption of stable conditions or constant probabilities. Seen from this perspective, the ideas and emotions of the gambler—the belief in fortune and luck, the clash of fierce but evanescent passions, the telescoping of time to the present point—were all of a piece. They represented a consistent rejection not only of calculation and calculability, but of all that made these possible: the indissoluble bonds between cause and effect, between interest and action, between past and future. The voluntary risks of the lottery symbolized the hope of escaping the stable, orderly world of the probabilists and the calculating reason they sought to model.

3.4.2 Life Insurance

By the first decades of the nineteenth century, gambling and insurance had come to be seen as antithetical approaches to risk taking: the one risk-seeking and the other risk-averse. But this distinction was none too clear during the seventeenth and much of the eighteenth century. Gamblers and insurers (particularly life insurers) were often the same people; insurance

---

[191] Ashton, *Gambling*, p. 238; Leonnet, *Loteries*, pp. 81 ff.; Coste, *Loteries*, p. 31.

[192] Harrison, "Religion"; McKibbin, "Gambling."

[193] An 1830 report on the French lottery claimed that bets for less than 3 francs accounted for only one-fifth of the total Parisian receipts, suggesting that the participation of the poor may have been exaggerated: Leonnet, *Loteries*, p. 86; cp. McKibbin, "Gambling," p. 156.

[194] McKibbin, "Gambling," p. 167.

offices doubled as betting centers; and stacks of learned tomes written first by jurists and then by mathematicians spelled out the structural similarity of the two approaches. Both the theory and practice of risk conspired to identify them, and only the slow emergence of new beliefs and values that closely paralleled those that eventually abolished state lotteries forced insurance and gambling asunder. In this section I shall argue that it was this divorce that made mathematically based insurance possible, hence the lag between the mathematical technology and its commercial application.[195]

Whatever we mean by modernity is in some way linked with new attitudes toward the control of the future and the possibility of a life relatively secure from the disruptions of chance. Hence, Keith Thomas sees no more revealing indicator of the decline in magical beliefs in favor of more rational ones than the rise of the insurance industry in early eighteenth-century England.[196] The argument seems, on the face of it, convincing. Maritime insurance expanded under the auspices of individual brokers who congregated at coffeehouses like Lloyd's; fire insurance first emerged in London in 1680; the Amicable Society for mutual insurance of lives was established in 1706; and the Royal Exchange and London assurance offices, both of which insured lives, were incorporated in 1720. Many other insurance schemes were launched and folded in short order.[197] In the feverish London insurance market of the mid-eighteenth century it was possible to buy insurance against cuckoldry, lying, and even losing at the lottery, the latter sold by none other than John Law himself.[198] Defoe envisioned an umbrella of mutual insurance as protection against all manner of disasters: "All the Contingencies of Life might be fenc'd against by this Method (as Fire is already) as Thieves, Floods by Land, Storms by Sea, Losses of all

---

[195] Parts of this section are taken from my "The domestication of risk: Mathematical probability and insurance, 1650–1830," in *The Probabilistic Revolution*, vol. I, *Ideas in History*, Lorenz Krüger, Lorraine J. Daston, and Michael Heidelberger, eds. (Cambridge, Mass.: MIT Press, 1987), pp. 237–260.

[196] Keith Thomas, *Religion and the Decline of Magic* (New York: Charles Scribner's Sons, 1971), pp. 651–656.

[197] For an overview of the eighteenth-century British insurance scene, see H.A.L. Cockerell and Edwin Green, *The British Insurance Business, 1547–1970* (London: Heinemann, 1976).

[198] John Francis, *Annals, Anecdotes, and Legends: A Chronicle of Life Assurance* (London, 1853), pp. 140 ff.; Nicolas Magens, *An Essay on Insurances* (London, 1755), vol. 1, p. 33; Ashton, *Lotteries*, p. 120.

Sorts, and Death it self, in a manner, by making it up to the Survivor."[199] However, *pace* Thomas, the vogue for insurance seems to have been less prudential than reckless, fueled more by the spirit of gambling than of foresight. As for the insurers, the more reputable relied on the traditional methods described in Section 3.2; the less reputable were frankly speculators. Both insurance offices and their customers were for the most part betting on the future, not planning for it.

The craze for the lottery and gaming described in the previous section did not exhaust the will to risk; betting on the outcome of every imaginable event was also much in vogue. Gambling clubs like White's of London (immortalized in the sixth picture of Hogarth's *Rake's Progress*) kept books to register bets between patrons, like this one of 28 February 1770:

> A bet was laid by a noble earl that he would procure a man to ride to Edinburgh from London, and back, in less time than another noble earl could make a million scores, or distinct dots, in the most expeditious manner that he could contrive.[200]

However, insurance offices were the real center for wagering in eighteenth-century London. Underwriters issued policies on the lives of celebrities like Sir Robert Walpole, the success of battles, the succession of Louis XV's mistresses, the outcome of sensational trials, the fate of eight hundred German immigrants who arrived in 1765 without food and shelter, and in short served as bookmakers for all and sundry bets. Lloyd's Coffeehouse was as much a resort for betting as for underwriting; the bets riding on the true sex of the French diplomat Charles d'Eon de Beaumont approached £70,000 in 1761–62, by far the majority of these placed at insurance offices.[201] Defoe noted with disapproval that these wagering offices attracted much the same clientele as gambling dens, being "throng'd with *Sharpers* and *Setters* as much as the Groom-Porter's or any Gaming-Ordinary in Town."[202] Life insurance, illegal in almost every European country except England, almost always consisted of a wager on the life of

---

[199] Defoe, *Projects*, p. 123.

[200] Quoted in Ashton, *Gambling*, p. 158; see Smollett, *Adventures*, pp. 88–89, for a parody of aristocratic betting.

[201] Francis, *Annals*, pp. 144–145; Ashton, *Gambling*, pp. 279, 152–155; Braun, *Geschichte*, p. 108.

[202] Defoe, *Projects*, p. 174.

a third person, commonly one who owed the policyholder money but often a complete stranger—a prince of the blood, a hale Scottish nonagenarian, the local bishop. So strong was the association between premium life insurance and wagering that Defoe, the great advocate of insurance projects, refused to countenance it altogether.[203]

Against the background of such practices, and of the legal connections created by the doctrine of aleatory contracts, it is hardly surprising that insurance (and other commercial risks) and gambling were often compared: Jean Le Clerc saw no difference between those who speculated heavily on the lottery and those who embarked on trade ventures in which one stood to either win or lose a great deal; an English pamphleteer made much the same point fifty years later in a moral defense of gambling.[204] Even those authors who believed the two to be distinct were by no means agreed as to the exact grounds for the distinction. Nicolas Magens thought it would be sufficient for merchants to put the public good before private self-interest to prevent "so many strange inventions of unnatural and gaming insurance."[205] Johann Tetens compared life insurance to a game of chance in which one bets that the premiums and more will be paid off to one's widow, but argued that here the lucky ones were the losers.[206] In France, both the detractors and supporters of insurance thought of it in gambling terms. J. P. Brissot de Warville attacked a proposal for a Parisian fire insurance company as an enterprise where "all is left to chance";[207] T. Vernier defended the government lottery during the Revolution on the grounds that annuities, tontines, and fire insurance were also "games of chance based on the expectation and the probability of events."[208] Thomas

---

[203] Defoe, *Projects*, p. 117.

[204] Le Clerc, *Réflexions*, p. 148; *Modest Defence*, p. 18.

[205] Magens, *Essay*, p. iv.

[206] Johann Tetens, *Einleitung zur Berechnung der Leibrenten und Anwartschaften die vom Leben und Tode einer oder mehrerer Personen abhängen* (Leipzig, 1785), p. v. Tetens, however, regularly described the amount paid into a widow's fund as "gewagt, oder aufs Spiel gesetzet" (p. 160).

[207] J. P. Brissot de Warville, *Seconde lettre contre la Companie d'Assurance pour les Incendies à Paris, & contre l'agiotage en général* (London, 1786), p. 30. Brissot's motives may be somewhat suspect, since he was then in the employ of the financier Étienne Clavière, who had much to lose if the company was granted a royal patent, but his Rousseauian rhetoric played upon familiar themes; see Jean Bouchary, *Les manieurs d'argent à Paris à la fin du XVIIIe siècle* (Paris: Bibliothèque d'Histoire Économique, 1939), vol. I, pp. 70–71.

[208] Quoted in Leonnet, *Loteries*, p. 44.

Erskine was even driven to challenge the integrity of the category of aleatory contracts, traditionally distinguished from usury by the element of risk: "For although the risque be ten times greater in lending to the gamester in driving to Whites, than to the merchant sailing to the Indies, yet the principle of public and mutual advantage being lost . . . the contract instantly changes its nature [i.e., becomes usurious]."[209] As Erskine evidently realized, as long as risk remained the sine qua non of such contracts, it was extremely difficult to distinguish between the innocent and the dubious on legal principle alone.

There seems to have been a positive thirst for risk during the South Sea Bubble era that found outlets not only in gambling, "projecting," speculating, and insuring, but also in bizarre hybrid enterprises. For example, mutual aid societies, variants on the medieval confraternities designed to protect members from misfortunes, sometimes voluntarily incorporated further chance elements into their workings. The Amicable Society was founded in 1706 to preserve the widows and children primarily of clergymen from sudden poverty should the household provider die, and was subscribed by equal contributions into a common pot. Its promoters and members understood the Society as a shield against uncertainty and insecurity, yet they were also attracted by a lotterylike scheme of payment: the annual pot was distributed equally among the beneficiaries of members who had happened to die that year, regardless of their number or previous contributions. In an age in which tontines[210] were more popular than life insurance, it was not unusual to find annuity and insurance schemes that deliberately included an element of outright gambling.[211] Even the original plans for the first mathematically based company included provisions for such a tontinelike arrangement to reward the survivors among the original subscribers.[212] Some mutual aid societies were perverted entirely to gambling ends, like the Equitable Society of 1712 that charged each of its

---

[209] Erskine, *Reflections*, p. 24.

[210] Tontines, named after their Neopolitan inventor Lorenzo Tonti, are financial agreements in which all members make an initial contribution to a common fund, which eventually goes to the last surviving member, or to the last surviving members after a specified date; see Schmitt-Lermann, *Versicherungsgedanke*, pp. 64–65, for their early history.

[211] See Deparcieux, *Essai*, pp. 50 ff., on tontines; also Société Générale Néerlandaise, *Mémoires*, pp. 223 ff., for the original annuity-tontine scheme offered by the Dutch town of Kampen in 1670.

[212] Maurice Edward Ogborn, *Equitable Assurances* (London: George Allen and Unwin, 1962), p. 32.

2,000 members of all ages and sexes four installments of £10 each and invested the proceeds in lottery tickets, any winnings to be paid out to the families of members who died within the year.[213] Straightforward commercial ventures also lured investors with tontine clauses—should any investor die before the plan's fruition, his share would be transferred to those remaining—that heaped one kind of risk upon another.[214] Risk was apparently not only tolerated, but relished.

In this climate of frank and often outrageous speculation, it is not surprising that a good many of the newly created insurance companies turned out to be swindles preying upon credulous and risk-happy customers.[215] Such enterprises represented the seamier side of a business that also boasted more reliable firms. But these latter depended on the application of mathematical probability no more than the fraudulent "bubbles" did. The vast bulk of the practice was maritime, and although premiums responded to decreases in risk (the disappearance of marauding Turks made insuring voyages to the Levant, Spain, and Portugal considerably cheaper), statistics played no role in pricing.[216] Traditional methods geared to the individual case still predominated; Lloyd's, for example, kept a "Captain's Register" with information on the careers of over 30,000 ship captains.[217] Fire insurance was too new to be burdened with the weight of tradition, and clients were offered graduated premiums depending on the kind of building (brick versus wood) and trade housed therein (sugar bakers, for example, paid especially stiff rates). Yet fire offices apparently never collected statistics on the subject,[218] and their classification schemes did not preclude variable premiums depending on particulars like kind of chimney or the coverage.[219] In these two areas, insurers would have received little guidance from the mathematical manuals even if they had consulted them.

However, the problems of annuities and life insurance had attracted considerable mathematical attention; yet the terms upon which they were bought and sold had little, if anything, to do with mortality statistics and probability. The actuary was originally a clerical rather than a mathemat-

---

[213] Rosin, *Lebensversicherung*, pp. 52–57.

[214] Rosin, *Lebensversicherung*, pp. 43–44; Walford, *Guide*, p. 31.

[215] Walford, *Guide*, p. 29.

[216] Magens, *Essay*, vol. 1, p. 84; Samuel Marshall, *Treatise on the Law of Insurance* (Boston, 1805), Book I.

[217] Braun, *Geschichte*, p. 110.   [218] Cockerell and Green, *British Insurance*, p. 27.

[219] Dickson, *Sun*, pp. 84–86.

ical position, a combination of secretary and bookkeeper,[220] and with this audience in mind the manuals of De Moivre, Simpson, and Dodson barely required more than arithmetic. Every such book included numerous tables of the values of annuities calculated by age, number of heads, and interest rates to spare the reader calculation. Nonetheless, their impact upon practice appears to have been minimal prior to the establishment of the Equitable Society for the Assurance of Lives in 1762, and even then, the dictates of mathematical theory were greatly tempered by other considerations.

Most eighteenth-century annuity schemes were as innocent of probability theory and statistics as life insurance was. The only partial exceptions were the societies founded to provide annuities to widows in the 1750s and 1760s, first in the Netherlands and then in Britain.[221] Most of these societies did scale premiums to age, but they failed at such a distressing rate that they can hardly have been a good advertisement for such procedures. Richard Price ascribed their failure to an insufficient amount of mathematics, for the organizers did not understand that claims would inevitably increase as the population of members aged, and they fixed premiums more by guesswork than by the tables.[222]

Moreover, there seem to have been other reasons for sellers of commercial annuities to ignore the mathematical manuals, for it appears from the eighteenth-century rolls that most of these annuities were essentially usurious loans disguised as aleatory contracts, particularly after the usury legislation of 1777 made any loan for interest greater than 5 percent illegal unless it involved some genuine risk. Borrowers circumvented this law by selling annuities on the lives of the *seller* rather than the buyer, at very low rates: for example, an annuity of £1,000 per annum might be sold at six years' purchase, £6,000, contingent on the life of the seller. Minus the cost of the life insurance, this amounted to a loan at an effective interest rate of about 12 percent. The overwhelming majority of annuities registered were of this form, and clearly required no recourse to the life tables for the annuity per se.[223] As late as 1793, when the Royal Exchange secured the

---

[220] Ogborn, *Equitable*, p. 48.

[221] Société Générale Néerlandaise, *Mémoires*, pp. 236 ff.

[222] Price, *Observations*, pp. 2 ff.

[223] Sybil Campbell, "Usury and annuities of the eighteenth century," *Law Quarterly Review* 44 (1928): 473–491; Campbell, "The economic and social effect of the usury laws in the eighteenth century," *Transactions of the Royal Historical Society*, 4th series, 16 (1933): 197–210.

right from Parliament to grant annuities on the grounds that the business as transacted by private individuals was rife with fraud and bankruptcy, no attempt was made to calibrate prices by mortality statistics.[224]

How were the premiums of eighteenth-century life insurance policies and annuities in fact determined? The three most prominent eighteenth-century British institutions dealing in life insurance were the Amicable Society and the Royal Exchange and London Assurances. Of these, only the Amicable offered long-term insurance, and it operated more as a friendly society than as a business.[225] Founded in 1706 by Sir Thomas Allen, Bishop of Oxford, the Amicable admitted anyone in good health between the ages of twelve and forty-five, up to 2,000 members, and charged each member an entrance fee of 10 s. and a fixed amount of (£6, 4 s.) each year. The annual income was to be divided equally among the beneficiaries of all those who died in that particular year.[226] Not only did the Amicable take no account of age, except to exclude the periods of greatest mortality (far too cautiously judging from the extant tables); it also had something of the lottery about it, as contemporary observers noted.[227]

Defoe's plan for a friendly society with aims similar to the Amicable gives some insight into the reasoning behind such organizations. Defoe was considerably more cautious than the founders of the Amicable, and would have excluded sailors and soldiers and December/May marriages, recognizing that mortality of wives reduced the society's obligations. His society also differed from the Amicable by raising a fixed sum (£500) by calls upon the members upon each death, in contrast to the Amicable's fixed premiums, so that it could be a bad bargain for long-lived members. As reassurance in the face of such uncertainties, Defoe advanced two arguments: first, members can quit the society at any time; and second, the probable annual contribution for 2,000 members would be about £12, 10 s., "yet this wou'd not be a Hazard beyond reason too great for the Gain." Among the organizers of such schemes, Defoe is almost unique in having based this estimate on the mortality data, or at least on another estimate

[224] Barry Supple, *The Royal Exchange Assurance* (Cambridge, Eng.: Cambridge University Press, 1970), p. 67.

[225] There were distinct national preferences in mortality-dependent financial schemes. The British favored premium insurance and annuities; in France annuities and tontines dominated the market; German principalities instituted many friendly societies; see Deparcieux, *Essai*; Schmitt-Lermann, *Versicherungsgedanke*.

[226] Magens, *Essay*, vol. I, p. 34.   [227] Price, *Observations*, p. 121.

approximated from that data, namely William Petty's claim that one out of forty Londoners dies each year.[228] Yet Defoe could not resist halving this secondhand figure to one out of eighty, to take account of the especially robust specimens he hoped would enroll in his society. He never suggested that entrance fees should be scaled by age, and apparently believed only selectively in the regularity of mortality among "the Midling Age of the People, which is the only Age wherein Life is any thing steady."[229] Informed as he was about the new demography, Defoe could not bring himself to trust its message in practice.

The Royal Exchange and London offices offered short-term (usually one-year) policies at a flat rate of 5 percent for every £100 insured, regardless of age. Life insurance made up only a small fraction of their trade during the eighteenth century, and those who bought policies could be divided into three categories, judging from the company's records: creditors insuring the lives of their debtors for the amount owed; gamblers betting on the life of some third person; and clergymen who insured their own lives, one year at a time. (The Amicable was also designed primarily for the benefit of the clergy; its 1713 prospectus also addressed physicans, surgeons, lawyers, and tradesmen.)[230] State of health is rarely mentioned (the records make occasional references as to whether the insured has had smallpox), but the extraordinary risks (e.g., a voyage "beyond the Cape of Good Hope") commanded much higher premiums (here, 15 percent). Indeed, the life insurance registers, like the maritime ones, provide a rough mental map of risk worldwide as seen through the eyes of an eighteenth-century Englishman: for example, one might travel almost anywhere in Europe at the domestic 5 percent rate, but a trip to the Bahamas increased the premium to 7.1 percent, and to North Carolina, 9 percent.[231]

Why was the practice of eighteenth-century insurance and annuities so resistant to the influence of mathematical theory? It should be noted that not only businessmen but also jurists took almost no account of how the

[228] Petty at times put the number at thirty; see William Petty, *An Essay Concerning the Multiplication of Mankind* (London, 1682), pp. 13–17.

[229] Defoe, *Projects*, pp. 132–140.

[230] Supple, *Royal Exchange*, p. 56; Royal Assurance Company, *Assurance Book on Lives*, vol. 1 (1733–1737), London Guildhall Library, MS. 8740; prospectus quoted in Rosin, *Lebensversicherung*, p. 51.

[231] Royal Assurance Company, *Assurance Book on Lives*, vol. 5 (1758–1771), London Guildhall Library, MS. 8740.

theory of aleatory contracts had been modified by mathematical probability. Robert Pothier's 1775 treatise on aleatory contracts characteristically declined to go into details about fixing premiums, aside from stipulating that they should be equitable, for "as it is not easy to determine what this just price is, one must give this just price much latitude, and hold the just price to be that which the parties have agreed to among themselves."[232] Nicholas Bernoulli's treatise on probability and the law was apparently unknown to him. For maritime and insurance annuities, one might argue that the inertia of an entrenched, successful practice based on nonmathematical estimates worked against the application of the new mathematical and statistical methods. This is no doubt part of the answer, but it cannot explain why new forms of insurance against fire and death did not take on a more statistical cast. The case of life insurance is particularly baffling, because of the availability of mortality statistics drawn from several locales and the growing belief that they revealed firm regularities. Part of the explanation perhaps lies in the lack of unanimity among the mathematicians as to the definition of life expectancy, the validity of certain simplifying assumptions, and the relative reliability of the various life tables.[233] Such controversies were far less important in England than in France, where a government-regulated economy and an official scientific body made for a smoother mesh between mathematical theory and economic practice.[234] But it was London, not Paris, that was the capital of the insurance world until the mid-nineteenth century, and given the elementary level of the English mathematical manuals, it is difficult to believe that London insurance merchants could fathom the deeper mathematical issues.

It took new beliefs and new values, not just the availability of new techniques, to make mathematically based life insurance attractive to buyers and sellers. Insurance sellers had to be persuaded that long-term regularities counted more than short-term perturbations in order to calculate

[232] Robert Pothier, *Traité des contracts aléatoires* (Paris, 1775), p. 75.

[233] Concerning the problem of determining the true mortality curve, see Laplace and Legendre's report to the Paris Académie des Sciences on the work of Kramp, presented 16 May 1789. Archives de l'Académie des Sciences, MS. *Procès-Verbaux* (1789), 108, pp. 137 ff.

[234] This meant that all insurance schemes were submitted to the Académie des Sciences for expert evaluation by mathematicians. See, for example, the reports of Laplace and Condorcet on an annuity project (6 March 1790), of Nicole and Buffon on Deparcieux (21 July 1745), and of Condorcet and Laplace on a proposed life insurance company (16 May 1787), all in the MS. *Procès-Verbaux*.

premiums by the mortality tables; they had to put their faith and money in general rules rather than individual cases. Buyers had to believe all this and more: they had to extend the virtues of prudence and foresight from provision for self to provision for family, and to dread a sudden catastrophe more than they hoped for an unforeseen windfall. None of these shifts was compatible with the time-honored link between gambling and life insurance. Gamblers, eighteenth-century style, fixed upon the short term, actively sought out risk, and forgot about social obligations, as their critics never tired of repeating. It is no coincidence that these critics and the urban professional middle class they represented were also early and vocal proponents of the new-style life insurance, for the critique of gambling and the advocacy of long-term life insurance were reverse and obverse of the same coin. These were the spokesmen for a meritocracy where hard work and talent, not winning the lottery, were the surest way to advancement; for a secure social order in which the well-off today would not find themselves poor tomorrow; for a sense of familial responsibility so strong that it reached beyond the grave; and for planning for the future in small increments, either in the form of insurance or savings. All of their cardinal virtues—prudence, foresight, self-discipline—were oriented toward the long term, toward the future that did not exist for the gambler. They preached control of one's life through industry and planning, but it would be a mistake to infer that they were optimists. In contrast to the lottery players, uncertainty to them meant not hope but fear. They saved for a rainy day and insured against disaster, ever mindful of the improbable but devastating accident that could undo all that merit had accomplished. Calculations of self-interest, economic and psychological, were a daily habit, and they exalted calculability into the virtue of reliability. Theirs were the values and beliefs that broke the link between gambling and insurance, and that made probability and statistics a permanent part of the insurance trade.

The legal doctrine of aleatory contracts was one obstacle to separating gambling from insurance, and therefore to quantifying risk. The key element of an aleatory contract was risk, conceived as an exchange of certain for uncertain goods. Gambling was the paradigm aleatory contract in spirit as well as in fact. Insurers who essentially used life offices to place bets on the lives of third parties were no more interested in the probabilities than the average purchaser of a lottery ticket. Quantifying uncertainty by means of probability theory may have diluted the risk that prevented, for example, a legal annuity from becoming an illegal usurious loan.

Could, for example, a gambling casino that calculated the odds and long-term returns really be said to be in a position of uncertainty equal to that of the individual player? This is not to say that insurers did not have some sense of the regularities governing the events upon which their trade depended—to judge from the stabilization of premiums, fire and maritime offices ironically seem to have been more sensitive to these than the early life offices were—but rather that the quantification of such risk seemed to presume too much certainty for the venture to be genuinely risky. The mathematical approach to such exchanges based on probabilistic expectation also amalgamated gambling and insurance problems. Notably, the mathematicians who attempted to find probabilistic grounds for distinguishing between gambling and insurance (1) abandoned the original definition of mathematical expectation, which was the shared method for pricing all aleatory contracts, from lottery tickets to annuities; (2) made an explicit appeal to the values of respectability and fiscal prudence; and (3) were roundly ignored by everyone except other mathematicians.[235]

The traditional orientation toward the individual case in insurance must also have slowed the acceptance of statistical methods. Actual experience with mortality in a period largely free from wars, famine, and (after the waning of smallpox) great epidemics may have promoted belief in regularity, but it was a mathematician, not an insurer, who founded the Equitable Society for the Assurance of Lives, the first company to take some notice of these regularities. Experienced practitioners initially opposed the idea, and even the great success of the Equitable did not inspire any genuine imitators for several decades. In any case, more than mathematics was needed to create the sort of life insurance that benefited the family of the deceased. Even if mathematically minded, the investor without family sentiment could and did choose to put his money into annuities that provided for his own comfort, as a rather regretful observation to this effect in the Equitable prospectus and also the preeminence of annuities over all other forms of reversionary payments in the mathematical treatises make clear. (The Amicable Society and the widows' funds are examples of the sentiment without the mathematics.) At least two sets of values, not necessarily related, had to converge in order to make the new-style life insurance appealing: first, a heightened sense of familial responsibility that made life insurance preferable to annuities; and second, an aversion to risk

[235] See n. 123, above.

conceived along the lines of a gamble that made the vaunted mathematical certainty of the Equitable reassuring.

The extraordinary success of the Equitable is the result not only of its exploitation of the regularity of the mortality statistics and the mathematics of probability to fix premiums (which were in any case much padded by fiscal considerations), but also of its creation of an image of life insurance diametrically opposed to that of gambling. The prospectuses of the Equitable and the companies that later imitated it made the regularity of the statistics and the certainty of the mathematics emblematic for the orderly, thrifty, prudent, farsighted *père de famille*, in contrast to the wastrel, improvident, selfish gambler. Long-term life insurance was aimed at a growing middle class of salaried professionals—clergymen, doctors, lawyers, skilled artisans who were respectable but not of independent means. In a world where apparently even clergymen could not count upon communal charity, the sudden death of the provider could topple the family from the middling ranks of society to the very bottom. Such reversals of fortune were the proper fate of the gambler, not the good bourgeois, and the new life insurance companies set about domesticating risk in the service of the domestic virtues: "family life and parsimony, frugality and orderliness."[236]

Late eighteenth-century critics of gambling and insurance writers paved the way for the Equitable by insisting on the distinction between necessary and unnecessary risks.[237] In England, legislation of 1774 (the so-called Gambling Act) made "interest" the distinction between legitimate insurance and gambling: with respect to life insurance, the holder of the policy was obliged to show a legitimate interest in the life insured that squared with the amount insured; otherwise the policy would be declared null and void.[238] Although the precise meaning of legitimate interest remained a matter of some controversy,[239] the intent of the law was clearly to distin-

---

[236] E. A. Masius, *Lehre der Versicherung und statistische Nachweisung aller Versicherungsanstalten in Deutschland* (Leipzig, 1846), p. 476.

[237] For example, Coudrette, *Dissertation*, p. 335; Dusaulx, *Lettre*, p. 12.

[238] John Raithby, ed., *The Statutes at Large, of England and of Great Britain* (London, 1811), vol. 13, p. 685 (Anno 14 Georg ii. III. c. 48).

[239] The case of William Pitt's coachmaker illustrates some of the problems in interpreting the law. The coachmaker insured Pitt's life for £500 as security for Pitt's debts to him. When Pitt died in 1806 owing an amount in excess of £1,000, the premiums were paid up. However, when Parliament appropriated funds to pay off the entire debt, the insurance company refused to pay up on the grounds that it was the debt that was insured, not the

guish between "Gaming or Wagering" and "the true Intent and Meaning" of insurance, and as such it marked a turning point in English conceptions of life insurance. Changes in practice were slower in coming, for they depended both on new values concerning familial responsibility and the stability of the social order, and a new interpretation of the mathematical "doctrine of chances" that was consistent with those values. The early career of the Equitable Society shows that this transition occurred only gradually, and that the Equitable's phenomenal financial success owed as much to the neglect of probability and statistics as to their use.

The early records of the Society for Equitable Assurance on Lives and Survivorships (established 1762) reveal the extent to which the first full-dress attempt to apply mathematical probability and statistics to the practice of insurance both shaped and was shaped by new values that promoted family over individual welfare, by an emphasis on the predictability versus the contingency of mortality, and, above all, by a policy of fiscal prudence that at times threatened to make the mathematical basis of the premiums irrelevant. The history of the Equitable is a rich and intricate one; I shall here be concerned only with those aspects that relate directly to these issues.[240]

The moving spirit behind the Equitable was the mathematician James Dodson (c. 1710–57), Fellow of the Royal Society, Master of the Royal Mathematical School, and author of several works on practical mathematics, including annuities. Denied admission to the Amicable Society on grounds of age (the Amicable admitted no one over the age of forty-five), he formed his own project for a life insurance company in 1756 and composed *First Lecture on Insurances* (unpublished) in the same year.[241] Although Dodson's death in 1757 cut short his calculations of premiums, which were to be based on the London table of mortality,[242] other backers of the project carried forth his plan. Their petition for a Royal Charter was rejected by the Privy Council on 1 May 1760, and although the opposition

---

life, since only the debt could be construed as legitimate insurable interest. The court decision given by Lord Ellenborough upheld the insurance company on the grounds that "This assurance . . . is in its nature a contract of indemnity, as distinguished from a contract of gaming or wagering." Quoted in Ogborn, *Equitable*, p. 148.

[240] For a complete history of the Equitable, see Ogborn, *Equitable*.

[241] The Equitable Society still possesses a copy of these lectures in manuscript, along with many other documents relating to its early history.

[242] Corbyn Morris, *Observations on the Past Growth and Present State of the City of London* (London, 1750).

of rivals such as the Amicable, the London Assurance, and the Royal Exchange no doubt played a role in this decision, the reasons given by the Privy Council shed some light on extant insurance practices and the novelty of a mathematical approach. In its decision, the Privy Council worried that the Equitable's premiums were too low (in fact, they were too high, as later experience would show) and that the starting capital was inadequate to launch such a venture, for the Council was wholly unpersuaded by the Equitable's argument that premiums alone would suffice. When it rejected the Equitable's petition for a second time on 14 July 1761, the Council expressed outright suspicion of the company's mathematical basis, "whereby the chance of mortality is attempted to be reduced to a certain standard: this is a mere speculation, never yet tried in practice, and consequently subject, like all other experiments, to various chances in the execution."[243]

Undeterred, the directors of the Equitable rented rooms and published a prospectus that explained the new insurance plan to the public. The Equitable was to be a mutual society, "the assured being mutually assurers one to the others," with the members entitled to dividends in the case of surplus, and subject to calls for extra contributions in case of deficit. Of the eight benefits to insurers listed, only one concerned provision for widows and children; the others reflected the actual insurance market, being mostly concerned with security on loans. However, the prospectus emphasized the importance of such benefits for the "families of clergymen, counsellors, physicians, surgeons, attornies; those who have places in public and private offices; and more frequently of artificers, manufacturers, and others who support themselves by their labour"—that is, those in "a middling station of life" who "would strenuously endeavour to avert that most sensible of all distresses, which must necessarily attend their families, should they be at once reduced from a plentiful and respectable, to an indigent and deplorable situation." Moreover, life insurance could supplement or replace public charity, "and many parishes may hereby be eased of burthens, which would otherwise have fallen on them." Yet the author of the prospectus realized that such provident views were still somewhat of a rarity even among the salaried middle class, and hastened to add that the Equitable could serve other ends as well, since "it hath been found by experience, that a future provision for family is, in the opinion of the

---

[243] Quoted in Ogborn, *Equitable*, p. 35; also Walford, *Guide*, p. 43.

generality of these persons of less importance than a provision for themselves in sickness, or old age, or at a time, when they may be disabled from labour."[244] The prospectus was at pains to distinguish its premiums from the flat rates other companies charged, "be the life ever so young and healthy," and included sample premiums so that the reader might make the comparison himself. Above all, the prospectus stressed the certainty of the underlying principle of the new scheme, which was "grounded upon the expectancy of the continuance of life; which, although the lives of men separately taken, are uncertain, yet in an aggregate of lives is reducible to a certainty."

Provision for family versus provision for oneself; private foresight versus public charity; the uncertainty of any individual death versus the certainty of mortality en masse—the kind of life insurance offered by the Equitable threw these contrasts, real and perceived, into relief. The early prospectus could not take for granted the attitudes toward familial responsibility beyond the grave and toward the stability of the social and natural orders upon which the attractiveness of such life insurance depended. Indeed, they were as much briefs for as appeals to these attitudes. Earlier annuity and tontine schemes had beguiled subscribers with the lottery player's dream of sudden upward social mobility from a bourgeois life to a princely one,[245] while the new life insurance played upon the specter of sudden downward social mobility. Earlier insurance schemes had deliberately emphasized the elements of uncertainty that were the essence of an aleatory contract and which had given them the allure of a gamble; the proponents of the new life insurance minimized the chance aspects.

The 1788 prospectus for the French Compagnie Royale d'Assurance, which was explicitly patterned on the Equitable, is a paean to these new attitudes unadulterated by the need to conform to accepted commercial practice, for life insurance had been heretofore illegal in France. The author of the prospectus praised the moral effects of life insurance as opposed to those of annuities, for the former provides "security against misfortune without discouraging either industry or activity. On the contrary, it encourages labor and economy . . . the facilities that it offers to the benefit of friendship, of filial piety, paternal tenderness, conjugal union, in a

---

[244] *A Short Account of the Society for Equitable Assurances on Lives and Survivorships* (London, 2 August 1764).

[245] See, for example, the 1671 prospectus quoted in Société Générale Néerlandaise, *Mémoires*, p. 230.

word, to generous sentiments can only tend to multiply the practice of all virtues." Moreover, it was preferable to charity, which tended to lead to sloth and disgrace.[246] The Compagnie Royale had taken care to employ a "profound mathematician," for it understood that such enterprises rested upon calculations: "Such indeed is the certainty of the various calculations upon which insurance is based that one can undertake it without capital, and by the simple amassing of the premiums."[247] Richard Price's evaluation of the mathematical basis of the Amicable Society makes clear the degree to which risk as uncertainty clashed with the new insurance sensibility, for he objected to the annual distribution of benefits on the basis of who had happened to die that year as *"a contingency"* that did not depend on the individual member's contribution. The regularity of mass mortality statistics had apparently made the contingency of individual deaths intolerable, even though, of course, they remained the basis for life insurance.[248]

Richard Price played an important part in the early affairs of the Equitable as a mathematical consultant, for Dodson's death had left the directors of the Equitable with a set of incomplete calculations. There was some disagreement among the directors as to which method to use for joint lives, "greatest" or "mean hazards," and judging from the early account books, the directors seemed to have adapted the premiums rather freely to what they saw as the individual exigencies of the case.[249] In 1768 the Equitable turned to Price for help in calculating reversionary payments, which may have led his interests in that direction, for his first publications on the subject followed soon thereafter. Price's treatise, *Observations on Reversionary Payments* (1771), contained a full and admiring account of the Equitable's practice, which he held up as a rare example of sound planning and solvency in the dubious business of annuities and insurance. We learn from Price that the Equitable had been cautious in every respect: it had calculated interest at the lowest rate (3 percent); it had used the mortality

---

[246] French opponents of insurance like Brissot de Warville worried that this might weaken the spirit of altruism and communal aid; see Brissot de Warville, *Dénonciation au public d'un nouveau project d'agiotage, ou Lettre à M. le Comte de S**** (London, 1786), pp. 31–34.

[247] Compagnie Royale d'Assurances, *Prospectus de l'établissement des assurances sur la vie* (Paris, 1788). The mathematician was Duvillard.

[248] Price, *Observations*, p. 121.

[249] Ogborn, *Equitable*, pp. 53, 81; *Rough Minutes of the Weekly Courts*, vol. 2 (3 January 1764–26 March 1765), MS. volume of the Equitable Society.

table that gave the shortest lifespans (Corbyn Morris's London table); it took the further precaution of insuring only healthy lives; and finally, it added a flat percentage (6 percent) to all of these premiums. In words that could have been made the Equitable's motto, Price exhorted its directors to proceed "frugally, carefully, and prudently," for despite the certainty of the calculations, "at particular periods, and in particular instances, great deviations will often happen."[250] When it came to practice, even Price admitted the force of the contingent.

The directors of the Equitable preserved certain elements of the older practices oriented toward the individual case. Every candidate for a life insurance policy was interviewed in person by the directors, made to swear that he had had smallpox and was not given to intemperance, and asked to give an account on any special risks run. For these latter, added premiums ranging from 11 to 22 percent were summarily charged at the discretion of the directors. Some policies on these matters emerged: for example on 21 April 1779, the directors resolved to charge travelers to the West Indies an extra 5 percent, but exceptions and modifications were made to suit individual cases.[251] The calculations of premiums provided them with a guide, which they regularly overruled as practice dictated.

In 1775 Richard Price's nephew William Morgan became actuary of the Equitable after a two-year study of insurance mathematics under his uncle's tutelage. Morgan's appointment virtually transformed the position of actuary from one of secretary to one of mathematical expert, with ever-increasing power within insurance companies.[252] Morgan's long tenure (until 1830) at the Equitable reinforced the sometimes exaggerated tendencies toward prudence and caution praised by Price, to the point where the company's spectacular surpluses were an endless bone of contention between the members, who insisted upon a distribution of dividends and/or decrease in premiums, and Morgan, who warned that some unforeseen disaster might flood the company with claims. In 1775, Morgan calculated the company's liabilities and discovered that 60 percent of its assets could be considered surplus, but he sided with his uncle against any distribution of dividends from this amount, lest "extraordinary events or a season of uncommon mortality" catch the Equitable unawares.[253]

[250] Price, *Observations*, pp. 128–130.
[251] Equitable Society, *Orders of the Court of Directors* (1774–1848), MS. volume of the Equitable Society.
[252] Francis, *Annals*, pp. 272–273.
[253] Quoted in Ogborn, *Equitable*, p. 105.

As the membership increased to over 5,000 policies in force by 1796, so did the surplus, until it had reached almost embarrassing proportions, and with it, the pressure for a distribution of dividends. By the fifth edition of *Observations on Reversionary Payments* (published posthumously by Morgan in 1803) even Price had relented, wondering whether it might not be better for the Equitable to use mortality tables "more adapted for the general state of mortality among mankind" and to calculate prices straight from the tables, thus reducing premiums by 20 percent, the interest still being computed at half the actual investment rate.[254] Morgan, however, held firm. Not even the salutary effects of the Law of Large Numbers on the regularity of the actual mortality experience of the Equitable would sway him, for "from the great difference in the sums assured in each life, the amount of the claims is so uncertain, that it shall often happen that events prove peculiarly unfavorable to the Society in a year which has been attended with no uncommon degree of mortality."[255]

The extent to which Morgan's legendary caution was justified by the calculations and the data was sharply challenged by mathematicians of the next generation such as Charles Babbage and Augustus De Morgan, who accused Morgan of ignoring the best available mortality data and more realistic interest rates in the name of an almost pathological prudence. De Morgan (who was Dodson's great grandson) exonerated the Equitable from any intent to defraud, but still maintained that its premiums were "enormous" due to an overestimation of mortality and margin of safety: "We should write upon the door of every mutual office but one be *wary*; but upon that one should be written *be not too wary* and over it *Equitable Society*."[256] The Equitable prospered under Morgan's regime, but its prosperity seems to have been less connected with its mathematical basis, although the prospectuses made much of this aspect, than to its willingness to "always modify the exact calculations of mathematics by those of prudence," in the words of the French prospectus.[257] Although its premiums may have been inflated in light of the actual mortality statistics, the Equitable attracted members. Indeed, when competitors like the Royal Exchange

---

[254] Price, *Observations*, 5th edition (London, 1803), pp. 175 ff.

[255] Quoted in Ogborn, *Equitable*, pp. 124–125.

[256] Quoted in Ogborn, *Equitable*, p. 206. See also Augustus De Morgan, "Review of Théorie Analytique des Probabilités, par M. le Marquis de Laplace," *Dublin Review* 2 (1837): 338–354, especially pp. 341 ff.

[257] Compagnie Royale, *Prospectus*, p. 53. See also Nicolas Fuss, *Éclairissements sur les établissemens public en faveur tant des veuves que des morts* (St. Petersburg, 1776), p. 31.

threatened the Equitable's effective monopoly on whole life insurance in the 1790s, they imitated the Equitable's methods but added the stipulation that premiums be at least 20 percent higher than those of the Equitable and were nonetheless immediately successful.[258] With only slight exaggeration, one might claim that these new-style life insurance companies flourished in spite of mathematical probability and mortality statistics. Not only did these concerns attempt to eliminate the element of risk that had previously been synonymous with insurance for their prudent middle-class clientele; they also attempted to eliminate any effective risk from the venture itself. In the latter case, they very nearly eliminated the mathematics and statistics that had been their claim to regularity and reliability in the former.

3.5 Conclusion

By the second decade of the nineteenth century, the divorce between insurance and gambling was almost complete. Pierre Simon Laplace placed the full weight of his mathematical prestige behind insurance as "advantageous to morals, in favoring the gentlest tendencies of nature."[259] Adolphe Quetelet echoed his master, comparing government insurance schemes favorably to lotteries;[260] and life insurance became the mathematician's favorite example for the utility of probability theory. The relationship between belief in statistical regularities and confidence in insurance was a symbiotic one: those who would persuade others of the existence of such regularities pointed to the financial success of insurance companies; insurance companies in their turn considered every new such regularity (e.g. between sunspots and epidemics) to be support for their practices.[261]

---

[258] Supple, *Royal Exchange*, p. 66. An 1805 overview of the London life insurance industry notes the remarkable success of the Equitable, but attributes it to the system of granting membership to all policyholders, who therefore are entitled to part of the profit. The mathematical basis of the society is not even mentioned; see Marshall, *Treatise*, p. 665.

[259] Pierre Simon Laplace, *Théorie analytique des probabilités*, in *Oeuvres* (Paris, 1847), vol. 7, p. 481; *Essai philosophique sur les probabilités*, 5th edition (Paris, 1825), pp. 192–194.

[260] Adolphe Quetelet, *Instructions populaires sur le calcul des probabilités* (Brussels, 1828), pp. 195–196.

[261] See, for example, Francis, *Annals*, p. 282; Underwriter's Agent of New York, *The Agent* (July 1872), p. 13.

The mathematical theory of risk had triumphed, and with it the belief that whole classes of phenomena previously taken to be the very model of the unpredictable, from hail storms to suicides, were in fact governed by statistical regularities. These regularities took the form of distributions rather than functional relationships, but they were hailed as regularities all the same. We associate statistical laws with indeterminism, but for much of the nineteenth century they were synonymous with determinism of the strictest sort.[262] Order was to be found in the mass and over the long run, in large numbers, no longer in the individual case. Too many particulars distracted rather than enlightened: astronomers eliminated judgments about that night's seeing conditions and this instrument's reliability from their reports; life insurance companies stopped interviewing prospective candidates personally. Rules increasingly replaced judgments, from the method of least squares in astronomy to the treatment of fevers in medicine.[263]

Was this simply the discovery of regularities that had existed all along, or had the world itself changed, become more stable and predictable? This question can be answered only piecemeal, with respect to this or that domain of phenomena. Longer periods of peace, the disappearance of pirates from many routes, and improvements in navigation no doubt made sea voyages a surer business; more stone and brick and less wood and thatch may have decreased the incidence of fires in cities like London, although new risks like cotton mills brought the numbers up again; mortality on the whole decreased, but there were so many local fluctuations due to smallpox, malaria, bad harvest, and so on that the average gain in years lived was not obvious.[264] Of course, lowered risks, even when they were real, did not necessarily imply regularities in the statistical sense: mortality figures may have been just as regular, dismally so, when one out of every two infants died as when almost none did. Indeed, if life genuinely

[262] See Theodore M. Porter, "A statistical survey of gases: Maxwell's social physics," *Historical Studies in the Physical Sciences* 12 (1981): 77–116, especially pp. 100–106.

[263] See Zeno G. Swijtink, "The objectification of observation: Measurement and statistical methods in the nineteenth century," in *The Probabilistic Revolution*, vol. 1, *Ideas in History*, Lorenz Krüger et al., eds., pp. 261–286; Stanley J. Reiser, *Medicine and the Reign of Technology* (Cambridge, Eng.: Cambridge University Press, 1978), concerning the introduction and influence of instruments like the thermometer.

[264] See n. 216, above; Cockerell and Green, *British Insurance*, p. 20; John McManners, *Death and the Enlightenment* (Oxford and New York: Oxford University Press, 1985), pp. 89–94.

improved in these areas over the course of the eighteenth century, these heartening changes might well have played havoc with earlier grim regularities or cycles. However, the psychological impact of these improvements, real and perceived, need not have followed the statistics so closely. A safer life could have been experienced as a more stable, predictable one, for it encouraged planning for the future in a way that periodic misfortunes did not. Children that usually survived past infancy; ships that usually returned from exotic destinations; dwellings that usually withstood fires for generations: in a mathematical sense these patterns were no more regular than the worst consistently coming to pass, or even cycles of prosperity and want, but they promoted a sense of security that the other equally well-defined patterns did not.

We must also specify the "who" as well as the "what" in order to answer sensibly our questions about a world changed in fact or in perception. Even within the same country at the same time, class and local conditions palpably affected fundamental variables such as mortality. It was axiomatic in the eighteenth century that country folk outlived city folk; that the children of the rich had a better chance of surviving than those of the poor; and that swamps were a health hazard. Second-order variables must also have shaped the daily experience of regularity or its opposite: the extent to which one's livelihood depended on the vicissitudes of the weather; the reliability as well as the amount of one's income; control over one's work and its fruits. For the urban middle class of professionals and tradesmen, these variables and others may indeed have fostered an ethos of control and predictability. In their lives, the ideology of the meritocracy was to some extent realized, and they imposed regularity on social and economic arrangements wherever possible. They kept household as well as business budgets, reckoned self-interest, praised punctuality and reliability, and perhaps even read popularizations, such as Buffon's, of the new demography.[265] And as Lawrence Stone has argued for England, they also led the way in making families smaller, more affectionate, and more egalitarian, thereby strengthening their emotional claims.[266] The anti-gambling and pro-life–insurance literature from late eighteenth-century France and Germany suggests similar developments.[267] They were the first to withdraw from the gambling scene and the first to purchase the new-style life insur-

---

[265] McManners, *Death*, pp. 94–104.   [266] Stone, *The Family*, chapter 6.

[267] Dupont de Nemours, *Opinion*, p. 3; Dusaulx, *Lettre*, p. 31; Tetens, *Einleitung*, p. v; Compagnie Royale d'Assurances, *Prospectus*, pp. 14–15.

ance, and it seems fair to assume that it was their values that ultimately drove a wedge between the two approaches to risk taking so long conflated in theory and practice. It is more speculative but still plausible to claim that it was their peculiar experience of new-found control, relative security from accidents, and self-imposed regularity that resonated to the mathematical regularities that were purportedly the basis of insurance schemes like the Equitable's. Their world at least really had become more stable and predictable.

But stability and predictability did not free them from fear. If the gambler was irrationally obsessed with the hope of a big win with a minute probability, the insurance customer was equally haunted by the fear of a big loss of comparably small probability. When interest-yielding investments are an alternative to paying out insurance premiums, and the probability of sudden death in the prime of life is very small (and even quantifiable from the mortality tables), it is not clear that life insurance is an economically rational strategy. Buffon took this probability as his minimal psychological unit of risk, since almost no one gave more than a moment's thought to the possibility.[268] But fifty years later, life insurance prospectuses were exploiting this fear for all it was worth, and their financial success suggests that it was worth a good deal. The very possibility of insurance apparently created "a new basis for anxiety," lest one suffer not only from misfortune but also from the appearance of not having exercised "the kind of intelligent control, the kind of 'care', that allows reasonable persons to minimize danger and avoid remorse."[269] A curious passage in Hume's *A Treatise of Human Nature* (1739) hints that this shift in sensibility was already underway. Hume analyzes the emotions of hope and fear, both stemming from uncertainty:

> Probability is of two kinds, either when the object is really in itself uncertain, and to be determin'd by chance; or when, tho' the object be already certain, yet 'tis uncertain to our judgment, which finds a number of proofs on each side of the question. Both these kinds of probabilities cause fear and hope; which can only proceed from that property, in which they agree, *viz.* the uncertainty and fluctuation they bestow in the imagination by that contrariety of views, which is common to both.[270]

[268] Buffon, "Essais," pp. 56–58.   [269] Goffman, "Action," p. 176.
[270] David Hume, *A Treatise of Human Nature* (1739), L. A. Selby-Bigg, ed. (London: Oxford University Press, 1968), II.iii.9, p. 444.

Fear and hope are thus symmetric, the one preponderating in proportion to the desirability and probability of the outcome. And yet Hume also claimed that "any doubt" will transmute into fear, "even tho' it presents nothing to us on any side but what is good and desireable."[271] Fear has the upper hand, even in situations which on the strength of Hume's own analysis should instead produce hope. This is the frame of mind that magnifies misfortunes, and shuns risk, the exact inverse of the lottery player's mentality.

It seems paradoxical that a more secure life should breed more fear than hope. Most historians have assumed just the reverse, following Descartes' view that surprise is the chief cause of fear.[272] But Montaigne may have been closer to the mark when he suggested that the opposite was also true, that foresight bred its own exquisite form of fear, anticipations made all the more menacing by a sheltered life and a well-cultivated imagination for future eventualities.[273] For those who must constantly face risks—the soldier in battle, the doctor in the midst of an epidemic, the peasant always a short step ahead of starvation—awareness of the full magnitude of what they stand to lose is mercifully blunted. A refusal or inability to contemplate the future has much the same effect. But for those who have succeeded in largely eliminating such risks from daily life, and for whom the future is as real or more so than the present, fears of even remote possibilities take on vivid colors. Add to this the knowledge that one's gains, social and economic, were both recent and precarious, as would have been the case for many members of the urban professional and trading class, and their preoccupation with improbable misfortunes becomes more comprehensible, if not more reasonable.

By the middle of the nineteenth century these attitudes had become so pervasive that they even made inroads into gambling: Houdini counseled would-be high rollers to emulate the merchant's "calm and cool . . . for the demon of bad luck invariably pursues a passionate player."[274] Earlier comparisons between the merchant and the gambler had turned on the willingness of both to take big risks, but now even fortune was on the side

[271] Hume, *Treatise*, II.iii.9, p. 447.

[272] René Descartes, *Les passions de l'âme* (1649), Art. 176, in *Oeuvres de Descartes*, Charles Adam and Paul Tannery, eds. (Paris: J. Vrin, 1967), vol. 11, p. 463.

[273] Michel de Montaigne, *Essais*, II.xi, Maurice Rat, ed. (Paris: Éditions Garnier Frères, 1962), vol. 1, p. 468; quoted in Delumeau, *Peur*, p. 8.

[274] Quoted in Steinmetz, *Gaming Table*, vol. 2, pp. 255–256.

of the prudent. In insurance, however, the ever-increasing emphasis on statistical regularities—in work-place accidents, in medical malpractice, in boiler explosions—threatened to undermine the very virtues of foresight and responsibility that had made the mathematical approach attractive in the first place.[275] Did the concepts of personal responsibility and blame for, say, Parisian carriage drivers make any sense if the number of traffic accidents hardly varied from year to year? Macroscopic statistical regularities had appealed to people who prized microscopic social regularities, but at bottom the two were very different things. The microscopic regularities were self-imposed by an act of will: reasonable men were calculable because they calculated; they chose to regulate their behavior out of rational self-interest. But the new macroscopic regularities discovered order in the large where chaos or caprice reigned in the small. Not only prudent men of affairs but also suicides yielded astonishing regularities when studied en masse. Individual rationality, taken in sum, was no longer a precondition for collective regularity.[276]

The fate of the argument from design in the probabilistic literature closely parallels this transition. Whereas De Moivre took the order revealed in stable statistical frequencies as incontrovertible evidence that an intelligent agent was at work in the world, Poisson argued that such order was only to be expected; we should suspect divine tinkering only when it was absent.[277] For the mathematicians, the clock no longer implied a clockmaker. The ascent of statistical regularities ultimately marked the decline of the reasonable man, as probability theory shifted its sights from the psychology of the rational individual to the sociology of the irrational masses.

[275] Daston, "Risk," pp. 254–255.

[276] Daston, "Rational individuals versus laws of society: From probability to statistics," in *The Probabilistic Revolution*, vol. I, *Ideas in History*, Lorenz Krüger et al., eds., pp. 295–304.

[277] De Moivre, *Doctrine of Chances*, pp. 251–254; Siméon-Denis Poisson, *Recherches sur la probabilité des jugements en matière criminelle et en matière civile* (Paris, 1837), p. 13.

CHAPTER FOUR

Associationism and
the Meaning of Probability

4.1 Introduction

By the turn of the eighteenth century, mathematicians had come to think about probabilities in at least four different ways: as deriving from the physical construction of certain objects (for example, the symmetry and uniform density of a fair coin or die); as the frequency with which certain events happened (how many people of a given age died annually); as the measure of the strength of an argument (how evidence weighed in for or against a judicial verdict); and as the intensity of belief (the firmness of a judge's conviction of the guilt or innocence of the accused). Since the mid-nineteenth century, probabilists have distinguished sharply between at least the first two "objective" and the latter two "subjective" senses of probability, and the controversy still rages over which is the "true" meaning of mathematical probabilities.[1] But their predecessors slid from one interpretation to another with an ease that bewilders latter-day commentators acutely aware of the conceptual gap that separates each from the others.

As we have seen in earlier chapters, the early probabilists had few scruples about extending their calculus from dice to mortality statistics, and in the case of Jakob Bernoulli and Gottfried Wilhelm Leibniz, to evidence and to belief. This philosophical insouciance, even on Leibniz's part, has led Ian Hacking to argue that the concept of probability was Janus-faced

---

[1] See, for example, John Maynard Keynes, *A Treatise on Probability* (London: Macmillan, 1921); Richard Von Mises, *Probability, Statistics and Truth*, 2nd English edition (London: George Allen & Unwin, 1957); L. J. Savage, *The Foundations of Statistics*, 2nd edition (New York: Dover, 1972), for a sampling of interpretations.

from the outset,[2] although its multiplicity was in fact more fourfold than double. However, later writers could and did at least distinguish subjective and objective faces of probability, if still unable to resolve which (if either) should guide research and applications. In contrast, classical probabilists from Jakob Bernoulli through Pierre Simon Laplace apparently saw no opposition, no choice to be made. In this chapter, I shall argue that it was the rise and fall of a particular version of associationist psychology that made this characteristic blurring between subjective and objective probabilities tolerable for classical probabilists, and ultimately intolerable for their successors.

The fact that this compromise, shorn of its psychological justification, is no longer possible for mathematicians, statisticians, and philosophers accounts for their bemused view of the classical interpretation. From a twentieth-century perspective, the classical interpretation combines an epistemic "art of conjecture" and a frequentist "doctrine of chances" with a cavalier—or healthy, depending on one's point of view—disregard for philosophical distinctions. Classical probabilists such as Jakob Bernoulli or Laplace appear to shift positions opportunistically as the occasion demands.[3]

These modern commentators, still urgently interested in the relation of probability to belief, evidence, and inductive inference and confirmation, have couched their analyses of classical responses to these problems in twentieth-century terms. The categories of subjectivist, logical, and frequentist probabilities postdate the works of the classical probabilists by over a century. Although these terms often illuminate both classical works and current dilemmas, the conceptual framework inherited from A. A. Cournot, Harold Jeffreys, Rudolf Carnap, John Maynard Keynes, Richard von Mises, Bruno de Finetti, and others often seems procrustean when imposed upon the work of an eighteenth-century thinker. In Ian Hacking's thoughtful discussion of the *Ars conjectandi*, for example, Bernoulli emerges as both more prescient and more quaint than a less anachronistic reading would warrant. On the one hand, Hacking credits Bernoulli with

---

[2] Ian Hacking, *The Emergence of Probability* (Cambridge, Eng.: Cambridge University Press, 1975), chapter 2.

[3] See M. G. Kendall, "The beginnings of a probability calculus," in *Studies in the History of Statistics and Probability*, E. S. Pearson and M. G. Kendall, eds. (Darien, Conn.: Hafner, 1970), pp. 19–34, on p. 31, who blames Jakob Bernoulli for the confusion; also Hacking, *Emergence*, chapter 2, who regards the amalgam as more useful, if no more consistent.

anticipating a frequentist "security level" for inductive influence (although he admits that Bernoulli himself did not exploit its potential within his mathematical treatment), and on the other, he saddles Bernoulli with a "useful equivocation" between *de re* and *de dicto* senses of possibility and corresponding epistemic and physical senses of probability.[4]

It is only with 20/20 hindsight, however, that Bernoulli and other classical probabilists through Laplace appear to vacillate between the objective and subjective interpretations of probability distinguished by Cournot.[5] Because the debate between these two camps has generated so much heat among mathematicians, statisticians, and philosophers since the early decades of this century, it is natural for the disputants to have sought historical antecedents for both their own positions and the confusions which they believe to have befuddled their opponents.[6] Frequently, these forays into the history of probability and statistics have sharpened the focus of current arguments over, for example, statistical inference, and at the same time viewed historical problems from a fresh conceptual angle.[7] Nonetheless, these historical studies have been conducted within a set of twentieth-century conceptual tools and, as often as not, with a pronounced bias on the part of the investigator mindful of the still-raging controversy. Hacking's remarks on the *Ars conjectandi* apply generally: "Since we still lack an adequate understanding of how to apply probability mathematics, no one writes dispassionately about Bernoulli."[8] This presentist orientation, although valuable for practicing probabilists, be they of a mathematical, statistical, or philosophical stripe, leads to serious historical distortions. In the case of the classical probabilists, it is only when we attempt to classify their work along objective versus subjective lines that the interpretations of Jakob Bernoulli or Laplace emerge as equivocal or poorly articulated.

[4] Ian Hacking, "Jakob Bernoulli's Art of Conjecturing," *British Journal of the Philosophy of Science* 22 (1971): 209–229, on p. 225; also Hacking, "Equipossibility theories of probability," *British Journal for the Philosophy of Science* 22 (1971): 339–355, on p. 341.

[5] Antoine Augustin Cournot, *Exposition de la théorie des chances et des probabilités* (Paris, 1843), pp. 81–82.

[6] See, for example, Hacking's attempt to discredit the twentieth-century logical probabilists through an analysis of Leibniz: Hacking, "The Leibniz-Carnap program for inductive logic," *Journal of Philosophy* 68 (1971): 597–610, on p. 599.

[7] The above-cited articles and book by Hacking on the history of probability display both features of this philosophical approach to best advantage.

[8] Hacking, "Jakob Bernoulli," p. 209.

Classical probabilists were briefly able to reconcile the subjective and objective facets of probability through an appeal to the theories of psychology and epistemology advanced by philosophers such as John Locke, David Hartley, and David Hume. Laplace, spurred by Jean d'Alembert's critique of conventional probability theory, was the first mathematician explicitly to distinguish between subjective and objective aspects of probability,[9] and only in the 1830s and 1840s did Siméon-Denis Poisson and especially Cournot begin to explore the full implications of the distinction. For the greater part of the eighteenth century, probabilists accepted the qualified equation of experience and belief set forth by a sensationalist epistemology.

Philosophical efforts to justify the equation culminated in M.J.A.N. Condorcet's attempt to identify the *motif de croire* with both the facts of experience and mathematical probabilities by assuming that nature unswervingly followed invariable laws and that these laws could be discerned from observable phenomena. Even these two cardinal principles warranted belief only because experience ceaselessly confirmed them: "The constant experience that facts conform to these principles is our sole reason to believe them."[10] Experience, belief, and probability were three aspects of a single psychological operation.

Section 4.2 of this chapter examines how the philosophy *cum* psychology of belief developed by Locke, Hume, Hartley, and Étienne Condillac incorporated and shaped ideas of probability. Section 4.3 shows how mathematicians used these ideas to understand what their calculus was about, and how the alliance between associationism and the classical theory of probability first joined and then severed objective and subjective probabilities.

4.2 Probability, Experience, and Belief

Since the 1840s, when Cournot and Robert Ellis reassessed the foundations of mathematical probability in terms of relative frequencies, the distinction between, in Cournot's words, "objective possibility," which

---

[9] Pierre Simon Laplace, "Recherches sur l'intégration des équations différentielles aux différences finies et sur leur usage dans la théorie des hasards" (1776), in *Oeuvres complètes de Laplace*, Académie des Sciences (Paris, 1891), vol. 8, pp. 69–200, especially p. 147.

[10] Marie-Jean-Antoine-Nicolas Caritat de Condorcet, *Essai sur l'application de l'analyse à la probabilité des décisions rendues à la pluralité des voix* (Paris, 1785), pp. x–xi.

denotes "the existence of a relation which subsists between things themselves," and "subjective probability," which concerns "our manner of judging or feeling, varying from one individual to the next," has been the departure point for all discussions concerning the interpretation of probability theory.[11] For the greater part of the eighteenth century, however, probabilists would have found such a distinction alien. Their work accommodated both objective and subjective senses of probability with an ease that has bemused later commentators.

The interpretation of probability theory derived from the legal hierarchy of proofs had been epistemic: degrees of probability or proof stemming from the evidence of things and of testimony corresponded to degrees of certainty in the mind of the judge. The intensity of conviction was correlated with the quality rather than the quantity of evidence. Jurists would not have counted up the number of times that someone seen to flee the scene of a murder with an unsheathed bloody sword was eventually proved guilty of the crime. The strength of the presumption of guilt created by such testimony sprang from the casual inferences connecting guilt and the observed circumstances, not the frequency with which the correlation held. Probabilities and concomitant degrees of conviction had been formally correlated in the hierarchy of proofs with types rather than with amount of evidence. The practice of *counting* instances inaugurated by the political arithmeticians was thus a novel, extra-legal approach to evidence, one which exerted a lasting influence on both philosophy and probability theory during the eighteenth century.

As we have seen in the case of insurance, the shift from a qualitative to a quantitative approach to evidence was not a simple one.[12] It required homogeneous, stable categories composed of identical units: one shipwreck, one death, one fire was like all the others for the statistician. Nothing about the construction and even the existence of such categories could be taken for granted. Was the proper way to classify sea voyages by route, season of the year, cargo, or experience of the captain? Should the bills of mortality be organized by cause of death, age, sex, or prevailing atmospheric conditions? What were the key variables and where did the suspected regularities lie, if anywhere?

Even where these questions could be answered, as in observational astronomy, qualitative concerns still influenced quantitative procedures for

---

[11] Cournot, *Exposition*, p. 82.   [12] See Section 3.1, above.

analyzing data: everyone agreed that it was the position of a comet at a given time that mattered, but there was for a long while an equally firm consensus that imponderables such as seeing conditions, the experience and alertness of the observer, and the quality of that particular telescope or chronometer did as well.[13] When Nicholas Bernoulli proposed that judges measure the veracity of witnesses by counting up how many times they had told the truth or lied in the past, the objections were not only on grounds of feasibility.[14] At a stroke Bernoulli had obliterated a set of distinctions evolved over centuries (and still in use today) expressly designed to extract the particulars of a case. In this system, neither the witness nor the circumstances of testimony could be assumed to duplicate themselves closely enough for Bernoulli's proportions to make sense. The increasing reliance upon and even preference for quantitative over qualitative evidence was almost as gradual in Enlightenment philosophy as in the practice of the same period.

Locke's *Essay on Human Understanding* (1689) mingled the old and new senses of evidence, and linked both to the legalistic probability of degrees of assent. On the one hand, Locke listed the traditional legal grounds for belief or probability: the number of witnesses, their integrity and skill, possible ulterior motives, internal consistency, and contrary testimony. On the other hand, he presented the equally traditional evidence of things in a new light. The probability of a proposition varied "as the conformity of our knowledge, as the certainty of observations, as the frequency and constancy of experience," as well as "the number and credibility of testimonies."[15] According to Locke, probability affirmed two types of propositions, corresponding to inductive and analogical inferences: "matters of fact," such as the color and texture of gold; and matters concerning things beyond the reach of the senses, such as the existence of atoms or angels. The "constant observation" of unvarying relations—for example, that iron has always been observed to sink in water—lent a very high probability,

---

[13] Zeno G. Swijtink, "The objectification of observation: Measurement and statistical methods in the nineteenth century," in *The Probabilistic Revolution*, vol. 1 *Ideas in History*, Lorenz Krüger, Lorraine J. Daston, and Michael Heidelberger, eds. (Cambridge, Mass.: MIT Press, 1987), pp. 261–286.

[14] Nicholas Bernoulli, *De usu artis conjectandi in iure* (1709), in *Werke von Jakob Bernoulli*, Naturforschende Gesellschaft in Basel (Basel: Birkhäuser, 1975), vol. 3, pp. 324–325.

[15] John Locke, *An Essay Concerning Human Understanding* (1689; all editions dated 1690), Alexander C. Fraser, ed. (New York: Dover, 1959), Book IV, chapter 15.

and hence "confidence," to inductive generalizations. Analogical general-izations founded on the "gradual connexion of all that great variety of things we see in the world" generated a weaker probability, but Locke advised that "a wary reasoning from analogy leads us often into the discov-ery of truths and useful productions, which would otherwise be con-cealed."[16]

Philosophical rather than mathematical probabilities dominated Locke's discussion, despite the suggestive references to frequencies. Locke still understood probability as an inferior form of knowledge, or rather as a make-do substitute for the genuine knowledge of intuition and demon-stration, "because the highest probability amounts not to certainty; with-out which there can be no true knowledge."[17] Probability was in fact a kind of penance, which God made "suitable, I presume, to that state of mediocrity and probationership he has been pleased to place us in here; wherein, to check our over-confidence and presumption, we might, by every day's experience, be made sensible of our shortsightedness, and lia-bleness to error."[18] Locke's traditional understanding of probability re-vealed itself in his emphatic distinction between "the highest probability" and certainty, in contrast to the continuum of mathematical probabili-ties.[19] He also compared probability to the "appearance" of truth and made it depend on authority, contrasting the true knowledge of the math-ematician who has actually demonstrated that the sum of the angles in a triangle is equal to two right angles with the "probability" of the man who merely takes the mathematician's ("a man of credit") word for it.

Yet Locke never lost sight of the objective and potentially quantitative grounds for the probabilities of appearance and authority: the former de-pends on "proofs, whose connexion is not constant and immutable . . . [but] appears for the most part to be so"; the latter on past experience of the mathematician's "not being wont to affirm anything contrary to, or besides, his knowledge, especially in matters of this kind."[20] However, these allusions are only one strand in a skein of probabilistic ideas that also included a great deal on evidence of the legal as well as the experimental sort, and on subjective "degrees of assent, from full assurance and confi-dence, quite down to conjecture, doubt, and distrust."[21]

---

[16] Locke, *Essay*, Book IV, chapter 16.   [17] Locke, *Essay*, Book IV, chapter 3.

[18] Locke, *Essay*, Book IV, chapter 14.

[19] See also the careful wording in Locke, *Essay*, Book IV, chapter 15.

[20] Locke, *Essay*, Book IV, chapter 15.   [21] Locke, *Essay*, Book IV, chapter 15.

Locke's principal goal was to anchor these states of mind to the evidence of experience, both direct and vicarious (i.e., that conveyed by testimony), so as to help rational judgment find its way in the "twilight of probabilities." His guidelines were a mixture of the semiquantitative ("constant and never-failing experience") and the irreducibly qualitative (the credibility of "particular testimonies"). In the most common case where both testimony and experience can be enlisted on either side of a question, Locke concluded "that it is impossible to reduce to precise rules, the various degrees wherein men give their assent,"[22] thus effectively siding with those who trusted to judgment exercised in the individual case rather than rules applied universally.

He also recognized the importance of qualitative concerns such as breadth as well as constancy of experience: the validity of an assertion about "matters of facts" hinged on the number and kind of facts from which it was derived. In Locke's example, the King of Siam, having never ventured forth from his native tropical climate, would understandably greet the Dutch ambassador's account of ice with disbelief. Judged solely in light of his constant experience of liquid water, the king was right to be incredulous, but the moral of Locke's parable was to avoid rash generalizations on the basis of limited experience. More than a century later, Laplace made much the same point regarding the application of probability theory to demographic questions: the accuracy of the results depended on the diversity and bulk of the data.[23] But the pressures of daily life inevitably forced even the most cautious into making decisions based on necessarily partial evidence. Locke warned against prematurely forming judgments and then holding them dogmatically, but this was more a plea for tolerance than a call to postpone action:

> Who almost is there that hath the leisure, patience, and means to collect together all the proofs concerning most of the opinions he has, so as safely to conclude, that he hath a clear and full view, and that there is no more to be alleged for his better information? and yet we are forced to determine ourselves on the one side or other. The conduct of our lives, and the management of our great concerns, will not bear delay; for those depend, for the most part, on the determination

[22] Locke, *Essay*, Book IV, chapter 16.

[23] Laplace, *Théorie analytique des probabilités* (1820; 3rd edition), in *Oeuvres*, vol. 7, pp. 398–399.

of our judgment in points wherein we are not capable of certain and demonstrative knowledge, and wherein it is necessary for us to embrace the one side or the other.[24]

This was not a counsel of despair, for Locke was enough of a probabilist of the new stamp to believe that it was possible to be rational without certainty. This rationality depended crucially on the connection between objective (both qualitative and quantitative) evidence and subjective belief or "assent" to an opinion. Locke's initial epistemological treatment of probability spelled out how the two *should* be linked; his later psychology of association explained how they *must* be.

Locke's seminal chapter "Of the Association of Ideas" was not added to Book II until the fourth (1700) edition of the essay, although earlier chapters had anticipated many of its points. Real ideas, for example, could be distinguished from fantasies by the constancy of their effects and "that steady correspondence they have with the distinct constitution of real beings."[25] However, the earlier editions of the *Essay* had offered no mechanism to explain why constant experience should produce conviction. Rather, Locke had simply asserted that rationality consisted in weighing this and all other sources of probability before forming opinions: ". . . the mind, if it *will proceed rationally*, ought to examine all the grounds of probability, and see how they make more or less for or against any proposition, before it assents or dissents from it."[26] His account was more a prescription for would-be rational thinkers than a description of natural mental operations. The new theory of the association of ideas purported to explain the psychological processes underlying reason: "Some of our ideas have a natural correspondence and connexion with another: it is the office and excellency of our reason to trace these, and hold them together in that union and correspondence which is founded in their peculiar beings."[27]

The well-regulated mind was thus naturally rational, where rationality meant keeping mental tally of "constant effects" and "exact resemblances" and apportioning probabilities and concomitant degrees of assent accordingly. Locke's rationality was synonymous with a judicious weighing of probabilities and expectations, as it had been for the Royal Society apolo-

[24] Locke, *Essay*, Book IV, chapter 16.
[25] Locke, *Essay*, Book II, chapter 33.
[26] Locke, *Essay*, Book IV, chapter 15; emphasis in the original.
[27] Locke, *Essay*, Book II, chapter 33.

gists. Happily, this rationality derived from the very way in which the mind operated upon experience to form judgments: normal psychology was both inherently probabilistic and empirical in its working. Experience generated belief and probability by the repeated correlation of sensations which the mind reproduced in associations of ideas. The more constant and frequent the observed correlation, the stronger the mental association, which in turn intensified probability and belief. Hence, the objective probabilities of experience and the subjective probabilities of belief were, in a well-ordered mind, mirror images of one another. This was why intuitive judgments based on broad experience could be trusted. If classical probabilists took the reasonable man as their standard, it was partially because his reasonableness was intrinsically probabilistic.

But if the mind naturally computed and compared probabilities, of what use was the calculus of probabilities? Once again, probabilists found their implicit reply in Locke, this time in his discussion of the pathology of reason. Although probabilistic rationality came naturally to the healthy mind, few minds survived the corruptions of custom and education. Artificial associations replaced natural ones and became entrenched through habit rather than constant experience: "This strong combination of ideas, not allied by nature, the mind makes in itself either voluntarily or by chance; and hence it comes in different men to be different, according to their different inclinations, education, interests."[28] Reasonable minds guided only by ideas "allied by nature" would, Locke implied, concur in their conclusions.

It was the artificial associations forged by early education, habit, convention, and prejudice which fragmented consensus. Similar forces impeded the mental reckoning of probabilities, leading to "wrong assent, or error." Dubious propositions instilled by childhood education, the undertow of passions or personal interest, the tyranny of unchallenged authority or custom, ignorance, and intellectual sloth all contributed to the decay of common reason. Locke believed that the mind, healthy or corrupt, was incapable of withholding assent from the preponderant probability, but warned that weaker minds could neglect to weigh in the full measure of relevant experience in their reckoning. In short, common reason was not all that common.

Only the strongest intellects could resist the distortions of custom and

---

[28] Locke, *Essay*, Book II, chapter 33.

habit. Sensationalist philosophers and psychologists such as Hartley and Condillac followed Locke's example in paying as much attention to the diseases of reason and their possible remedies as to its healthy functioning. Condillac warned that "we make for ourselves different rules of probability according to the interests which dominate us": an urgent interest strengthened a slight probability, and the ability to judge according to "well-founded probabilities" usually came only with indifference to the outcome.[29] Although reason was the natural result of reflection upon experience, it was not the typical one. Hence, the calculus of probabilities was a necessary aid to all but the most indomitable minds. Probabilists could—and did—claim that although their mathematical results were nothing but "good sense reduced to calculation," this did not render their calculus redundant. Reason was too often overwhelmed by sophistry, prejudice, habit, and the sheer bulk of experience to be trusted implicitly.

This resolution of the apparent contradiction between the tenet that mathematical probability should always confirm the dictates of common reason, as in the St. Petersburg problem, and yet at the same time serve to correct that common reason, was a recurring theme in the writings of the classical probabilists. Abraham De Moivre defended his demonstration and discussion of Bernoulli's theorem on the grounds that even though its results were "level to the lowest understanding, and falling in with the common sense of mankind," common sense required the assistance of "formal demonstration" because of "the scholastic subtleties with which it may be perplexed" and the confusing "abuse of certain words and phrases."[30] Condorcet offered much the same justification for his probabilistic treatment of decisions and judgments: "Almost all of the results [of the *Essai*] will be found to conform to that which the simplest reason would have dictated; but it is so easy to obscure reason by sophistry and vain subtleties that I thought myself fortunate to have been able to support [even] a single useful truth with the authority of a mathematical demonstration."[31]

The interplay, not to say tension, between the descriptive and prescriptive elements in mathematical probability theory closely paralleled that between epistemological and psychological elements in contemporary philosophical accounts of probability. Hume's treatment of the subject is

[29] Étienne Bonnot de Condillac, *Traité des sensations* (1754), in *Oeuvres de Condillac* (Paris, An VI/1798), vol. 3, p. 95n.

[30] Abraham De Moivre, *The Doctrine of Chances*, 3rd edition (London, 1756), p. 253.

[31] Condorcet, *Essai*, p. ii.

the clearest and, for our purposes, most instructive example of these parallels, for Hume assimilated philosophical to mathematical probabilities with an unprecedented thoroughness. He bent Locke's empiricist, associationist framework in two directions, the one psychological and skeptical, and the other mathematical and quantitative. No Enlightenment thinker drew the distinction between how we come to believe and the rational grounds for that belief more sharply than Hume, and probability was a central example (one might almost say, victim) of this skeptical opposition of habit and reason.

Yet even Hume in the end had recourse to probabilities of the reasonable sort to argue against belief in miracles. Psychology never broke entirely free of epistemology, any more than descriptive did from prescriptive elements in classical probability theory. Indeed, their interdependence was an article of faith for philosophers who followed Locke's genetic approach. As Hume put it,

'Tis impossible to reason justly, without understanding perfectly the idea concerning which we reason; and 'tis impossible perfectly to understand any idea, without tracing it up to its origin, and examining that primary impression, from which it arises. The examination of the impression bestows a clearness of the idea; and the examination of the idea bestows a like clearness on all our reasoning.[32]

In his *Treatise on Human Nature* (1739), Hume distinguished between two sources of probability: those based on chance and those based on causes. Chance corresponded to no real idea, but rather to "the negation of cause" which suspended the "perfectly indifferent" imagination between alternative outcomes. Unless disrupted by the action of a constant cause, this absolute indifference was the natural mental state and the guarantee of equiprobability in all calculations of probabilities.[33] Hume denied that a preponderance of equipossible combinations or "chances" necessarily entailed belief in the outcome so favored. Neither demonstration—we cannot know for certain which outcome will occur in any given case—nor probability—this would be begging the question—commands assent: "The question is, by what means a superior number of equal chances operates upon

[32] David Hume, *A Treatise of Human Nature* (1739), L. A. Selby-Bigge, ed. (Oxford: Clarendon Press, 1975), pp. 74–75.
[33] Hume, *Treatise*, p. 125.

the mind, and produces belief or assent; since it appears, that 'tis neither by arguments deriv'd from demonstration nor from probability."[34]

Hume's solution to this psychological query shows how closely interwoven subjective and objective elements of probability could be in the eighteenth-century philosophical discussions. Translating the (objective) physical homogeneity of a die and proportion of markings into the (subjective) indifference among the faces into balancing mental impulses, Hume concluded that the "vivacity of the idea is always proportionable to the degrees of the impulse . . . and belief is the same with the vivacity of the idea, according to the precedent doctrine." The ultimate causes of belief differed little from the bases for reckoning mathematical probabilities: the symmetry of the die and our indifference among alternative outcomes. However, Hume insisted upon the intermediate psychological link of summed mental impulses. Mathematical probabilities alone could not command belief.

The probabilities of cause (true probability, for "what the vulgar call chance is nothing but a secret and conceal'd cause") derived from the Lockean "association of ideas to a present impression." As in Locke's associationism, Hume's psychological mechanism strengthened belief with frequent repetitions:

> As the habit, which produces the association, arises from the frequent conjunction of objects, it must arrive at its perfection by degrees, and must acquire new force from each instance, that falls under our observation. The first instance has little or no force: The second makes some addition to it: The third becomes still more sensible; and 'tis by these slow steps, that our judgment arrives at a full assurance.[35]

Hume described in graphic detail how the mind brought past experience to bear on an uncertain future event and how the mind naturally proportioned belief (synonymous with the "vivacity" of the idea in Hume's philosophy) according to frequencies of past observations: "Each new experiment is as a new stroke of the pencil, which bestows an additional vivacity on the colours {of the idea}, without either multiplying or enlarging the figure."[36] This psychological enhancement of mental images by frequent repetition insured that "vivacity" or belief was always proportional to mathematical probability.

---

[34] Hume, *Treatise*, p. 127.
[35] Hume, *Treatise*, p. 130.
[36] Hume, *Treatise*, p. 135.

In contrast to Locke, Hume regarded both kinds of probability in purely quantitative terms. Chances were all equal (and therefore could be handled like arithmetic units) because the psychological state of absolute indifferences to which they corresponded was always identical to itself. The causes revealed by experience also depended only on their numbers: "every past experiment has the same weight, and that 'tis only a superior number of them, which can throw the balance on any side."[37] Custom and education can produce the same effect of belief, albeit only with more frequent and prolonged repetition than firsthand experience to compensate for lost vivacity of impressions. Although the mind tallied these frequencies and proportionate degree of belief unconsciously and automatically, Hume insisted upon the exquisite sensitivity of the reckoning:

> When the chances or experiments on one side amount to ten thousand, and on the other to ten thousand and one, the judgment gives the preference to the latter, upon account of that superiority; tho' 'tis plainly impossible for the mind to run over every particular view, and distinguish the superior vivacity of the image arising from the superior number, where the difference is so inconsiderable.[38]

Habit may not be strictly rational, but it is exact. Gone are Locke's qualitative considerations, and even his refusal to join probability with certainty at the extreme point of the continuum. Hume subscribed to the mathematical version of probabilities fully enough to interpolate a third category of "proofs" between demonstrative knowledge and uncertain probabilities, which are "those arguments, which are derived from the relation of cause and effect, and which are [nonetheless] entirely free from doubt and uncertainty."[39] With Hume, philosophical and mathematical probabilities became almost indistinguishable.

Hume's associationist account of probabilities fell squarely within the Lockean tradition that equated probabilities, experience, and belief, despite Hume's novel definition of belief in terms of the "vivacity" of mental images. Upon a cursory reading, a contemporary well versed in the fourth edition of Locke's *Essay* would have been familiar with the main line of argument, if not with the detail of the psychological development. However, Hume diverged from the other methodological exponents of probability on at least one critical issue. While Hume contended that the mind

[37] Hume, *Treatise*, pp. 125, 136.   [38] Hume, *Treatise*, p. 141.
[39] Hume, *Treatise*, p. 124.

naturally proceeded according to probabilities, he denied that this process was reasonable. Hume's treatment of probabilities was scrupulously descriptive: probabilistic thinking was an inescapable fact of mental life, but it was not necessarily rational. The attack on the validity of inductive inference applied with equal force to probabilistic reasoning. Although Hume did distinguish "philosophical" from "unphilosophical" probability, the former being defined as that "allow'd to be reasonable foundations of belief and opinion," he reminded his readers that the psychological mechanisms responsible for both types of probability were identical.[40] The wise man knew to correct for the vagaries of imagination, the immediacy of impressions, or the "general rules" of prejudice, all of which deceptively heightened the vivacity of a given idea. Yet (paradoxically), he could correct for such distortions only by applying still other general rules.[41]

One such overriding general rule invoked by Hume was the principle of the uniformity of nature. Hume was at pains to point out that this principle enjoyed neither deductive nor inductive support. Nonetheless, he argued that it was a psychological necessity, an almost involuntary precept implanted by beneficent nature to compensate for the shortcomings of human reason. Once accepted, this principle permitted Hume to take a more conventional position on the connection between probabilistic thinking and reasonableness, although he turned both principle and reasonableness to the unorthodox task of attacking the plausibility of all reports of miracles. A "wise man," Hume asserted, "proportions his belief to the evidence." In cases of unexceptionably uniform experience, he attains "the last degree of assurance and regards his past experience as full *proof* of the future existence of that event." Lacking such unremitting regularity, he weighs experience pro and con: "All probability, then, supposes an opposition of experiments and observations where the one side is found to over balance the other and to produce a degree of evidence proportioned to the superiority." Since all experience points toward the existence of inviolable natural laws, and the principle of the uniformity of nature sanctioned this inference, "the proof against a miracle, from the very nature of the fact, is as entire as any argument from experience can possibly be imagined," where a miracle was defined as a "violation of the laws of nature."[42]

Thus, despite his reservations concerning the validity of induction and

[40] Hume, *Treatise*, p. 143.

[41] Hume, *Treatise*, p. 150.

[42] Hume, *An Enquiry Concerning Human Understanding* (1758), Charles W. Hendel, ed. (Indianapolis: Liberal Arts Press, 1955), pp. 118–119, 123.

probability, Hume made probabilistic thinking the *de facto*, if not the *de jure*, standard of reasonableness. Hume's views on probability belonged in large part to the mainstream of empiricist epistemology; indeed, his account of the relationship between the frequency of experience, the measure of probability, and the force of belief, with its attention to psychological detail and striking imagery, was perhaps the most forceful of all the eighteenth-century discussions of this topic. However, Hume's contention that inductive inference rested on habit rather than reason did not go entirely unnoticed or unchallenged by mathematical probabilists. Richard Price certainly knew Hume's arguments well enough to disapprove of their skeptical tendencies, even if he did not fully appreciate their force, and it is plausible to assume that Price hoped that his commentary on Bayes' theorem might serve as a rebuttal. Price's commentary and its bearing on Hume's critique will be discussed in Chapter Five. In the context of this discussion, it is sufficient to remark the ways in which Hume reinforced the associationist interpretation of probability as a Janus-like concept with inseparable objective and subjective faces.

Hume's philosophical and psychological account of the probabilities of experience and belief was certainly more thoroughgoingly quantitative than Locke's, but it contained almost no reference to the mathematical theory of probability per se. This omission was repaired by David Hartley's *Observations on Man, His Frame, His Duty and His Expectations* (1749), which drew heavily upon the latest mathematical findings to support his more physiological brand of associationism. Hartley's contribution to associationism was more synthetic than original. Combining elements of Locke's sketch of associationism, Newton's physiological speculations concerning the vibratory basis of sensation advanced in the "Queries" to the *Opticks*,[43] and the Reverend John Gay's pleasure/pain theory of morality set forth in his *Preliminary Dissertation Concerning the Fundamental Principle of Virtue and Morality* (1731), Hartley produced a naturalistic account of morality. Although Hartley intended his treatise primarily as a contribution to natural theology, his supporting chapters on psychology, epistemology, and methodology had a far greater impact on eighteenth- and early nineteenth-century thought: Joseph Priestley, Erasmus Darwin, William Goodwin, James and John Stuart Mill were among his intellectual heirs.[44]

[43] Issac Newton, *Opticks*, based on the 4th edition (1730) (New York: Dover, 1952), Queries 14, 23, and 31.

[44] See Robert M. Young, "David Hartley," *Dictionary of Scientific Biography*, Charles C. Gillispie, editor-in-chief (New York: Charles Scribner's Sons, 1972), vol. 6, pp. 138–140.

Hartley's full-blown associationism underscored the kinship between "frequently renewed" sensations, the doctrine of chances, and assent. All of these depended on the repetition of associations and concomitant cerebral vibrations until grooves of mental habit were etched in the brain. Hartley greeted De Moivre's proof of Bernoulli's theorem that in the long run observed frequencies approached stable a priori probabilities as a mathematical affirmation of the "Order and Proportion, which we everywhere see in the Phenomena of Nature," mentioning that an unnamed friend—perhaps Bayes—had shown him the solution to the inverse problem of finding a priori probabilities from observed frequencies. Despite the ingenuity and elegance of the mathematical methods, Hartley claimed that they were nonetheless "evident to attentive Persons, in a gross general way, from the common Methods of Reasoning."[45]

The very workings of the human understanding, when undeflected by strong emotion or uncritical custom, imitated the law of large numbers. Every observed repetition of the same or similar events ingrained the association of corresponding sensations more deeply by triggering the "corresponding miniature vibrations" of "medullary substance."[46] Assent and expectation were strengthened accordingly. "Practical assent," or the readiness to act, sprang from such involuntary psycho-physiological associations. The further linkage of these associations with the "Idea, or Internal Feeling, belonging to the word Truth" produced "rational assent." The method of induction and analogy generalized the rational assent accorded propositions about past experience to propositions about similar future events. Evidence for future predictions was "of the same kind with that for the Propositions concerning natural Bodies, being, like it, taken from Induction and Analogy. This is the Cause of the rational Assent. The Practical depends on the Recurrency of the Ideas and the Degree of Agitation produced by them in the Mind."[47]

Hartley's *Observations* was heavily influenced by Joseph Butler's enormously popular *Analogy of Religion, Natural and Revealed, to the Constitution of Nature* (1736).[48] Butler's *Analogy* was the culmination of the probabilis-

---

[45] David Hartley, *Observations on Man, His Frame, His Duty, and His Expectations* (London, 1749), vol. 1, p. 331. See Stephen M. Stigler, "Who discovered Bayes' theorem?" *The American Statistician* 37 (1983): 290–296, on the identity of Hartley's "ingenious Friend."

[46] Hartley, *Observations*, vol. 1, pp. 59–67.

[47] Hartley, *Observations*, vol. 1, pp. 324–325, 332.   [48] Young, "Hartley," p. 139.

tic tradition of natural theology inaugurated by Grotius and elaborated by John Tillotson, Robert Boyle, and the circle of Royal Society apologists. Butler's treatise also incorporated inductive and analogical reasoning, conceived in explicitly frequentist terms, within the extant epistemic interpretation of probability and the religious scaffolding built around it. Butler retraced the by now familiar arguments that contrasted the omniscience of an "infinite Intelligence" with the perpetual uncertainty of merely human understanding, to which probabilities must be "the very Guide of Life," and argued that "in point of Prudence and of Interest" we are obliged to act in accordance with the highest expectation.

Following Locke, Butler explained that probability differed from demonstration in admitting of degrees "from the highest moral Certainty; to the very lowest Presumption," which matched the "Degree of Conviction." Degrees of probability/conviction rose and fell with experience. Butler argued that a single observation of the phenomenon of the tides imparted "the lowest imaginable" presumption that it would recur the next day. However, the combined observations of a lifetime, and of all mankind, "gives us a full assurance that it will." We have only this probability, based on a vast number of observations, that the sun will rise tomorrow.[49] Butler's sunrise example, along with his claim that "this general way of argument is evidently natural, just and conclusive," was later adopted by Price in his commentary on Bayes' theorem, and became a commonplace of the mathematical literature on inverse probabilities.

As in Locke's *Essay*, Hartley's discussion of assent, both rational and practical, was inseparable from probability. However, while Locke's treatment preserved the traditional qualitative sense of probability, Hartley made full use of the mathematical theory of probability. Hartley was well acquainted with De Moivre's *Doctrine of Chances* as well as with the natural theological argument from expectation and its determination of practical action.[50] Hartley's assimilation of induction and analogy to mathematical probability owed a good deal to De Moivre's discussion of the implications of Bernoulli's theorem in the second edition (1738) of the *Doctrine of*

---

[49] Joseph Butler, *The Analogy of Religion, Natural and Revealed, to the Constitution and Course of Nature* (London, 1736), Introduction. Hume was also much impressed, though not persuaded, by Butler's work, and hoped to solicit Butler's opinion of the *Treatise* before publication; see James Noxon, *Hume's Philosophical Development* (Oxford: Clarendon Press, 1973), p. 77.

[50] Hartley, *Observations*, vol. 1, p. 338.

Chances.[51] Like De Moivre, Hartley emphasized that the mathematical results only confirmed the results of ordinary reason and adduced the theorem as further evidence for divinely instituted order in nature. He enlarged upon the comparison between mathematical and inductive methods in a discussion of the *"Newtonian* differential method" of fitting curves to collections of discrete points. The results of both methods—curve fitting and inductive analogy—at best enjoyed only a high probability. The mathematical method was "liable to the same Uncertainties, both in Kind and Degree, as the general Maxims of Natural Philosophy drawn from Natural History, Experiments, & c."[52] Induction and analogy, like conviction and assent, were intrinsically probabilistic. At root, both were the products of identical psycho-physiological operations.

Perhaps the most influential Continental exponent of this bond between belief, induction and analogy, and probability was George Buffon, whose *Essais d'arithmétique morale* (1777), included among the supplements of his encyclopedic *Histoire naturelle*, was read and commented upon by d'Alembert, Condorcet, and Laplace. Buffon introduced his remarks on probabilities with the familiar hierarchy of mathematical, physical, and moral certainties, which derived respectively from demonstrative, inductive, and analogical arguments. Buffon concerned himself exclusively with the probabilistic moral and physical certainties, dismissing the mathematical variety as simply "truths of definition." Physical certainty, such as the unshakable conviction that the sun will rise tomorrow, resulted from a "constant series of observations, which constitute what one might call the *experience of all time."* The inferior probability of moral certainty "presumed only a certain number of analogies with what we already know."[53]

Buffon set out to quantify the probabilities represented by this pyramid of mathematical, physical, and moral certainties. Physical certainty—"an immense sum of probabilities which forces us to believe"—commanded involuntary assent; moral certainty, while persuasive, was only a "greater or smaller probability." Buffon illustrated physical certainty with a sunrise example strikingly similar to that used by Price, although Buffon made no use of inverse probabilities. Using the conceit of a man seeing and

---

[51] See the exchange of letters between R. C. Archibald and Karl Pearson in *Nature* 117 (1926): 551–552, regarding the publication history of this passage in De Moivre.

[52] Hartley, *Observations*, vol. 1, pp. 338–340.

[53] George Leclerc Buffon, "Essais d'arithmétique morale," in *Histoire naturelle. Supplément* (Paris, 1777), vol. 4, p. 52.

hearing for the first time, Buffon attempted to show how belief and doubt arose from experience. Probability theory quantified the degree of physical and moral certainty thus attained. In the sunrise example, Buffon computed the probability that the sun would rise on the morrow on the assumption that both outcomes—the sun rising or not—were equiprobable. Buffon reckoned the probability that the sun would rise tomorrow from the knowledge that it had done so without exception every day for the past 6,000 years (the canonical age of the earth established by the chronology of Bishop Ussher) to be $1 - \frac{1}{2^{n-1}}$, where n was the total number of observed sunrises since creation. Such overwhelming probabilities constituted physical certainty.

In order to evaluate moral certainty, Buffon resorted to the psychological measure of the least tremor of fear, equated with the probability that a man fifty-six years old would die within the next twenty-four hours. No reasonable, healthy man of that age, Buffon argued, doubted his survival for so short a period, despite the fact that there existed a small but measurable probability that he might die: approximately .0001, according to the mortality tables. Hence, moral certainty accrued to any proposition with probability at least equal to .9999. Because $2^{13} < 10^4 < 2^{14}$, Buffon contended that an event need only happen thirteen to fourteen times to accrue moral certainty that it will occur again.

Buffon did not regard such inductively derived probabilities as universally applicable. The method was valid only when the underlying causes were, on the one hand, constant, and on the other, inaccessible to direct inspection. The recurrence of an event of unknown cause repeated fourteen times merited moral certainty only if it were not due to "the effect of chance." Buffon did not admit the existence of any genuinely random events: chance effects, like all others, were governed by natural causes. However, the causes that regulated chance effects were intermittent and heterogeneous, "necessarily variable and versatile in so far as this is possible. Thus by the notion of chance, it is evident that there is no link, no dependence between its effects; so that consequently the past cannot in any way influence the future."[54]

Nor were probabilities appropriate where certainty could be had. In cases in which the causes were manifest (Buffon cited the interlocking gears and weights that drive clockwork), there was no need for cumulative

[54] Buffon, "Essais," p. 61.

observations to amass physical certainty. But in the intermediate case between the independent effects of chance and the exposed mechanism of determinate causes, probabilities could be fruitfully applied to events presumed to result from constant, if hidden, causes. Moreover, analogy could replace 6,000 years of patient observation. Once moral certainty had been achieved with the requisite fourteen repetitions, Buffon claimed that the observer was entitled by analogy with past experience of nature's global uniformity to "leap over an immense interval, and to conclude by analogy that this effect depends on the general laws of Nature, that it is consequently as ancient as all of the other effects, and that there is a physical certainty that it will always happen as it always has happened."[55]

Condorcet, as we will see in Chapter Five, attempted to quantify this enormous boost to belief imparted by past experience of nature's uniformity by using inverse probabilities. Armed with Laplace's version of Bayes' theorem, Condorcet based a philosophy of scientific method and belief on probability theory. Although Condorcet was perhaps the most ambitious proponent of the methodological potential of the calculus of probabilities, his combination of experience, belief, and mathematical probabilities was by no means original, and probably owed a large debt to Buffon's essay, as well as to Laplace's 1774 and 1776 memoirs.[56] Condorcet was also influenced by the descriptions of the aberrations of reason found in Locke and emphasized by Condillac. These discussions of how the mind erred in assessing probabilities and in proportioning belief to experience strengthened Condorcet's arguments for the calculus of probabilities as an auxiliary to all decision making. If the natural reckoning of probabilities could be so easily disrupted by self-interest, imagination, or prejudice, the need for a mathematical corrective was all the more urgent.

In his *Essai sur l'origine des connaissance humaines* (1746) and *Traité des sensations* (1754), Etienne Bonnot de Condillac propounded a radical form of sensationalist psychology. Whereas Locke had acknowledged two sources of ideas, experience and reflection, Condillac attempted to reduce epistemology to a monistic system based on experience alone. Condillac emphasized that the way in which the mind organizes experience depends on our needs and wants. It was attention that forged associations between

[55] Buffon, "Essais," p. 64.

[56] Condorcet also mentioned the Bayes/Price essay in his introductory *Histoire* to Laplace's memoir "Sur les probabilités," *Histoire et mémoires de l'Académie royale des Sciences. Année 1778* (Paris, 1781), p. 43.

ideas presented together by sensation, memory, or imagination, and attention in turn varied according to "our temperament, our passions, our state, or to sum everything up in a word, our needs."[57] Even more than Locke, Condillac was preoccupied with pathological associations instilled by prejudice, early education, overly vivid imagination, or brain consistency. (For example, he believed that young girls were far more prone to confuse "chimeras for realities" because even faint associations left permanent impressions on their soft brains.) Everyone was to some degree debilitated by "disordered" association of ideas; mad minds simply formed such erratic associations more frequently and about more important matters than sound minds did.

The mental estimation of probabilities suffered from similar sorts of distortion, according to Condillac. Chains of association between ideas grew stronger through habitual reiteration in memory as well as through the repetition of observed correlations. Depending on the degree of pleasure or pain attached to the constituent associations, the chains could be further fortified by habit. In Condillac's famous conceit of the partially sentient statue, hope and fear arose from such habits: "Our statue's habit of experiencing agreeable and disagreeable sensations makes him expect to experience them again. If this judgment is joined to the love of a pleasant sensation, it produces hope; and if it is joined to hatred of an unpleasant sensation, it forms fear."[58] Thus not only frequency, but personal interest colored expectation. Condillac applied this insight to the intuitive evaluation of probabilities: "We conduct ourselves according to experience, and we make different rules of probability according to our dominant interest . . . where we are wise enough to make decision only on well-founded probabilities, it is often only because we have little interest in acting."[59]

Condillac's promotion of personal needs and interests to equal status with observational frequency in determining the strength of the associations between ideas threatened the identification of normal mental operations and reasonableness. According to Condillac, normal psychology dictated that needs and interests be averaged in along with the tally of a past experience in assessing future expectations: "If [the interest] was large, the slightest degree of probability ordinarily suffices" to impel action. He did not condone this lopsided reckoning; he only asserted its universality. Condillac

---

[57] Condillac, *Essai sur l'origine des connaissances humaines* (1749), in *Oeuvres*, vol. 1, p. 67.
[58] Condillac, *Traité*, pp. 93–94.   [59] Condillac, *Traité*, p. 95n.

also heeded the obstacles to reason, particularly the effects of early education and overwrought imagination, more than his predecessors had.

The emphasis on personal needs and interests, and on the widespread pathologies of uncritical habit and unbridled imagination, undermined the more optimistic psychologies of Locke, Hartley, and even Hume. While Hume had denied the ultimate validity of inductive and probabilistic reasoning, he had nonetheless accepted its practical necessity and, when suitably fortified by metaphysical assumptions of the uniformity of nature, philosophical desirability as a criterion of rational belief, as in the case of miracles. The British philosophers had concurred that probabilities and belief not only should be, but actually were, proportioned to one another through the associative operations of sound minds. Although they had taken due notice of possible distortions of these operations, they had done so in the belief that these were corrigible aberrations. Condillac, in contrast, claimed that at least one source of error in the intuitive weighing of probabilities, the pressure of needs and interests, was as natural as balancing frequencies of past events, and that the illusions perpetrated by imagination were almost as common. The reliability of subjective probabilities diminished accordingly.

4.3 The Dissociation of Objective and Subjective Probabilities

Condillac's misgivings about the reliability of the link between the probabilities of experience and belief were the beginning of the end for the classical interpretation of probability. Once wishful thinking became the rule rather than the exception in the psychology of hopes and fears, subjective and objective probabilities could no longer be so automatically equated. Associationist psychology first made this connection possible, and later undid it almost altogether, with ultimately fatal consequences for the classical theory. However, at first Condillac's analysis of mental errors in computing probabilities strengthened the hand of the mathematicians. They could now point to an urgent need for their theory as prophylactic and corrective to these all too common errors, as opposed to the far weaker justification that a watertight mathematical demonstration for what everyone had known all along could not hurt. Of course, this new strategy, most evident in the work of Condorcet and Laplace, did shift the emphasis within the classical theory from descriptive to normative. Without ceasing to be the reasonable calculus, probability theory became ever more a prescription of how we should reason, rather than a description of how we in fact do. Yet the descriptive element never wholly vanished, for the classical probabilists kept returning to the intuitions of an enlightened elite as the source and check of their findings. It took the trauma of the French Revolution and its aftermath—Condorcet and Laplace both lived in the thick of these events—to unseat this minority of *hommes éclairés* and the ideal of rationality they represented in the minds of the probabilists.

Condorcet did not challenge the link between reasonableness and the assessment of probabilities, but he echoed Condillac's doubts concerning the validity of purely intuitive reckoning. Good sense still meant reasoning by expectation, but the calculus of probabilities became the necessary guide to such reasoning. Condorcet championed mathematical probability as the only trustworthy index of rational belief, and as the cornerstone of the scientific study of human society and conduct. My purpose here is not to examine Condorcet's program for a *mathématique sociale*, which has already attracted a good deal of scholarly attention,[60] but rather to situate Condorcet's interpretation of probabilities within a tradition that embraced the efforts of philosophers, psychologists, and mathematicians alike.

Until d'Alembert launched his thirty-year attack on probability theory, probabilists accepted their mandate to codify reasonableness in mathematical form as a straightforward, if difficult, task. D'Alembert's criticisms exposed the ambiguities in the concept of "reasonableness" and the consequent complexity (and perhaps impossibility) of expressing it mathematically. However, d'Alembert never went so far as to exhort probabilists to redefine the objectives of the theory: for d'Alembert, the failure to mathematicize reasonableness was the failure of the calculus of probabilities per se. Mathematical probability and reasonableness remained inseparable.

Similarly, the associationist psychology of Locke and his disciples had affirmed the bond between probability theory and reasonableness by linking subjective and objective aspects of probability. Depending on the physiological mechanism espoused, oft-repeated vibrations or deeply en-

---

[60] See Gilles-Gaston Granger, *La mathématique sociale du Marquis de Condorcet* (Paris: Presses Universitaires de France, 1956); Roshdi Rashed, *Condorcet: Mathématique et société* (Paris: Hermann, 1974); and Keith M. Baker, *Condorcet: From Natural Philosophy to Social Mathematics* (Chicago: University of Chicago Press, 1975).

graved impressions in the brain created involuntary associations between ideas frequently paired in sensation. The associations in turn generated belief. The notion of probability encompassed both the objective elements of experience and the subjective elements of belief. By positing psychological reasons for this parallelism of objective and subjective probabilities, the associationists *explained* the connections between reasonableness and probability theory that the authors of the Port Royal *Logique* and the Royal Society theologians had merely asserted.

Locke, Hartley, and Hume admitted that psychological pathologies sometimes disrupted the natural operations of reason, but denied that these pathologies were inherent. Condillac's claim that the distortion of interest and imagination inevitably warped the formation of associations and thus slanted beliefs and judgments had much the same effect on Condorcet's view of mathematical probability that d'Alembert's critique had: Condorcet did not relinquish the bond between probability theory and reasonableness, but he did shift the probabilists' responsibility from describing to correcting good sense.

While Condorcet's admiration of Condillac's philosophy was restrained—he considered Condillac to have done no more than embellish Locke's discoveries[61]—he accepted Condillac's strictures against imagination as the antithesis of analysis and the source of illusion. Condillac had likened the imagination to a coquette who aims only to please, even at the price of inaccuracy, by flattering prejudice and passions.[62] Condorcet shared Condillac's distrust of imagination as a pleasant sedative to the analytic reason that penetrated to the origin of ideas, recombined their constituent elements, and directed judgment. The calculus of probabilities had to take over these functions for all but the most incorruptible intellects. This theme became increasingly prominent in Condorcet's mature writings on probability.

In a manuscript fragment written sometime between 1774 and 1781,[63] Condorcet presented his earliest views on the relationship between the objective probabilities of frequency and the subjective probabilities of belief. His approach and examples were heavily influenced by Locke's nominalism: we believe that a piece of heavy yellow metal soluble in aqua regis

[61] Baker, *Condorcet*, pp. 114–116.

[62] Condillac, *Essai*, pp. 135–136.

[63] Condorcet, Bibliothèque de l'Institut MS. 875, ff. 89–99. Condorcet mentions Laplace's 1774 memoir on inverse probabilities, but not the Bayes/Price essay, which he knew of by 1781 at the latest (see n. 56, above).

will also dissolve in a sulphur potassium compound (*foie de souffre*) because long observation testifies that whenever a piece of metal displays the first properties, it will also exhibit the second.[64] Although this knowledge remained only a probability, such an unbroken sequence of observations warranted belief, just as drawing white balls a hundred million times in a row warranted the expectation that the next drawing would be white. Even the fundamental proposition "that which I believe I touch exists" rested on potentially calculable probabilities gleaned from constant experience:

> . . . in reality this proposition means nothing else but that I have certain sensations, following which certain constancies are given. I will experience them again, and I even may experience them every moment of my existence. Thus there is only the experience that this has happened constantly which makes me believe in this proposition, thus one need only know this experience in order to evaluate the degree of this probability.[65]

According to Condorcet, probabilities varied from person to person, depending on their store of relevant experience. Citing the standard distinction between mathematical, physical, and moral certainties, Condorcet argued that physical and moral certainty admitted different degrees, since "when different people say that they are sure, each has a different probability derived from his experience." By certainty, Condorcet meant psychological certainty: in the physical sciences, wider experience confirmed conviction because of the constant order exhibited by the phenomena, but in the moral sciences, only those with experience too narrow to have encountered exceptions professed certainty. Nonetheless, Condorcet took future expectations founded upon past experiences to be "the unique rule of our opinions and of our actions" and claimed to be able to compute the increment in probability added to this rule in particular cases by cumulative observation. While it would be "a chimera" to attempt these mathematical computations for most probabilities in either the sciences or the "conduct of life," Condorcet nonetheless hoped that quantitative terms of comparison would aid in decision making: "Am I as sure that this new experiment, this discovery which I announce is true as I am of the proposition, gold forms amalgam with mercury, and how many more experiments would be necessary to achieve this degree of probability?"[66]

---

[64] Condorcet, MS. 875, f. 95.   [65] Condorcet, MS. 875, f. 98.
[66] Condorcet, MS. 875, f. 98.

In this early fragment, Condorcet underscored the distinction between the intuitive certainty of mathematical demonstration and the probabilities produced by experience. In later works, however, he asserted that even the truths of mathematical demonstrations were probabilistic, based on the experience that "if I have once demonstrated a truth, I constantly find this same truth every time I wish to follow the demonstration." Only the intuitive self-evidence of the idea or sensation immediately present to consciousness yielded certainty.[67] But in the earlier manuscript, Condorcet insisted that only the vulgar confused the belief inspired by mathematics and the belief prompted by probabilities derived from experience. Probabilistic belief, the only type available in the physical and moral sciences, was not derived from the nature of reality—Condorcet denied any intrinsic connection between probabilities and the "real state of things"—nor from the workings of the mind alone, but rather from the subjective representation of objective experience:

> To believe, in the physical and moral sciences, is nothing else but to represent things as existing in a certain manner; thus when many experiences have represented to us a certain combination of things, it is always this same combination which represents itself to us, and consequently we believe that that which has happened, will happen again.

Belief, then, sprang from frequently repeated mental associations impressed by experience. Thus far, Condorcet's account of the connections between experience, belief, and probability was quite compatible with Hartley's or Hume's associationist theory of natural reason. However, Condorcet went on to add a qualification closer to the spirit of Condillac's psychology of imagination and interests:

> This explains how a vivid imagination makes us believe the falsest things with such force as we believe according to the inclinations of our passions. And we conduct ourselves according to the probability for the same reason, because we only do that which we desire as advantageous, that which we represent to ourselves as advantageous.[68]

Condorcet's distrust of psychological assessments of probability deepened in later works, and the fervor with which he campaigned for a

---

[67] Condorcet, *Essai*, p. xiii.　　　　[68] Condorcet, MS. 875, f. 99.

broader-based, more widely disseminated calculus of probabilities rose accordingly. In the Preliminary Discourse to his *Essai sur l'application de l'analyse à la probabilité des décisions* (1785), Condorcet hedged his admission that the results of applying the calculus of probabilities to the decisions of deliberative assemblies concurred with what "the simplest reason would have dictated" with many caveats. Because "simple reason" was so frequently plagued by "sophistry" and "vague subtleties," the clarifications of mathematical demonstration were always welcome.[69]

At a deeper epistemological level, probabilities assessed the accuracy and sufficiency of observations, as well as the rectitude of the reasoning applied to them. Arguing that our understanding must be satisfied with "average," as opposed to "real" probabilities, and surrendering all hope of certainty, Condorcet tempered the skeptical implications of his argument by holding out the prospect of sounding out the probabilistic bases of our knowledge with "a sort of precision." Far from succumbing to the "discouragement and indolence" that accompanied the pyrrhonist doctrines of the ancients, men guided by the calculus of probabilities would be freed from the tyranny of "vague impressions," a freedom heretofore enjoyed only by a tiny elite of "enlightened men." Mathematical probabilities would permit us "to judge and conduct ourselves, not according to a vague, unconscious impression, but rather according to an impression submitted to calculation, and whose relation to other impressions of the same sort is known to us."[70]

Defending the utility of his results concerning tribunals and assemblies against the criticism that common sense would have arrived at the same conclusions without the elaborate mathematical machinery, Condorcet reiterated his arguments concerning the fallibility of simple reason. Beyond its vulnerability to sophistry and subtlety, common reason also balked when circumstances required fine distinctions: "Reason suffices so long as one needs only a vague observation of events; the calculus becomes necessary as soon as truth depends on exact observation."[71] Condorcet appealed to the universal experience of changing one's mind without being able to pinpoint the reasons for doing so: without the calculus of probabilities to weigh the impact of new experience, we become merely the "slaves of our impressions." Impressions subvert reason because belief is proportional to

---

[69] Condorcet, *Essai*, p. ii.　　　　　[70] Condorcet, *Essai*, p. xciii.

[71] Condorcet, *Essai*, p. clxxxv.

215

reiteration, and the same natural tendency that leads us to believe in the constancy of the natural order ascribed equal assurance to mental associations retraced by habit or self-interest. Imagination and temperament perverted the passive mental tally of impressions to unreasonable ends:

> Reason and the Calculus tell us that probability increases with the number of constant observations which is the basis of our belief; but does not the force of the natural tendency which makes us believe depend at least as much on the force of the impression that these objects make on us? Thus if reason does not come to our aid, our opinions will actually be the work of our sensibility and our passions.[72]

Condorcet hoped that the calculus of probabilities would rescue the two groups most subject to these aberrations of reason, namely "children and the people," from error and prejudice. The elite of *hommes éclairés* could trust right reason to prevail over the distortions of imagination and interest, although Condorcet recommended that even they avail themselves of the calculus of probabilities in complicated cases. The less fortunate majority became the intended audience of Condorcet's public course on mathematics at the Paris Lycée and of his posthumously published elementary text on probabilities. In his introductory lecture to the Lycée's abbreviated course on mathematics (actually astronomy and the calculus of probabilities), Condorcet hammered away at the familiar theme of mathematical probability as a corrective to intuitive "motives for belief." Promising his listeners a survey of the "spirit of the methods," albeit one which omitted their mathematical content, of mathematical probability, Condorcet maintained that the calculus of probabilities would free them from involuntary mental forces:

> We will prove that the motive for belief in these real truths to which we are led by the calculus of probabilities differs from that which determines us in all our judgments, in all our actions, only because the calculus gives us the measure of this motive; and that we submit, by the enlightened assent of reason, to a force whose power we have calculated, instead of submitting mechanically to an unknown force.[73]

---

[72] Condorcet, *Essai*, p. cxci; compare Hume, *Treatise*, p. 119. On Hume's influence on Condorcet, see Baker, *Condorcet*, pp. 138–140.

[73] Condorcet, "Discours sur l'astronomie et le calcul des probabilités, lu au Lycée"

The motive for belief produced by the calculus of probabilities did not differ in kind from that which guided everyday life, only in amplitude and precision. For Condorcet, the calculus of probabilities acted as a kind of mental magnifying glass, which sharpened, without qualitatively modifying, inward perception.

By 1793, the French Revolution had imparted new urgency to Condorcet's campaigns for public instruction and "social mathematics."[74] In an article entitled "Tableau général de la science qui a pour objet l'application du calcul aux sciences politiques et morales," originally published in the *Journal d'Instruction Sociale* (22 June and 6 July 1793) and later reprinted as an appendix to Condorcet's *Eléments du calcul des probabilités* (An XIII/ 1805), Condorcet gave two major reasons why the cultivation of social mathematics was a prerequisite to the general happiness and improvement of the human race. First, all of our judgments, including those upon which we act, are based on an intuitive reckoning of probabilities by "vague and almost mechanical feelings." Although social mathematics did not claim to be able to quantify all such opinions and judgments, its estimates, Condorcet maintained, were superior to those of "instinct and habit." Second, absolute truths that transcended calculation and degree often resisted practical application, especially in matters of fact susceptible to measure or numerous combinations. In such cases, errors proliferated unless mathematics intervened.[75]

Four years of political tumult prompted Condorcet to add demagoguery and bad faith to the list of psychological agents of distortion to be countered by the calculus of probabilities. Social mathematics was to supply the antidote to the corrosive effects of mob politics on enlightened consensus. In order to reestablish public order and prosperity,

> . . . stronger combinations, more precise means of calculation are necessary, and these can only be adopted by proofs which, like the results of calculations, impose silence on bad faith, as well as upon prejudice . . . since all of the truths recognized by enlightened men have been confused in the mass of uncertain changing opinions, we

---

(1787), in *Oeuvres de Condorcet*, F. Arago and A. Condorcet-O'Connor, eds. (Paris, 1847), vol. 1, p. 499.

[74] Granger, *Mathématique*, chapter 4.

[75] Condorcet, "Tableau général de la science qui a pour object l'application du calcul aux sciences politiques et morales" (1793), in *Oeuvres*, vol. 1, p. 541.

must fetter men to reason by the precision of ideas, by the rigor of proofs, [in order] to set truths presented to them beyond sophistry of interest.[76]

Condorcet believed that the calculus of probabilities would make ordinary minds proof against the "seductive influences of imagination, interest, and the passions" and force them to affirm the opinions of enlightened men.

In Condorcet's schema, the calculus of probabilities applied equally well to the operations of the mind as to the study of men and things, and particularly to the determination of threshold probabilities for rational belief and action. Although Condorcet contended that social mathematics ultimately reinforced the views of the enlightened elite and called upon skilled mathematicians to apply themselves to the development of the theory, he nonetheless maintained that social mathematics was not the exclusive purview of experts, "an occult science, whose secret is restricted to a few adepts" but rather "a common, ordinary science."[77] Just as sailors could determine longitudes from lunar tables without any technical knowledge of celestial mechanics, the use of the calculus of probabilities could be taught to everyone regardless of mathematical preparation.

The calculus of probabilities could be applied to the "operations of the human mind" because it was conceived as a mathematical model of those operations. Condorcet assumed that enlightened men who naturally reckoned by inverse probabilities would always agree with one another: right reason admitted only one answer to any question. By public instruction in the results, if not the mathematical theory, of the calculus of probabilities, the circle of reasonable consensus could be widened to include everyone. "The empire usurped by words over reason, by the passions over truth, by active ignorance over enlightenment" would be destroyed.[78]

Laplace shared Condorcet's reservations concerning the ability of most people to assess probabilities intuitively. While Laplace affirmed that mathematical probability was nothing more than "good sense reduced to a calculus," he carefully distinguished between the objective sources of probability and subjective estimations in one of his earliest papers on the subject, declaring that mathematicians must deal only with the former: "In all these physico-mathematical researches the physical causes of our sensations, and not the sensations themselves, are the object of Analy-

---

[76] Condorcet, "Tableau," p. 542. [77] Condorcet, "Tableau," p. 550.
[78] Condorcet, "Tableau," p. 542.

sis."[79] Although Laplace's work has usually been regarded as the best illustration of the subjective tendencies inherent within the classical interpretation of probability, his philosophical orientation was often explicitly objective, as in this passage. His well-known statements on probability as the measure of "ignorance concerning the different causes which contribute to the production of events, and their complexity, combined with the imperfection of analysis"[80] have overshadowed his repeated assertions that mathematical probability, rather than mental states, treated events in the external world.

Laplace was not necessarily equivocating; he believed that incomplete knowledge about nature and society was knowledge nonetheless, derived from "the method founded on observation and calculation," which Laplace recommended so strongly to the moral sciences on the basis of its enormous success in the natural sciences.[81] Although he accepted and even elaborated the connections between associationist and mathematical probability, he also took Condorcet's cautions concerning the psychological sources of probabilistic error seriously enough to emphasize repeatedly the objective aspect of probabilities. When Poisson broke up the classical triumverate of experience, probability, and belief by separating objective frequencies and subjective beliefs, he was pursuing Laplace's misgivings about probabilistic illusions to their logical conclusion.[82]

Significantly, Laplace's discussion of the relationship between associationist psychology and mathematical probability fell under the heading of "Des illusions dans l'éstimation des probabilités" in his *Essai philosophique*. Just as the sense of touch corrected optical illusions, "reflection and calculation" righted probabilistic illusions, according to Laplace.[83] Laplace's principal motive for examining the "laws of intellectual organization" was not to demonstrate the triple parallelism between sensory evidence, mathematical probability, and rational belief but rather to expose the psycho-

[79] Laplace, "Recherches," p. 147.

[80] Laplace, "Recherches," pp. 144–145. Laplace repeated this passage almost verbatim in his *Essai philosophique sur les probabilities* (1814), which became the introduction to the *Théorie analytique des probabilités*. See also Roger Hahn, "Laplace's first formulation of scientific determinism in 1773," *Actes du XIe Congrès International d'Histoire des Sciences* (Cracow, 1968), vol. 2, pp. 167–171.

[81] Laplace, *Essai philosophique*, in *Oeuvres*, vol. 7, pp. lxxviii.

[82] Siméon-Denis Poisson, *Recherches sur la probabilité des jugements en matière criminelle et en matière civile* (Paris, 1837), p. 31.

[83] Laplace, *Essai*, p. cxii.

logical roots of one class of probabilistic illusions, which included the superstitions of astrology, divination, and auguries. Laplace envisioned psychology as the continuation of the "visible" physiology of nerves that transmitted sensations produced by external objects to the brain, where they left "permanent impressions which modified the *sensorium* or seat of thought in an unknown manner."[84] While Laplace adopted both the vocabulary and disclaimers that the posited physiological mechanisms were only hypothetical from Newton's "Queries" to the *Opticks*, his detailed description of the physiological processes belied his caution. He also incorporated important elements of pleasure/pain psychology and followed Condillac's emphasis on deviations from reason.

Although Laplace's psychology was physicalist, it was not necessarily reductionist in the usual sense. Laplace drew heavily upon physical phenomena for analogies: the variety and complexity of the internal senses were comparable to the myriad phenomena of light and electricity; the "sympathy" exhibited by animate beings of the same or similar species for one another was comparable to the resonance of two pendulums hung from the same beam; complex ideas accreted from simple ones as the partial tides produced by sun and moon combined to cause full tides and so forth. The boundary between analogy and reality was often blurred in Laplace's account. For example, were the "vibrations" excited by the principle of sympathy in the sensorium figurative or literal? As Laplace himself observed, "almost all the comparisons that we draw to material objects, in order to render intellectual things palpable, are at bottom identities."[85]

The fundamental principle of association was the subject of Laplace's most vivid physicalist descriptions. Praising the associationist theories that explained the connections between contiguous sensations, between ideas and their signs, as the "real part of metaphysics," Laplace sketched the aspects of the theory that bore on probabilities in the language of "traces," "impressions," "images," and "vibrations" that dramatized the passive, automatic nature of perception and belief. According to Laplace, the "interior image" constructed from sensations of external objects resulted from the superposition of impressions received from the various senses and those associated with that object in memory.[86] By some unknown psychological mechanism, we are able to distinguish these remem-

---

[84] Laplace, *Essai*, p. cxxiii.
[86] Laplace, *Essai*, pp. cxxv–cxxviii.

[85] Laplace, *Essai*, p. cxxxviii.

bered impressions of past sensations from the fancies of imagination, and it is this ability, Laplace maintained, which permits us to posit causal connections. The intensity of impressions varied with age and circumstances: impressions received in early childhood, for example, left more indelible traces in the sensorium than those left in adulthood. Frequent repetition facilitated all mental actions, forming habits so ingrained that they were often indistinguishable from instinct. Following Condillac,[87] Laplace remarked that attention heightened the "vivacity" of the impressions under scrutiny at the expense of all others, magnifying both sensitivity to, and intensity of, similar impression.

More important from the standpoint of reckoning probabilities was the psychological tendency to reinforce or even create frequently repeated "modifications of the sensorium." Belief was such a modification, one which often persisted long after the proofs for it—"founded on evidence, the testimony of the senses, or probabilities"—had been forgotten. The psychological forces of habit, sympathy, and pleasure seeking conspired to make credulity irresistible. Laplace contended that this was the psychological explanation for Pascal's insight that the infidel could come to faith simply by going through the external motions of the devout and quoted Pascal's argument at length: "How few things are demonstrated? Proofs convince only the mind; custom makes our strongest proofs. Who has demonstrated that a new day will dawn tomorrow or that we die? And what is more universally believed?"[88]

Habit, interests, and taste both created and maintained these modifications of belief, entrenching them (perhaps literally) in the sensorium by frequent repetition. Consequently, false estimations of probabilities were common: "Accustomed to judge and to comport ourselves according to a certain class of probabilities, we give these probabilities our assent, and they determine us with more force than probabilities which are substantially greater, [being] the results of reflection or of calculation."[89] The passions exaggerated mental estimations of probabilities by enhancing the impression of the feared or desired object: "Its image, deeply etched in the sensorium, weakens the impression of the contrary probabilities."[90] The accurate estimation of probabilities required a calm mind unruffled by strong emotions. Laplace proposed to fight fire with fire: supplant preju-

---

[87] Condillac, *Essai*, p. 67.

[88] Laplace, *Essai*, p. cxxxiv.

[89] Laplace, *Essai*, p. cxxxvi.

[90] Laplace, *Essai*, p. cxxxvii.

dices rigidified in habit with new habits; appeal to the imagination in order to erase the deceptive vivacity of emotionally colored probabilities by envisioning the true magnitude of probabilities as lines of different lengths.

Despite his belief that the aberrations introduced by habit, prejudice, and interest into the associationist machinery of the mind were corrigible, Laplace's emphasis on probabilistic illusion and his preference for a physical or statistical interpretation of probabilities greatly influenced his disciple Poisson. In his major work on probability, *Recherches sur la probabilité des jugements* (1837), Poisson drew a sharp distinction between the objective and subjective aspects of probability. He maintained that while the words "chance" and "probability" were generally used as synonyms, it was essential to differentiate between them: "In this work, the word chance will refer to events in themselves, independent of our knowledge of them, and we will keep the word probability . . . for the reason we have to believe" that an event will occur.[91] Poisson did not reject the latter, epistemic sense of probability, but he scrupulously distinguished between objective chance and subjective probability throughout his exposition. Although he still worked within the framework of classical probability, Poisson no longer accepted the equivalence of objective and subjective probabilities as self-evident.

Significantly, Poisson also questioned the associationist theories that had consolidated that equivalence. Although Poisson remained faithful to the classical assumption that mathematical probability and the psychology of rational belief were two sides of the same coin, he questioned the particular psychological theory that had connected the two since Locke's speculations on the association of ideas. In some cases, at least, reasonable men did not apportion belief to the frequencies of past events. Poisson cited the example of scientists who endorsed experimental results after only a few or even a single trial. Indeed, Poisson implied that the associationist tallies hypothesized by Hume held only for habitual, as opposed to reasonable, expectations and applied only to the unenlightened masses. Scientists, reasonable men par excellence, did not asssociate cause and effect by sheer repetition, in the way that words and objects become mentally linked.[92]

---

[91] Poisson, *Recherches*, p. 31.
[92] Poisson, *Recherches*, p. 162.

Poisson had evidently taken to heart Laplace's warnings against illusory probabilities created by habits born of convention. With Poisson, mental associations had completed the transition from being the safeguard of rational belief, proportioned to experientially derived probabilities, to becoming the source of distortions that estranged belief and true probability.

Although Poisson distinguished between objective and subjective aspects of probability, he did not reject the subjective version. His work on inverse probabilities, described in the next chapter, employed many of the epistemological assumptions invoked by Bayes, Laplace, and Condorcet, although Poisson was more acutely aware of their limitations. Cournot, perhaps better known as a philosopher and political economist than as a probabilist, was among the first to break completely with the classical tradition. Along with the British critics of the subjective aspects of the classical interpretation, notably Ellis and George Boole,[93] Cournot asserted in his *Exposition de la théorie des chances et les Probabilités* (1843) that mathematical probability was not a measure of belief or even the "reason to believe," but "the measure of physical possibility."[94]

Cournot insisted upon Poisson's distinction between "probability" and "chance," between "a measure of our knowledge and . . . a measure of the possibility of things, independently of our knowledge," in the hopes that its scrupulous application would exonerate probability theory from the criticisms that even "good minds" had leveled against it.[95] Cournot emphasized the relation between mathematical probability and the nature of things, which he termed "possibility," or "the existence of a relation which subsists among things themselves, and which has nothing to do with our manner of judging or feeling, which varies from one individual to another." Subjective "probability," on the other hand, referred to just these individual judgments and was responsible for a "crowd of equivoca-

---

[93] George Boole, *An Investigation of the Laws of Thought* (1854) (New York: Dover, 1958); R. L. Ellis, "On the foundations of the theory of probabilities," *Transactions of the Cambridge Philosophical Society* 8 (1849): 1–6.

[94] Cournot, *Exposition*, p. 81. Cournot claimed to have arrived at the distinction between objective "chances" and subjective "probabilities" independently of Poisson, and cited a letter (p. vi) from Poisson of 26 January 1836, which affirmed the equivalence and importance of their joint distinction.

[95] Cournot, *Exposition*, p. iv.

tions," having "falsified the idea which one should have of the theory of chances and of mathematical probabilities."[96]

4.4 Conclusion

Cournot's work marks the advent of a new interpretation of mathematical probability exclusively in terms of objective frequencies. Cournot was also among the first to recognize that the classical approach to probability had been an *interpretation*, distinct from mathematical probability per se. It is beyond the scope of this book to examine Cournot's alternative formulation and its ramifications. However, Cournot's critique of the classical interpretation did establish the terms by which the theory has been evaluated ever since, as a wholly subjective view of probabilities. Cournot posed as the central question of his own treatise whether probability theory "was only a *jeu d'esprit*, a curious speculation, or whether it on the contrary studies very important and very general laws which govern the real world."[97] Once judgment and belief had been severed from the real world, interpretations of probability theory had to be classified into one of two mutually exclusive categories, as "subjective" or "objective." Classical probability theory, because of its explicitly epistemic orientation, has been assigned to the former category by writers since Cournot, who have dismissed the objective elements within the work of the classical probabilists as "equivocal" or "inconsistent."

Indeed, without the psychological theories that bound objective and subjective probabilities together, the classical probabilists do appear to vacillate. The doctrine of the association of ideas smoothed transitions from one sense of probability to another by providing a psychological basis for their equivalence. Not only philosophers like Hartley and Hume but mathematicians like Condorcet and Laplace treated associationist psychology and mathematical probability as kindred topics. This alliance between psychological and mathematical theories was a natural consequence of the central objective of classical probability theory: to provide a mathematical model for a certain kind of psychological process, of rational decision under uncertainty.

So long as associationist mechanisms, both psychological and physio-

[96] Cournot, *Exposition*, p. 82.   [97] Cournot, *Exposition*, p. 70.

logical, guaranteed the proportioning of belief to experience, and to experience conceived in ever more quantitative terms, so long could mathematicians conflate subjective and objective probabilities in useful and interesting ways. But what psychology giveth, psychology taketh away: by the early nineteenth century the associationists had supplied the mathematicians with their best arguments for severing subjective and objective probabilities, and, ironically enough, for severing associationist psychology from mathematical probability theory.

CHAPTER FIVE

The Probability of Causes

5.1 Introduction

Natural philosophers of the late seventeenth and eighteenth centuries described their methods in terms of two reciprocal processes: reasoning from causes to effects, and from effects to causes. The former was the surer and therefore the preferable route, corresponding to the synthetic method in geometry. Rationalists like Descartes hoped to deduce the phenomena from the fundamental laws and structure of nature as theorems were deduced from the axioms and definitions of geometry, or, to borrow a then popular analogy, as a watchmaker might "deduce" the workings of a clock from its internal mechanism of springs or weights. But in most cases the watchcase was sealed shut, and natural philosophers were forced to reason in reverse—from the movements of the hands to the "hidden springs and principles" within, from manifest effects to hidden causes.

Isaac Newton sometimes spoke of "deducing" explanations from the phenomena, but almost all of his contemporaries (and on other occasions, Newton himself) took a more guarded view of the possibility of achieving such certainty by means of this second, empirical method. Empirical inferences were irreparably uncertain, for they could not exclude alternative hypotheses that also explained the observations at hand. Watches run by spring movements had faces identical to those driven by weights. Even René Descartes admitted at the end of his *Principia philosophia* that mechanical hypothesis could only aspire to a very high probability, despite the many phenomena it could account for.[1] As Robert Boyle put it in one of his dialogues:

---

[1] René Descartes, *Principia philosophia* (1644), Art. CCV in *Oeuvres de Descartes*, Charles Adam and Paul Tannery, eds. (Paris: J. Vrin, 1964), vol. 8.1, pp. 327–328.

Perhaps you will wonder, Pyrophilus, that in almost every one of the following essays I should speak so doubtingly, and use so often, perhaps, it seems, it is not improbable, and such other expressions, as argue a diffidence of the truth of the opinions I incline to, and that I should be so shy of laying down principles, and sometimes of so much as venturing at explications. But I must freely confess to you, Pyrophilus, that I have met with many things of which I could give myself no one probable cause, and some things, of which several causes may be assigned so differing, as not to agree in any thing, unless in their being all of them probable enough.[2]

Thus scientific method entered the realm of probabilities and became a topic for the mathematicians. Natural philosophers, like the learned judge and the prudent merchant, were more often than not forced to decide under uncertainty, for human faculties were too limited and nature too devious to penetrate directly to what Locke called the "real essence" of things except in rare instances. Classical probabilists from Jakob Bernoulli through Siméon-Denis Poisson sought to quantify Boyle's "probable enough," and, more broadly, to provide a mathematical model for how this special group of reasonable men came to reliable conclusions on the basis of incomplete evidence. Empirical reasoning presented two major difficulties for its practitioners: first, by itself the method did not single out any unique hypothesis; and second, it provided no firm grounds for extrapolating from an inevitably circumscribed set of observations to a generalization about all cases, everywhere and always. Mathematicians hoped to solve both problems by making such methods precise. If the probabilities of competing hypotheses could be measured, they could also be compared, not only with one another but with some absolute standard of certainty, like Bernoulli's threshold of moral certainty.

Bernoulli's and Bayes' theorems were the chief mathematical results of the probability of causes, but the theorems alone were not sufficient to constitute such an application. In order to extend the mathematical theory of probability to empirical inference further assumptions, constraints, and

---

[2] Robert Boyle, *Some Considerations Touching Experimental Philosophy in General* (1663; 2nd part, 1671), in *The Works of the Honourable Robert Boyle* (London, 1772), vol. 1, p. 307. See also Barbara J. Shapiro, *Probability and Certainty in Seventeenth-Century England*, chapter 2.

analogies were required. Associationism supplied the mechanism that translated objective experience into subjective belief, but objective experience came in several varieties, not all of them readily quantifiable. In John Locke's example, the King of Siam was misled into denying the possibility of ice by trusting exclusively to the bare frequency of experience, without considering its breadth: centuries of observations confirming that water never froze, but all of them made in the tropics. Bacon's "shining instances" were single observations or experiments that displayed a phenomenon with such clarity that no repetitions were needed. The evidentiary force of eye-witness reports that the accused had been spotted fleeing the scene of the murder, bloody weapon in hand, did not derive from how many times such circumstances had led to convictions in the past. Imponderables like the astronomer's bad head cold or choppy air the night of the comet's perihelion were almost as important in evaluating an observation's reliability as the calibration of the chronometer.

These qualitative considerations were all impeccably empirical, but since they seemed mathematically intractable, the probabilists ignored them. Over the course of the eighteenth century, empiricist philosophers also shifted their emphasis from the qualitative to the quantitative aspects of experience. The result was a theory of empirical inference that applied only to events that were virtually identical, and that occurred (or failed to) over and over again. For philosophers and probabilists alike, induction meant summing these units of past experience into an expectation about the future. They debated whether such expectations were rational, but not over the summation process itself or the constraints on the kind of experience needed to make it work.

In general, the probabilists believed that inductive inferences were rational as well as useful. Indeed, even those who had read David Hume's arguments on causality, like Richard Price, argued that the probability of causes was the best refutation to such skeptical doubts. Neither Jakob Bernoulli nor Abraham De Moivre questioned the validity of inductive reasoning; nor did they distinguish between the mundane inference that leads us to believe that this loaf of bread will be nourishing because others like it have been, and the natural philosopher's conviction that a newly discovered planet will obey the law of gravitation because all other massive bodies have been observed to do so. The mathematicians endorsed the "common sense of mankind" that reasoned from past to future.

However, they did not trust common sense to judge accurately the rate at which the probability of a conjecture increased with the number of confirming instances, nor to single out the soundest conjecture from the pack of contenders. Only quantitative arguments could help the bewildered empiricist here. Bernoulli acknowledged that "even the stupidest man knows by some instinct of nature per se and by no previous instruction" that the greater the number of affirmative observations, the surer the conjecture, but this did not make his theorem about how many observations corresponded to what degree of security superfluous.[3] Similarly, De Moivre admitted that his argument from observations to causes "being level to the lowest understanding, and falling in with the common sense of mankind, needed no formal demonstration, but for the scholastic subtleties with which it may be perplexed; and for the abuse of certain words and phrases, which are sometimes imagined to have a meaning merely because they are often uttered." Mathematical probability would hone common sense to numerical precision, and codify induction "if not to force the Assent of others by strict Demonstration, at least to the Satisfaction of the Enquirer himself."[4]

However, even assumptions about the link between objective and subjective probabilities and constraints on the kind of objective experience that counted did not suffice to make the probability of causes a part of the mathematicians' "art of conjecture." Still further assumptions about the nature of chance and causation were required. Bernoulli's and Bayes' theorems specify the relationship between the observed frequency of an event and its underlying probability; the probability of causes was to be a mathematical description of reasoning from effects to causes. In order to equate causes with underlying probabilities, the probabilists had to create a new model of causation, made concrete in the ubiquitous closed urn filled with colored balls that unites almost all applications of the classical theory. Applications as diverse as the probability of causes and the probability of judgments stood or fell on the strength of the perceived analogy between the phenomena and the urn model. Since analogies can be ephemeral, it is possible for a mathematical theory to lose as well as to gain a domain of application, and this is indeed what happened in both these cases in the mid-nineteenth century. Just because such applications are thus historical

---

[3] Jakob Bernoulli to Leibniz, 3 October 1703, in *G. W. Leibniz: Mathematische Schriften*, C. I. Gerhardt, ed. (Hildesheim: Georg Olms, 1962; reprint of 1855 edition), vol. 3, part I, pp. 77–78.

[4] Abraham De Moivre, *The Doctrine of Chances*, 3rd edition (London, 1756), p. 254.

in character, it is particularly important to understand the context that made them plausible in the first place. In this chapter I describe the classical approach to the probability of causes within that context: Section 5.2 explains Bernoulli's own interpretation of his theorem and the introduction of the urn model of causation; Section 5.3 examines the impact of Bayes' theorem on eighteenth-century discussions of the problem of induction; and Section 5.4 describes how M.J.A.N. Condorcet, Pierre Simon Laplace, and Siméon-Denis Poisson attempted to mathematize scientific reasoning.

5.2 Bernoulli's Theorem and the Urn Model of Causation

In his introductory "Histoire" to Laplace's 1780 "Mémoire sur les probabilités," Condorcet reduced all problems in probability theory to a single model:

> All questions in the Calculus of Probabilities can be reduced to a single hypothesis, that of a certain quantity of balls of different colors mixed together, from which different balls are drawn by hazard in a certain order or in certain proportions.[5]

This urn model suggested two types of questions: if the number of balls of each type is known a priori, to determine the probable results of the drawings; or, if only the results of the drawings are known, to determine the probable mixture of balls. Condorcet observed that the first type of problem, that of direct probabilities, had been the sole concern of probabilists until recently, but that the recent work of Thomas Bayes, Price, and especially Laplace on the second class of inverse probabilities had inaugurated a new era in mathematical probability. Inverse probabilities extended the theory to a new class of applications "far more extensive and useful than those of the ordinary [direct] calculus; in fact, all of our physical and moral knowledge reduces to probabilities of this type." Yet despite the novelty and generality of the method of inverse probabilities, the urn filled with colored balls still provided an adequate model of the prob-

[5] M.-J.-A.-N. Condorcet, "Histoire: Sur les probabilités," *Histoire et mémoires de l'Académie royale des Sciences. Année 1778* (Paris, 1781), pp. 43, 45–46.

ability of causes, inverse as well as direct, in Condorcet's opinion. This section describes the genesis of the urn model of causation through the attempts of classical probabilists to create a mathematics of induction.

Jakob Bernoulli was well aware of the significance of the theorem that bears his name. Declaring that he had pondered over the problem for twenty years, Bernoulli advised his readers that its great novelty, utility, and difficulty made it "exceed in weight and value all the remaining chapters of this thesis." Bernoulli's successors seconded that judgment. Bernoulli's theorem represents the first mathematical attempt to relate probabilities to observed frequencies of events where equipossible cases could not be plausibly posited: what assurance do we have that the observed frequency of outcomes will closely approximate the underlying probabilities, and how much does this assurance improve with an ever greater number of observations or trials? Will repeated trials indefinitely increase the probability that the observed ratio of outcomes equals the "true" ratio "so that this probability finally exceeds any given degree of certainty," or is the probability bounded by "its own asymptote . . . some degree of certainty is given which one can never exceed" no matter how many observations we accumulate?[6] In modern notation, Bernoulli showed that

$$\lim_{n \to \infty} P(|p - m/n| < \epsilon) = 1, \text{ for } \epsilon \text{ as small as desired,}$$

$$\text{where } p = \text{"true" ratio}$$
$$n = \text{number of trials}$$
$$m/n = \text{observed ratio.}$$

Bernoulli's theorem opened up a whole new realm of problems for classical probabilists: the art of conjecture need no longer be restricted to games of chance or to the few other situations where equipossible outcomes could be assumed without doing violence to actual circumstances. Mathematical conjectures about far more complex and interesting situations like human diseases and the weather became possible. Moreover, Bernoulli also drew the attention of mathematicians to the relationship between probabilistic conjecture and inductive reasoning. The early probabilists had enlisted assumptions about physical symmetry and equity to justify equipossible outcomes, and had been unconcerned, with the exception of Cardano, with testing the extent to which these a priori probabilities were borne

---

[6] Jakob Bernoulli, *Ars conjectandi* (1713), in *Die Werke von Jakob Bernoulli*, Basel Naturforschende Gesellschaft (Basel: Birkhäuser Verlag, 1975), vol. 3, pp. 249–250.

out by a posteriori observations concerning, for example, the behavior of dice. Bernoulli, perhaps stimulated by the work of the political arithmeticians John Graunt and William Petty, sought a mathematical link between a priori probabilities and a posteriori observations. Because Bernoulli and his eighteenth-century successors equated the a priori probabilities with causes and the a posteriori observations with effects, his theorem became a tool for discovering the probability of causes from effects.

The significance of Bernoulli's theorem, particularly with respect to statistical inference, is the subject of a large literature. My intention here is not to contribute to that ongoing discussion, but rather to reconstruct Bernoulli's own understanding of this theorem and of a posteriori probabilities. However, there is one current issue concerning statistical inference that can legitimately be traced back to Bernoulli's theorem. Because this problem is central to classical, as well as to modern, discussions of a posteriori probabilities, it may be best to clarify it at the outset.

Bernoulli's theorem assumes that the a priori probability p is known. His computations seek the number of observations required to make the probability that m/n will fall within the specified interval $[p - \epsilon, p + \epsilon]$ 10, 100, 1,000, etc., times more likely than the probability that it will not. To make such calculations of n, Bernoulli required p and ϵ. A typical computation seeks the number of trials necessary to insure a .999 probability that the observed ratio falls between 31/50 and 29/50, given that $p = 3/5$. However, occasionally Bernoulli appears to have wanted to apply his theorem (illegitimately) to the inverse problem: given an observed ratio of m/n, with what probability can we conjecture that m/n equals an unknown p (instead of computing the probability that p lies in the interval $m/n \pm \epsilon$, given m/n)? The latter problem is generally known as that of statistical inference, described by De Moivre as "the hardest problem that can be proposed on the subject of chance." Bernoulli's theorem addressed the former question, although it suggests a way of roughly answering the latter by testing the probability that p lies within $[m/n - \epsilon, m/n + \epsilon]$ for various values of p. But it provides no direct way of finding a unique, unknown p given m/n for finite n.

Yet Bernoulli frequently wrote as if his theorem did supply a method. Recapitulating the topics to be covered in the unfinished fourth section of his planned treatise *Ars conjectandi* in a letter to Gottfried Wilhelm Leibniz, dated 3 October 1703, Bernoulli sketched his investigations regarding "whether what is hidden from us by chance a priori can at least be known by us a posteriori." Bernoulli's example makes it clear that he hoped to determine unknown probabilities from observations. He contrasted two situations: how much more likely is it that a seven rather than an eight will result from tossing two dice; and how much more likely is it that a young man of twenty years will survive an old one of sixty? In the first case, Bernoulli observed that since we know the possible combinations that would produce a seven or an eight, and the ratio of these to the total number of possible outcomes, we can determine the probabilities a priori.

However, neither the number of "fertile" nor total combinations can be so easily ascertained in most situations "which depend upon either the work of nature or the judgment of men." In the second case, there exist innumerable diseases that beset the human body, and in unknown proportion of deadliness. Bernoulli despaired of deducing the probabilities of, say, dying from dropsy instead of from the plague from the internal composition of human bodies, in the way that the probabilities of throwing a seven instead of an eight might be inferred from "the similarity of sides and the balanced weight of the die." Consequently, he turned to "experiments" or trials that might reveal the hidden probabilities empirically, in a way not fully justified by the results of his theorem. Returning to the example of the old and young men, Bernoulli claimed to have demonstrated the following sort of inference "accurately and geometrically":

> For had I observed it to have happened that a young man outlived his respective old man in one thousand cases, for example, and to have happened otherwise only five hundred times, I could safely enough conclude that is twice as probable that a young man outlives an old man as it is that the latter outlives the former.[7]

That Bernoulli's fellow-probabilists, beginning with Leibniz and continuing in the work of De Moivre, Bayes, Laplace, and Condorcet, were more dubious of this kind of claim is shown by their persistent attempt to prove that this reasoning was mathematically valid, despite Bernoulli's assertion that he had already done so. Their efforts met with equivocal success: probabilists continue to contest the solution put forth by Bayes and Laplace.[8] A.A. Cournot, writing in 1843, supported Bernoulli's

---

[7] Jakob Bernoulli to Leibniz, 3 October 1703, in Gerhardt, ed., [Leibniz] *Schriften*, vol. 3, part I, p. 77.

[8] See Max Black, "Probability," in *Encyclopedia of Philosophy*, Paul Edwards, editor-in-chief (New York: Macmillan & The Free Press, 1967), vol. 6, pp. 464–479.

contention that his theorem was sufficient to discover probabilities a posteriori, arguing that as m and n grew very large, Bayes' theorem approached Bernoulli's: "It is not, as many authors seem to assert, in this case that Bernoulli's rule becomes exact in approaching that of Bayes; it is Bayes' rule which becomes exact, or rather assumes an objective value, in approaching Bernoulli's rule."[9]

Some historians, notably Ian Hacking, have exonerated Bernoulli from any such equivocation between his theorem and statistical inference, on the grounds that if Bernoulli had truly meant his theorem as a tool for statistical inference in particular instances (rather than simply as a good approximation in the long run), "then it is plausible to urge that he should have followed the path discovered by Bayes."[10] However, this negative evidence is unconvincing. If Bernoulli did not pursue the line of reasoning which led Bayes to his theorem and yet clearly hoped to turn his own results to similar purposes, it was for the same reasons that classical probabilists continued to worry about the problem of a posteriori probabilities in the wake of multiple demonstrations designed to lay the matter to rest. Bernoulli's theorem and his interesting misapplication of that theorem raised issues central to the eighteenth-century understanding of causes, probabilities, and inductive reasoning from effects to cause.

Before proceeding to these issues, a discussion of the mathematical problems is in order. Bernoulli's theorem, the first "stepping stone between the theories of probability and statistics,"[11] provides a natural departure point. Although earlier writers, such as Gerolamo Cardano, Edmund Halley, and the author of the last chapters of the Port Royal *Logique*, had appealed to the principle that there was an approximate fit between observed frequencies and "true" probabilities which improved as the number of observations increased, Bernoulli was the first to attempt a mathematical treatment of this common-sensical precept. Bernoulli further sought the rate at which the probability that the observed ratio approxi-

[9] Antoine Augustin Cournot, *Exposition de la théorie des chances et des probabilités* (Paris, 1843), p. 166.

[10] Ian Hacking, "Jacques Bernoulli's *Art of Conjecturing*," *British Journal for the Philosophy of Science* 22 (1971): 209–229, on p. 225.

[11] O. B. Sheynin, "On the early history of the law of large numbers," in *Studies in the History of Statistics and Probability*, E. S. Pearson and M. G. Kendall, eds. (Darien, Conn.: Hafner, 1970), pp. 231–239, on p. 236.

mates the true probability to a given accuracy increases with the number of observations. Ultimately, he hoped to specify the conditions by which one might compute the number of trials necessary to insure a probability at least equal to moral certainty, which he had earlier presumed to be "amply sufficient in civil life, where what is morally certain is considered absolutely certain in order to form our conjectures in any situation that may arise."[12]

Bernoulli's proof drew upon his work on infinite series (a treatise on infinite series was appended to the *Ars conjectandi*) and the binomial expansion. In Part II of the *Ars conjectandi*, Bernoulli had set forth the principles of permutations and combinations, linked with mathematical probability from the outset of the theory and an area of active mathematical research in its own right during the seventeenth century.[13] Bernoulli admitted that the bulk of his results had been borrowed from the works of F. van Schooten, John Wallis, Leibniz, and Jean Prestet, reserving credit for a theorem on the sum of a series of figurate numbers to himself. Although much of Chapter 3 of Part II of *Ars conjectandi* duplicated the results of Pascal's *Traité du triangle arithmétique* (1665), Bernoulli appears not to have known of Pascal's work. However, Bernoulli produced a "table of combinations" that is essentially the arithmetic triangle, and proceeded to investigate its "truly curious and surprising" properties, including the familiar derivation of the coefficients of the binomial expansion. By identifying the terms of a binomial expansion,

$$(r + s)^m, t = r + s,$$

with the probabilities (when divided by nt) of a given ratio of successful and failed trials, Bernoulli was able to recast his problem as one of estimating the proportion of the central term of the expansion, and the $2n$ terms surrounding it, to the sum of all the terms of the expansion. Once Bernoulli could show that (a) the middle term was the largest in the expansion, (b) the ratio of the middle term to any other term a given distance l terms away was always smaller than the ratio of this second term to a term a further l terms away; and (c) the exponent n can be so chosen to

---

[12] Jakob Bernoulli, *Ars conjectandi*, p. 243.

[13] See Isaac Todhunter, *A History of the Mathematical Theory of Probability from the Time of Pascal to that of Laplace* (New York: Chelsea Publishing Company, 1949; reprint of 1865 edition), chapter 4.

make the ratio of the sum of the middle term and the n preceding and n following terms to the sum of all remaining terms away larger than any given ratio; then his task became one of approximating this ratio.[14]

Bernoulli's method of approximating by inequalities the required number of observations nt, such that the observed number of successes m/nt was c times more likely to fall within the interval

$$\left[\frac{r-1}{t}, \frac{r+1}{t}\right]$$

corresponding to the $nr - n$ and $nr + n$ terms around the middle term

$$\binom{nt}{ns} r^{nr} s^{ns}$$

was inspired by Archimedes's approximation of π, and designed to give conservative answers. Yet however overgenerous Bernoulli's tolerance levels for the required number of observations may have been, they did offer a rough guide to how the observed proportion of "fertile" to "sterile" cases, in Bernoulli's terminology, ever more closely approximated the true proportion of their respective probabilities. The computations depended on knowing the prior probabilities r and s, as well as the desired probability c of reliable observations.

Yet Bernoulli himself did not hesitate to invert his method, without proof or further justification, to find the probability with which r and s could be inferred from the *observed* ratio of fertile to sterile cases. Apparently, Bernoulli viewed this type of reasoning from effects to causes as fully justified, and not simply on the grounds of expediency. Bernoulli had recognized—and rejected as insufficiently supported by "accurate and geometrical" argument—the natural and eminently useful human instinct to trust the lessons of experience in proportion to the extent of that experience. Although Bernoulli hoped that his results would illuminate the daily decisions of civil and economic life, he explicitly chose not to base his arguments on common practice and manifest utility. In the final chapters of *Ars conjectandi*, Bernoulli sought a mathematically sound, quantitatively refined rendition of common-sense reasoning. He would thus not have intentionally doubled back on his argument to make universal practice the basis for reversing the direction of inference, that is, inference

---

[14] For a discussion of Bernoulli's method of approximation and its limitations, see Karl Pearson, "James Bernoulli's theorem," *Biometrika* 17 (1925): 201–210.

from known probabilities to unknown observed ratios to that from known observed ratios to unknown probabilities.

In order to explain Bernoulli's confidence in this inverse application of his theorem, we must probe more deeply into Bernoulli's conception of the relationship between causes, probabilities, observed effects, and the hidden nature of things. Fortunately for historians, Leibniz took Bernoulli to task on many of these points in their correspondence of 1703–1704, forcing the latter to clarify the assumptions which in his opinion legitimated the mathematics of a posteriori probabilities. These assumptions were largely metaphysical in nature, bearing on the questions of uniformity and simplicity of nature. By comparing Bernoulli's responses to Leibniz's challenges with the other reflections on probability and causation scattered throughout the *Ars conjectandi*, it is possible to reconstruct the metaphysics that prompted and sanctioned the partnership of classical probability theory and the philosophy of induction.

In a letter dated 3 December 1703, Leibniz pressed Bernoulli on the latter's use of a posteriori probabilities.[15] Leibniz's argument against Bernoulli's brand of reasoning from past observation was twofold: first, in cases as complex as the diseases which beset the human body, the causes might evolve with time, discrediting earlier observation; and second, the number of human diseases could conceivably be infinite, thus rendering comparative ratios of the sort postulated by Bernoulli meaningless. Leibniz affirmed the general constancy of nature, who "has her own habits," but did not rule out deviations from past rules in particular cases "because of the very mutabilities of things." Moreover, even if the causes and their ratios remained fixed for all time, the original number might be so vast, even infinite, as to elude the grasp of human understanding, at least insofar as computing ratios of possible combinations, and therefore the calculus of probabilities, were concerned.

Leibniz's criticisms implicitly questioned the adequacy of Bernoulli's urn model of probability and causation. Since urn models have since become a staple of the literature on probability, a stock way of conceptualizing more intricate problems involving chance, it is important to note that Bernoulli's use of the now familiar urn example to model the relation between underlying causes and observed effects was perhaps the first quantitative attempt to construe a chance mechanism metaphorically. Hereto-

---

[15] See n. 3, above.

fore, probabilists had treated lotteries, dice games, and coin tosses at the immediate level of practical problems, not as analogues for more general processes in nature. Bernoulli's appropriation of the urn example to describe the processes linking inaccessible causes to observed effects expanded not only the domain of problems upon which probabilists might test their skills, but also the conceptual tools for extending the range of the theory's applications still further. By likening situations as disparate as the diseases that afflict the human body and the decisions reached by a tribunal to drawing black and white balls at random from an urn, probabilists hoped to free their theory from its preoccupation with gambling puzzles. Bernoulli's urn model of causation set the pattern for other applications of classical probability theory.

The salient features of Bernoulli's urn model were few and easy to grasp. In the simplest case, an urn is filled with a certain number of black and white balls, from which drawings are made at random and with replacement. Any one ball is assumed to be as equally likely to be drawn as any other. If the ratio of black to white balls is known a priori, Bernoulli's theorem specifies the number of trials necessary to insure that the observed ratio of black-to-white balls drawn falls within a certain margin of the true ratio with a given probability. Here, the true ratio of black-to-white balls corresponds to the "cause"; the actual ratio manifested in the drawings, to the "observed effect." Bernoulli's theorem quantified the reliability of the relation between the two, depending on the number of trials. In the case of human diseases, Bernoulli compared the human body to a "tinder box" filled with diseases, combined in a determinate ratio like balls of various colors (each color corresponding to a different disease) which was related to the observed prevalence of dropsy, plague, and so on, as if the diseases were being grimly "drawn" at random from their opaque receptacle. Bernoulli speculated that the same model might be applied to physical phenomena (e.g., the weather) and civil life alike.

Leibniz attacked the urn model as an oversimplification. What if the mix of balls contained in the urn changed with time? What if the number of balls were infinite? What if no determinate ratio existed among the various types of balls? Bernoulli rejected the last possibility out of hand: meteorological perturbations and human diseases could only appear indeterminate with respect to limited human knowledge—God created only determinate entities. Moreover, Bernoulli doubted that the number of diseases was actually infinite, but even if it were, he insisted that this would not rule out a mathematical treatment of the sort he had advanced. Citing Archimedes's determination of π though a process of ever-narrowing inequalities as his precedent, Bernoulli argued that an infinite number (presumably here a number with an infinite expansion like π) might be reliably approximated. Similarly, he claimed that the ratio of two infinite numbers might be determined precisely enough by finite trials to be expressed by finite values.

Leibniz further remonstrated that Bernoulli's a posteriori estimation of probabilities depended on an infinite number of cases, although only a finite number of experiments will ever be at our disposal. Comparing the observed trials to a given set of points and the "fixed underlying outcomes" to the curve passing through these points, Leibniz pointed out that for any finite number of points, an infinite number of curves can be found to pass through all of them. Analogously, the inferred a posteriori probability was only one of an infinite number of estimates that fell within the margin defined by Bernoulli's theorem. Leibniz tempered his criticism by acknowledging that although a "perfect estimation" must elude such a posteriori reasoning, "an empirical estimate would nonetheless be useful and sufficient in practice."

Leibniz's objection struck at the heart of Bernoulli's strangely optimistic claim that his theorem could be used to discover a priori probabilities from observation. Imagine an urn containing black and white balls in an unknown ratio. After fifty drawings, with replacement, we record twenty-three black balls and twenty-seven white ones. Applied to this example, Bernoulli's theorem would tell us the probability of this observed ratio for fifty trials *if* the urn was indeed half filled with black balls and half with white. However, it could also yield the probability of an observed ratio of twenty-three black: twenty-seven white for an urn presumed to contain that proportion of black-to-white balls—or for that matter, for any proportion. How was one to choose among all possible estimates of the a priori probability? Bayes' theorem suggested a way of comparing these conditional probabilities which remains controversial to this day, but Bernoulli evidently did not believe that any such special method was needed. Bernoulli assumed that the "simplest" p—in this case, $1/2$—would also be the best estimate. Once a p had been selected, an ever-increasing number of trials n could test the probability that the observed ratio m/n would fall within a given interval around this value p.

Bernoulli defended his simplicity assumption to Leibniz on metaphysical grounds, admitting that all curve fitting, as in the determination of a cometary trajectory from a sprinkling of observations, "would be quite flimsy and uncertain if it were not conceded that the curve sought is one of the class of simple curves." But, Bernoulli insisted, "this indeed seems quite correct to me, since we see everywhere that nature follows the simplest paths."

As part of the mathematical theory of probability, Bernoulli's theorem is irrefragable. However, as a model for the relationship between underlying causes and observed effects, it rests upon a number of interpretive assumptions. Probabilists still dispute the verisimilitude of these assumptions and the difficulty of assuming convergence for a finite sequence of relative frequencies, although most of these discussions hinge upon measures of irregularity or randomness unknown to Bernoulli.[16] Even for Bernoulli and his contemporaries, though, these assumptions were potentially controversial, as Bernoulli's exchange with Leibniz indicates. Three assumptions in particular merit close examination in the context of early eighteenth-century thought, both for their bearing on Bernoulli's theorem and for the light they shed on the connections between the natural philosophy, scientific methodology, and probability theory of the period: first, the plausibility of the urn model itself as an analogy for investigations into nature's "hidden causes"; second, the combinatorial mechanism of causes that insured the stochastic independence of successive events; and third, the simplicity assumption that skirted the thorny problem of convergence. All three assumptions were deeply imbedded in the actual natural philosophy of Bernoulli's time.

Even if Leibniz had accepted Bernoulli's defense of the urn model on the issues he had raised, there was much which a thoroughgoing rationalist might still presumably find objectionable about such a "black-box" approach to causation. Why posit a chance mechanism to generate phenomena that, as both Bernoulli and Leibniz believed, could ultimately be deduced from their proper causes? The rationalist model of causation commenced from simple principles and reasoned downward to effects: causation was essentially a deductive, not a combinatorial, connection. In contrast, the urn model suggested that the choice among any number of causes competing to be realized was based on the proportion of combina-

---

[16] For a recent survey of these issues, see Terence L. Fine, *Theories of Probability: An Examination of the Foundations* (New York and London: Academic Press, 1973), pp. 89–97.

tions, all equally likely, favoring but not determining one outcome rather than another. Bernoulli's theorem demonstrated that even such a seemingly haphazard jostling of causes eventually produced regular effects. But only if "all events were to be continued throughout eternity" would they exhibit "a certain necessity and, so to speak, a certain fate"; that is, a determinism compatible with the rationalist program both Bernoulli and Leibniz espoused, at least from the perspective of an omniscient deity.

The hierarchical explanations of the rationalists assumed that the ultimate causes appeared to the intellect as clear and distinct ideas, the departure point for all reasoning. Descartes' famous rules of reasoning advised readers to accept "nothing more than what presented itself so clearly and so distinctly to my mind that I might have no occasion to place it in doubt." From these translucent principles supplied by intuition, the sound thinker would follow the chain of deductive inference "by the continuous and uninterrupted action of a mind that has a clear vision of each step in the process."[17] Bernoulli's urn model, particularly as applied to the inverse use of his theorem, hid from view both the causes and the links that joined them to effects—a strange candidate indeed for a rationalist model of causation.

Bernoulli's urn model was a plausible simulation of causation only when seen against the background of an alternative philosophy of nature, one which envisioned causes as inaccessible, hidden from human view. Bernoulli, and, to a lesser extent, Leibniz, were able to reconcile this image of a secretive nature with their rationalist professions of faith by widening the gap between the divine intellect, which inferred predicate from subject in even contingent propositions of the "Caesar crossed the Rubicon" type, and merely human minds, which could only grope dimly toward the truths of natural and moral philosophy. Deduction and certain knowledge had become the increasingly exclusive province of the deity, while more feeble minds were relegated to induction and probability. Probability came consequently to be the method of the rationalist *manqué*, the mathematical backbone of induction. Bernoulli's urn model derived from the more empirical, pessimistic crosscurrents of seventeenth-century scientific methodology. Its philosophical lineage traces back to the works of Locke, Boyle, Newton, and even earlier to Francis Bacon, who sought fundamental causes and principles beneath the level of appearances.

The gap between sensation and reality assumed numerous forms: Bacon's search for the "latent processes" of nature that were "infinitely small, or at least too small to strike the senses";[18] Locke's distinction between the secondary qualities which constitute sensations, and the primary qualities, which act upon the organs of perception to produce sensations through the motion and configuration of "insensible particles";[19] Boyle's redefinition of the scholastic notion of "form" in terms of the motions and arrangement of homogeneous, invisible corpuscles.[20] While Bacon portrayed nature as devious or at least elusive, to be interrogated by experiment and hounded in her wanderings, he did accept notions derived from the immediate data of sense as at least as valid as those that presumed to penetrate nature more deeply: specifics like "man" and "dog" were less likely to mislead than "substance," "essence," or "quality."

Yet Bacon despaired of the ability of the unguided human intellect, encumbered as it was by the idols of the tribe, cave, market, and theater, to probe beyond sensation: ". . . the discoveries which have been hitherto made in the arts and sciences are such as might be made by practice, meditation, observation, argumentation,— for they lay near to the senses, and immediately beneath common notions, but before we can reach the remote and more hidden parts of nature, it is necessary that a more perfect use and application of the human mind and intellect be introduced."[21] Later writers carried Bacon's pessimistic assessment of unaided human reason still further, insisting that method could only palliate, not cure, these inherent deficiencies. By the mid-seventeenth century, this pessimism had hardened into the commonplace distinctions made by Boyle, John Wilkins, John Tillotson, and others between the certain, demonstrative knowledge that stemmed from intuition, confined largely to mathematics and a few metaphysical principles, and the tentative probabilistic knowledge that obtained in natural, and a fortiori, in moral philosophy.

The limitations of human reason and the way to overcome them are recurring themes in seventeenth-century writings on philosophy and sci-

---

[17] Descartes, *Discours de la méthode* (1637), Parts I and II, and *Regulae ad directionem ingenii* (1684 Dutch edition; 1701 Latin text), Rules 3 and 7, in *Oeuvres*, vol. 6, pp. 1–11, 22–31; vol. 10, pp. 366–370, 387–392.

[18] Francis Bacon, *New Organon* (1620), in *The Works of Francis Bacon*, James Spedding, Douglas Heath, and Robert Leslie Ellis, eds. (Boston, 1863), vol. 8, pp. 126–129.

[19] John Locke, *An Essay Concerning Human Understanding* (1689; all editions dated 1690), Alexander Campbell Fraser, ed. (New York: Dover, 1959), vol. 1, pp. 172–173.

[20] Boyle, *Origin of Forms and Qualities* (1666), in *Works*, vol. 3, pp. 15–17, 28, 35–36.

[21] Bacon, "The Great Instauration," in *Works*, vol. 8, p. 33.

ence. Although Descartes believed that good sense was more or less equally distributed among all men, he also preached the corrective of a rigorous method for the defects of human understanding. The Port Royal Logique prefaced its instructions in the "art of thinking" with a diagnosis of the causes of bad judgment and unsound reasoning: chief among them was mistaking probable relations for certain ones. Among Continental writers, Marin Mersenne and Pierre Gassendi were the chief spokesmen for the study of phenomenal relationships as the best to which natural philosophy might aspire. While both attacked the extreme skepticism of the pyrrhonists, they conceded that knowledge of things-in-themselves lay beyond the reach of human understanding. In Mersenne's lengthy rebuttal to the radical skeptics, De la vérité des sciences (1675), he cautioned against intellectual hubris: "It should not be thought that we can penetrate to the nature of individuals, nor to what passes internally in these, for our senses, without which our understanding can know nothing, see only what is external."[22] Gassendi's major work, Syntagma (1659), echoed Mersenne's pragmatic retort to the pyrrhonists: while the real nature of things is inaccessible to human reason, we are capable of indubitable knowledge about appearances. Natural philosophy dealt with appearances, not essences.[23]

Seventeenth-century proponents of the new science espoused induction as a necessary, if imperfect, method, and did their best to make a virtue of necessity. Newton advised readers of the Opticks that "although the arguing from Experiments and Observations by Induction be no Demonstration of general Conclusions; yet it is the best way of arguing which the Nature of Things admits of, and may be looked upon as so much the stronger, by how much the Induction is more general."[24] In Query 31 and in the Rules of Reasoning in the second edition of the Principia, Newton further explained how the imperfect methods of induction might be best exploited in natural philosophy. While Newton ascribed the impossibility of fully demonstrative science to the "Nature of Things" rather than to the inadequacies of the human intellect, his methodological conclusions fall squarely within the seventeenth-century tradition of making do with what one has.

---

[22] Marin Mersenne, De la vérité des sciences (Paris, 1625), p. 212.

[23] See Richard Popkin, The History of Scepticism from Erasmus to Descartes (Assen: Van Gorcum, 1964), pp. 132–146.

[24] Isaac Newton, Opticks, based on the 4th edition (1730) (New York: Dover, 1952), p. 404.

Whether writers chose to stress the opacity of nature or the obtuseness of human reason or the bluntness of human senses to some extent did influence the remedy prescribed. Philosophers primarily concerned with human frailty, like Locke and the Port Royal authors, dwelled on degrees of probability and the commensurate proportioning of belief to evidence; those preoccupied with the hidden realms of nature paid more attention to techniques of induction. But the differences between the two orientations could easily be exaggerated. Since both groups believed that nature was objectively determined from a divine standpoint, the issue was reduced to an epistemological one: how to investigate the hidden deep-structure of nature with the maximum certainty.

Newton's celebrated Rules of Reasoning might be interpreted as so many bridges set down between immediate experience and remote reality. Nature's economy, simplicity, and homogeneity validated generalizations from the properties of macroscopic to those of microscopic bodies. Newton illustrated Rule III, regarding the "universal qualities of all bodies whatsoever," with an argument as to how the hardness, impenetrability, mobility, and inertia of the whole implied similar attributes for the parts. Rule II and Query 31 of the *Opticks* drew equally important methodological conclusions from the precept of nature's uniformity. Gravity could be compared to interparticulate forces on the corpuscular level because nature could be presumed "very consonant and conformable to herself," endlessly repeating the same motifs in all her works. The natural philosopher who understood one facet of nature therefore possessed a key to all others. Induction progressed horizontally, by collecting instances of the same type into an eventual generalization; analogy worked vertically, extrapolating the results of inductions at one level to the unobserved phenomena at another.

Rule III raised both of the issues that eventually assimilated induction and analogy to probabilistic reasoning. The first concerned inductions proper: to what extent must experience confirm a conjecture before it may be legitimately extended to cases beyond the scope of observation, either in time or in sensory accessibility? Newton replied that rules which held without exception "to all bodies within the reach of our experiments" could be safely generalized to past, future, and otherwise remote cases, without specifying how extensive the "reach of our experiments" must be. Later philosophers and mathematicians turned probability theory to the task of measuring the strength of an inductive generalization by its experiential "breadth." Bernoulli's theorem was among the earliest and most influential contributors to this program; twentieth-century philosophers of science continue the tradition in their efforts to construct a calculus of confirmation for scientific theories.[25]

The second question concerned analogical inferences, which gained plausibility with the degree of similarity already established among the entities under comparison, shading off into induction as similarity approached identity. Thus David Hartley, discussing the correspondence of the structure and function of organs belonging to different animal species, cautioned that "there will be a probable Hazard of being mistaken, proportional in general to the known Difference of the two Animals, as·well as a probable Evidence for the Truth of Part, at least, of what is advanced, proportional to the general Resemblance of the two Animals."[26] Although attempts to quantify the closeness of analogy and concomitant probability of inference never made much mathematical headway, the problem was set firmly within a probabilistic framework from the seventeenth century onward.

Hence, Bernoulli's urn model reproduced characteristic elements of late seventeenth-century natural philosophy. Like the hidden causes of nature, the true ratio of balls—even their total number—remained permanently inaccessible to the human observer with no way to pry open the urn. The curious fell back on drawings, just as natural philosophers denied insight into the inner workings of nature contented themselves with observations of appearances, or "phenomena." As the composition of the urn could be conjectured from the results of the trials, so causes could be tentatively inferred from effects. Although it did not explain the source of these conjectures (the province of analogy), Bernoulli's theorem measured how the conjecture improved as the number of observations increased—in other words, as the induction was made more general.

Other aspects of Bernoulli's urn model seem more difficult to square with prevailing views of causation, including Bernoulli's own. Although Bernoulli was a staunch determinist, his urn model posited only chance connections between the inaccessible causes and the observed effects. Causes "determined" effects only in a combinatorial fashion. The combi-

---

[25] See Max Black, "Induction and probability," in *Philosophy in the Mid-Century*, Raymond Klibansky, ed. (Florence: Nuovo Italia, 1958), pp. 154–163.

[26] David Hartley, *Observations on Man, His Frame, His Duty, and His Expectations* (London, 1749), vol. 1, p. 342.

natorial model of cause guaranteed the independence of successive trials, assumed by Bernoulli's theorem, an assumption apparently at odds with the philosophy of regular causation: wouldn't the causes responsible for one event continue to influence the results of the next one? Bernoulli's theorem proved that even such loose links between cause and effect would eventually produce stable results. However, it is not apparent from the theorem itself why Bernoulli's urn model could have been presented to his contemporaries as a plausible account of causation, even if nature were acknowledged to be as opaque to human scrutiny as the sealed urn.

The urn model represented natural causes by the fixed ratio of black-to-white balls, a ratio that was determinate and unchanging, as befitted a deterministic understanding of nature governed by immutable laws. Chance thus played no part in causation per se. Yet effects—the observed ratio of black-to-white balls—exhibited only an overall or approximate congruence to the underlying causes. The connections between cause and effects were mediated by proportions of combinations and permutations which made the outcome of any particular event a matter of probability. Only in the late nineteenth century did this brand of statistical explanation gain wide acceptance among physicists through the work of Maxwell, Boltzmann, Clausius, and others. For the natural philosophers of the early eighteenth century, to thus insert chance between cause and effect, even as an epistemological device to symbolize human ignorance of the precise links that joined the two, must have seemed foreign indeed.

There was, however, at least one area in which combinations and permutations had entered natural philosophy in a qualitative way. The seventeenth-century revival of Epicurean atomism in the works of Gassendi, Boyle, Walter Charleton, and others[27] made frequent appeal to combinatorial concepts in order to explain how homogeneous matter took such diverse forms. Configurations of corpuscles, as well as their size, shape, and motion, accounted for the enormous variety of material forms known through experience. In his *Origin of Forms and Qualities*, Boyle imagined that the homogeneous, indivisible *minima naturalia* might be inclined at different angles "in reference to the horizon" and ordered like soldiers "behind one another in ranks." At the macroscopic level, the corpuscles' "dis-

---

[27] See Robert Kargon, "Atomism in the seventeenth century," in *Dictionary of the History of Ideas*, Philip Wiener, editor-in-chief (New York: Charles Scribner's Sons, 1974), vol. 1, pp. 132–141.

position or contrivances as to posture and order" resulted in the distinctive texture of a kind of substance.[28] According to the corpuscular philosophy, the size, shape, motion, and arrangement of these corpuscles entirely determined the species of matter and the secondary qualities which defined it in practice.

Walter Charleton synthesized the moderated skepticism and refurbished corpuscular philosophy of Gassendi in his *Physologia Epicuro-Gassendo-Charltoniana* (1654). Both the "Obscurity of Nature" and the "Imperfections of our Understanding" hampered the progress of natural philosophy. The design of nature and of human senses were equally to blame for disagreements among philosophers: "If Nature were not invelloped in so dense a cloud of Abstrusity, but should unveil herself, and expose all her beauteous parts naked to our speculation: yet are not the Opticks of our mind either clear or strong enough to discern them."[29]

Yet Charleton was nonetheless confident that natural philosophy, assiduously cultivated, would yield true knowledge about the workings of nature, and that this knowledge would bear out the atomistic theory. Charleton also emphasized the importance of the permutation and combination of atoms. Just as the permutations and combinations of letters generated words, sentences, and whole treatises, "so likewise, if there are but twenty-four diverse Figures competent to Atoms, they alone by variety of Order, or transposition, would suffice to the Constitution as incomprehensible a diversity of Qualities."[30] Charleton illustrated his point with permutations of letters into anagrams or words within a Latin verse, examples commonly found in mathematical treatments of combinations.[31]

Boyle and Charleton insisted that the configuration of corpuscles that stamped gold as yellow, ductile, dissoluble in *aqua regis*, and so on (as opposed to the equally distinctive configurations of silver, wood, or stone corpuscles) was dictated by divine will. Although configurations of component corpuscles could be altered by natural processes (Boyle believed that alchemical transformations were hence entirely consonant with the

---

[28] Boyle, *Origin of Forms*, pp. 35–36.

[29] Walter Charleton, *Physologia Epicuro-Gassendo-Charltoniana* (1654), reprinted (New York and London: Johnson Reprint Corporation, 1966), p. 5.

[30] Charleton, *Physologia*, p. 132.

[31] See, for example, John Wallis, *A Discourse of Combinations, Alternations and Aliquot Parts* (1685), in *The Doctrine of Chances*, Francis Masères, ed. (London, 1795), pp. 286–291; Jean Prestet, *Elemens des mathématiques* (Paris, 1675), pp. 350–354.

corpuscular theory), the corpuscles recombined according to divinely set patterns. The Cambridge Platonists and other opponents of these latter-day atomists nonethless pointed to the atheistic, pagan sources of the philosophy, claiming that it eliminated divine guidance: the universe resulted from nothing more than the chance combination and permutations of the fundamental corpuscles. Orthodox proponents of the corpuscular philosophy such as Boyle, Charleton, Joseph Glanvill, and others took care to dissociate themselves from such heretical implications by appropriating the arguments of their adversaries. Glanvill's *Philosophia Pia* (1671) and Charleton's *Darkness of Atheism Dispelled by the Light of Nature* (1652) sharply distinguished the refurbished atomism of the Royal Society from atheistic Epicureanism by themselves attacking as ludicrous the belief that so intricately constructed a system as nature should have come about by a chance collocation of atoms, without providential intervention. The Royal Society natural philosophers caricatured Epicureanism in order to purge their own brand of atomism from any taint of atheism: it is unlikely that any of their contemporaries held the view that, in Boyle's words, "atoms, meeting together by chance in an infinite vacuum, are able of themselves to produce the world."[32]

Yet although they raised the specter of randomness, only to dismiss it as absurd in the next breath, the corpuscular philosophers did alter the framework within which mathematical probability was conceived and applied. The image of a totally haphazard universe became the backdrop for the efforts of John Arbuthnot, Abraham De Moivre, Richard Price, and others to fortify the argument from design with probabilistic reasoning. It also may have indirectly smoothed the way for Bernoulli's urn model of causation by providing an epistemological precedent: the combinations and permutations of the various sizes and shapes of atoms, like those of the variously colored balls in the urn, only appeared to produce their observed effects by chance. In reality, natural laws instituted by providence dispelled chaos.

More importantly, the corpuscular philosophers also provided a model for a combinatorial mechanism of causation. The analogy of the letters of the alphabet to the varieties of atoms appeared again and again in their treatises as the means by which a finite collection of fundamental particles

---

[32] Boyle, *Of the Excellency and Grounds of the Corpuscular of Mechanical Philosophy* (1665), in *Works*, vol. 4, p. 68.

might produce the enormous diversity of phenomena. Boyle answered those who claimed that the corpuscular philosophy could not be extended to "all the phenomena of things corporeal" by comparing their doubts to those "that should affirm, that by putting together the letters of the alphabet, one may indeed make up all the words to be found in a book, as in Euclid, or *Virgil*; or in one language, as Latin, or English; but that they can by no means suffice to supply words to all the books of a great library, much less to all the languages in the world."[33] Charleton expressed the same idea in much the same terms, and even computed the number of combinations engendered by twenty-four letters (i.e., 2^{24}) to show the combinatorial fertility of even twenty-four shapes of atoms. John Wallis used the permutations of twenty-four letters to point out the enormity of the numbers involved: translated into changes rung on bells, "Ten Thousand Thousand years" would not suffice to execute all permutations of twenty-four.[34]

Thus Bernoulli's paradoxical assumption that causes were at once determined and combinatorial, regular and stochastically independent, had a precedent in the corpuscular theories of Gassendi, Boyle, and Charleton. Combinations of atoms produced distinctive observable qualities; combinations of balls dictated the ratios observed in successive trials. In both cases, the proportions of one shape of atom (or one color of ball) to another were fixed, and so, presumably, was each resulting combination of elements for an omniscient spectator. The merely human investigator, however, had recourse to the combinatorial art of conjecture to represent the relation between effects and their causes. Although the latter day atomists drew no probabilistic conclusions from their discussions of the combinations and permutations of atoms, some, such as Charleton, explicitly linked their analyses to the mathematics of combinations. The alphabet analogies are suggestive: a major seventeenth-century application of combinatorics appears to have been the construction of anagrams, and cryptograms.[35] The associations between combinatorics and cause forged by the corpuscular philosophy prepared the way for Bernoulli's assimilation of cause-effect relations to mathematical probability.

Bernoulli's third assumption of simplicity also commanded an illustrious following among natural philosophers of the period. Despite his

---

[33] Boyle, *Excellency*, p. 71. [34] Wallis, *Discourse*, p. 284.

[35] See, for example Mersenne, *Vérité*, pp. 527–543; Bacon, *Works*, vol. 9, p. 117; vol. 2, pp. 499–503.

criticisms, Leibniz himself considered simplicity to be one important criterion of a scientific hypothesis.[36] Newton agreed, stipulating that "Nature is pleased with simplicity."[37] "Simplicity" was an elusive term with almost as many senses as uses, and natural philosophers capitalized on this ambiguity. In some cases, simplicity might mean following Occam's razor in explanation (i.e., invoking the smallest number of causes); in others it might refer to the nature of the explanation itself. Bernoulli's appeal to the simplest curve invoked this second, more nebulous sense of the word. Bernoulli's attempts to use his theorem to do the work of its own inverse depended on the intrinsic plausibility of the simplest estimate of the a priori probability. Although forced to render this assumption explicit by Leibniz's curve-fitting comparison, neither Bernoulli nor his contemporaries flinched at affirming nature's preference for simple solutions.

De Moivre's discussion and refinement of Bernoulli's theorem replaced Bernoulli's assumption of simplicity with the kindred assumption of order. Observing that Bernoulli's method summing the terms of the binomial expansion $(a + b)^n$ was "not so much an Approximation as the determining very wide limits, within which they demonstrated that the Sum of the Terms was contained," De Moivre set about recalculating the approximate sums when n was an "infinite Power" with the help of Stirling's formula—effectively approximating the area of sections of the normal curve.[38] De Moivre's method drastically reduced the number of trials required to boost the probability that the observed ratio fell within a given interval around the true probability, and showed that the size of the interval was inversely proportional to the square root of the number of trials. This last quantity would serve as "the *Modulus* by which we are to regulate our Estimation." De Moivre interpreted his results as a severe restriction upon the empire of chance: "And thus in all Cases it will be found, that *altho' Chance produces Irregularities, still the Odds will be infinitely great, that in process of Time, those*

[36] Louis Couturat, *La logique de Leibniz d'après des documents inédits* (Paris: Felix Alcan, 1901), p. 268.

[37] Newton, *Philosophiae naturalis principia mathematica*, 2nd edition (Cambridge, 1713), pp. 357–358.

[38] De Moivre originally published his treatment of Bernoulli's theorem in a pamphlet, dated 12 November 1733, which was privately circulated; see R. C. Archibald, "A rare phamphlet of Moivre and some of his discoveries," *Isis* 8 (1926): 671–684. De Moivre translated and expanded this piece in the 2nd (pp. 235–242) and 3rd (pp. 243–250) editions of the *Doctrine of Chances* (1738; 1756).

Irregularities will bear no proportion to the recurrency of that Order which Naturally results from ORIGINAL DESIGN."[39]

De Moivre presented his own version of the inverse of Bernoulli's theorem, but as a remark rather than as a proposition. Bernoulli's theorem demonstrated that, assuming a "a certain determinate Law," the ratio of observed events would "continually approach to that Law" as the number of observations increased. De Moivre reasoned that if the ratio of events were observed "to converge to a determinate quantity" $P{:}Q$, then this ratio revealed the "determinate Law" that governed the event. Otherwise, if the law were expressed by some other ratio $R{:}S$, Bernoulli's theorem dictated that events would converge to $R{:}S$. But this violates the hypothesis that the observed ratio converges to $P{:}Q$, and therefore De Moivre claimed to have proved the inverse theorem by contradiction.

De Moivre's "proof" was heavily laden with assumptions. First, as modern advocates of a relative frequency interpretation ruefully admit, convergence obtains only in the limit. In practice, the required infinite number of observations never obtains. Attempts to determine unique probability values from "condensation points" within finite sequences of observations have met with frustration; apparent convergence of finite relative frequencies can be shown to be compatible with, even caused by, the randomness of the observational data.[40] De Moivre, however, was untroubled by any of these monstrous possibilities. He recognized that his argument would be vitiated if no determinate law regulated the observed ratio of events—"for then the Events would converge to no fixt Ratio at all"—but rejected this hypothesis out-of-hand as an absurdity: the series *must* converge. But De Moivre did not take up the problem of how to identify the convergent ratio from a finite sequence of observations. Like Bernoulli, he evidently assumed that the "simplest" ratio compatible with the data would be the true one.

De Moivre's attempts to yoke Bernoulli's theorem and his version of its inverse to the argument from design ran in a circle. In order for his argu-

---

[39] De Moivre, *Doctrine*, p. 252. For the mathematical significance of De Moivre's work, see Ivo Schneider, "Der Mathematiker Abraham De Moivre," *Archive for History of Exact Sciences* 5 (1968–69): 177–317; also Stephen M. Stigler, *The History of Statistics: The Measurement of Uncertainty before 1900* (Cambridge, Mass.: Harvard University Press, 1986), pp. 70–88.

[40] See Thornton C. Fry, *Probability and Its Engineering Uses* (New York: Van Nostrand, 1928), pp. 88–91; Fine, *Theories*, p. 93.

ment on the inverse theorem to stand, De Moivre acknowledged that observations must converge to a "fixt Ratio." Events converged to a fixed ratio because they followed a determinate law. We know that all events follow determinate laws because of the inverse theorem. De Moivre embarked on this closed loop because he found genuine randomness to be inconceivable: hence the a priori necessity for determinate laws. Assuming the existence of such determinate laws, Bernoulli's theorem promised that observations would eventually conform to them with ever greater precision.

From this vindication of "Original Design," De Moivre launched into an argument for the beneficence of the cosmic order he believed he had succeeded in demonstrating mathematically:

> Again, as it is thus demonstrable that there are, in the constitution of things, certain Laws according to which Events happen, it is no less evident from Observation, that those Laws serve to wise, useful and beneficent purposes, to preserve the stedfast [sic] Order of the Universe, to propagate the several species of Beings, and furnish to the sentient Kind such degrees of happiness as are suited to their State.[41]

Although the existence of order was, De Moivre believed, mathematically "demonstrable," its divine origins and benign character required other kinds of arguments "from Observation." Reasoning along the well-worn lines traced by the Leibniz-Clarke correspondence and by Boyle lecturers William Derham and Richard Bentley, De Moivre contended that brute, passive, inert matter could not "modify its own essense, or give to itself, or to any thing else, an original determination of propensity." Therefore, the order must have been imposed on matter by an external, nonmaterial agent. Citing John Arbuthnot's probabilistic argument for divine providence from the proportion of male-to-female births,[42] De Moivre concluded that this agent was "all-wise, all-powerful and good."

The second part of De Moivre's argument "from *final Causes*" rested on probabilistic foundations that several eighteenth-century mathematicians, including Jean d'Alembert and Nicholas Bernoulli, attacked as shaky. However, such criticisms failed to dampen the enthusiasm with which their colleagues advanced this probabilistic argument for design, first in a

[41] De Moivre, *Doctrine*, p. 252.

[42] John Arbuthnot, "An Argument for Divine Providence, taken from the Regularity observ'd in the Birth of both Sexes," *Philosophical Transactions of the Royal Society of London* 27 (1710–12): 186–190.

theological and later in a secular context. The mathematical variants of this argument and the controversy it engendered among mathematicians from Arbuthnot to Poisson constitute a complex and fascinating chapter in the history of eighteenth-century probability theory.[43] Here, it is sufficient to note that De Moivre's work established Bernoulli's theorem as a more workable tool for estimating the numerical adequacy of a set of observations; tacitly endorsed Bernoulli's implicit model of combinatorial causation; rendered more explicit Bernoulli's linked assumptions concerning the simplicity of nature, convergence of finite sequences of relative frequencies, and the validity of the inverse theorem; and, finally, made Bernoulli's theorem and its inverse into weapons for the natural theologians.

5.3 Bayes' Theorem and the Problem of Induction

The philosophical import of Bernoulli's theorem had been epistemological: if nature is governed by determinate laws, patient observations will ultimately conform to those laws. The perturbations of chance will eventually give way to stable ratios. Joseph Glanvill's caveat to natural philosophers that "we cannot pry into the hidden things of Nature, nor observe the first Springs and Wheels that set the rest in motion"[44] retained its force, but the natural philosopher could take heart from Bernoulli's theorem that chance irregularities would not swamp the order in observed effects that betokened the underlying order of hidden causes. Bernoulli's theorem struck a blow against those radical skeptics who had maintained that regular causes need not produce regular effects.

In De Moivre's hands, Bernoulli's theorem, or rather its inverse, became an even more formidable weapon against the skeptics. Not only could reg-

[43] The argument centered on the problem of discerning—and evaluating the probability of—orderly arrangements in nature. Although the argument originated in a theological context, Daniel Bernoulli and Laplace devised secular versions. See also Willem 'sGravesande, *Démonstration mathématique de la direction de la providence divine*, in *Oeuvres philosophiques et mathématiques* (Amsterdam, 1774), vol. 2; and Nicholas Bernoulli's criticisms in a letter to Montmort, published in the latter's *Essai d'analyse sur les jeux de hazard*, 2nd edition (Paris, 1713), pp. 373–393.

[44] Joseph Glanvill, *Essays on Several Important Subjects in Philosophy and Religion* (London, 1676), pp. 1–2.

ular causes be expected to ultimately produce regular effects in the long run; observation of effects would disclose those causes with any desired degree of probability to the persevering observer. In order to prove the second, stronger, epistemological claim, De Moivre was obliged to introduce ontological assumptions to the effect that regular, simple causes ("determinate Laws") existed. Since De Moivre found chance per se to be "utterly insignificant"—that is, in the literal sense of being meaningless—this assumption cost him little uneasiness: nature was opaque, but not random.

Together these claims immeasurably strengthened the hand of the inductivist natural philosopher. Natural philosophy might be barred by the subtlety of nature, fallible senses, and flawed understanding from ever penetrating the true causes and essences of things directly, but Bernoulli's theorem and, especially, its inverse, guaranteed that induction could discover these by the longer, indirect route of observation and experiment. On the strength of De Moivre's results, David Hartley extended probabilistic reasoning from the sphere of practical matters to that of induction and analogy. Hartley was confident that "no Sceptic" could forebear from giving his "practical assent at least to the Doctrine of Chances," and invoked De Moivre's formulation of Bernoulli's theorem (Hartley apparently knew the theorem only from De Moivre) as assurance that if the causes of an event bore a fixed ratio to one another, then the observations of that event would approach the ratio if "the Number of Trials be sufficient":

> This may be considered as an elegant Method of accounting for that Order and Proportion, which we every-where see in the Phenomena of Nature. The determinate shapes, sizes, and mutual Actions of the constituent Particles of Matter, fix the Ratios between the Causes for the Happenings, and the Failures; and therefore it is highly probable, and even necessary, as one may say, that the Happenings and Failures should perpetually recur in the same Ratio to each other nearly, while the Circumstances are the same.[45]

Like De Moivre and Bernoulli, Hartley assumed implicitly that determinate, fixed ratios of causes did exist. However, he evidently did not consider De Moivre's "remark" a sufficient guarantee of the inverse theorem, for he went on to mention an unnamed "ingenious Friend" who had

---

[45] Hartley, *Observations*, vol. 1, p. 338.

shown him the solution to the "inverse Problem, in which he has shown what the Expectation is, when an Event has happened p times and failed q times, that the original Ratio of Causes for the Happening or Failing of an Event should deviate in any given Degree from that of p to q." Hartley likened this sort of inverse reasoning to what he termed the "Newtonian differential Method," essentially Leibniz's curve-fitting example: given the ordinates and abscissas for several points of an unknown curve, find the "law"—i.e., the equation—of the curve. The ordinates corresponded to the observed effects, the abscissa to the circumstances under which they were observed, and the equation to the "law of Action" that produced those effects under the given circumstances. Hartley considered the analogy between curve fitting, inverse probabilities, and induction to be exact: all derived general "laws" from particular effects. The comparison extended even to the limitations of all three methods:

> This Parallel is the more pertinent and instructive, in as much as the mathematical Conclusion drawn by the differential Method, though formed in a way that is strictly just, and so as to have the greatest possible Probability in its Favour, is, however, liable to the same Uncertainties, both in Kind and Degree, as the general Maxims of Natural Philosophy drawn from Natural History, Experiments, & c.[46]

Hartley was a philosopher and psychologist rather than a mathematician. Yet he was quick to perceive the connection between the mathematical results and the method of induction, and between both of these and associationist psychology.[47] Induction and belief were both a matter of counting for Hartley; Bernoulli's theorem and its inverse justified the hope that counting, carried on long enough, would "determine the Proportions, and, by degrees, the whole Nature, of unknown Causes, by a sufficient Observation of their Effects."

This grand hope of making mathematical probability the guarantee of induction depended on Hartley's anonymous "ingenious Friend." Hartley had not accepted De Moivre's "remark" as a sufficient substitute for the inverse theorem since it held rigorously only in the infinite case. It is possible that Hartley knew Bayes: both were Fellows of the Royal Society, but there exists no evidence of such a connection. Bayes' proof of the inverse

---

[46] Hartley, *Observations*, vol. 1, pp. 339, 340.
[47] See Section 4.2, above.

theorem was published posthumously in the *Philosophical Transactions* of 1763 by his literary executor Richard Price; Hartley's *Observations on Man* appeared in 1749. Hartley intimates that he knew of the result through a private communication but offers no other particulars. Hartley's statement of the theorem itself is strikingly similar to Bayes, down to the notation, but by 1749 the canonical use of certain symbols (e.g., p and q to represent indices of terms of the binomial expansion) appears to have been widespread. Although the connection between Hartley and Bayes remains at best shadowy, both interpreted the solutions to the inverse problem as the sine qua non of a rigorous theory of induction.[48]

Richard Price emphasized the same point in his covering letter to John Canton requesting him to communicate Bayes' essay to the Royal Society. Like De Moivre, Price believed that "common sense" sufficed to show that judgments about the future drew upon past outcomes, and "that the larger number of experiments we have to support a conclusion, so much the more reason we have to take it for granted." In Price's opinion, however, the promptings of common sense were often too vague to be useful in particular cases. Only the inverse theorem could determine "in what degree" observation confirmed a given conjecture and thus *measure* "the strength of *analogical* or *inductive reasoning*." Price complained that the psychological conviction of common sense could not be generalized to all cases of inductive reasoning, "concerning, which at present, we seem to know little more that it does sometimes in fact convince us, and at other times not." Echoing De Moivre's concern that sophistry, affections, and other sources of intellectual error might devalue inductive results, Price voiced his hope that these snares "might in some measure be avoided, if the force that this sort of reasoning ought to have with us were more distinctly and clearly understood." Inverse probabilities made common sense more exact and therefore more trustworthy.[49]

---

[48] For other possibilities concerning the identity of Hartley's "ingenious Friend," see Stephen M. Stigler, "Who discovered Bayes' theorem?" *The American Statistician* 37 (1933): 290–296. Stigler mainly considers evidence for connections among mathematicians, but other interests might have brought Bayes and Hartley together; compare Bayes' *Divine Benevolence, or an Attempt to Prove That the Principal End of Divine Providence and Government Is the Happiness of His Creatures* (1731), and Hartley's *The Progress of Happiness Deduced from Reason* (1734).

[49] Thomas Bayes, "An essay towards solving a problem in the doctrine of chances," communicated by Richard Price, *Philosophical Transactions of the Royal Society* 53 (1763): 370–418; 54 (1764): 296–325.

Price admired De Moivre's work in sharpening Bernoulli's theorem "to a greater degree of exactness" but rejected De Moivre's "Remark" on its inverse as useless in practice. Price pointed out that De Moivre's results could not claim to be "rigorously exact, except on supposition that the number of trials made are infinite," a rule that offered little guidance as to how many observations sufficed to make it "exact enough to be depended on in practice." Evidently, Price did not accept De Moivre's tacit assumption that a simple ratio would emerge from a finite number of trials. Applauding De Moivre's efforts to enlist his results in the service of the argument from design, Price nonetheless added that Bayes' inverse theorem would be better suited for this purpose, "for it shows us, with distinctness and precision in every case of any particular order or recurrency of events, what reason there is to think that such recurrency or order is derived from stable causes or regulations in nature, and not from any irregularities of chance."[50]

In cases where the number of trials was large, common sense needed little aid from the doctrine of chances to accept the ratio of events as a good approximation to the ratio of causes. But when the number of trials was small, the inverse theorem revealed the limits between which it was "reasonable" to believe that the true ratio must lie. The inverse theorem could, Price claimed, thereby supplement common sense by determining on the basis of a limited number of observations "the precise degree of assent which is due to any conclusions or assertions relating to them."[51] It connected objective and subjective probabilities where psychological ties were weakest. Bayes' essay, exclusive of Price's prefatory letter and appendices, was divided into two parts, introduced by the statement of the problem itself:

> *Given* the number of times in which an unknown event has happened and failed: *Required* the chance that the probability of its happening in a single trial lies somewhere between any two degrees of probability that can be named.

Section I set forth definitions of the elements of probability and proved the theorem in conditional probability which bears Bayes' name and is expressed in modern notations as

$$P(B|A) = \frac{P(A \cap B)}{P(A)}, \text{ where } A \text{ and } B \text{ are events.}$$

[50] Bayes, "Essay," p. 374.　　　　　[51] Bayes, "Essay," p. 418.

Section I ended with a derivation of the formula expressing the probability of an event A happening p times and not happening q times in $p + q$ trials is the $(q + 1)^{\text{th}}$ term of the binomial expansion,

$$(a + b)^{p+q}, \text{ where } a = P(A); b = P(A'); \text{ that is, } \left(\begin{array}{c} p+q \\ q \end{array}\right) a^p b^q.$$

Although the reasoning of Section I may strike modern readers as somewhat convoluted because of Bayes' preference for expectations and concomitant "reasonable" trades over probabilities, there is nothing controversial about the results presented there.[52]

Section II contained the scholium that apparently caused Bayes himself doubts grave enough to withhold the essay from publication. According to Price, Bayes had had misgivings about the postulate that permitted his theorem to be generalized to cases where nothing is known about the prior probabilities of A and B, thus solving the inverse problem. This postulate assumed that "the probability for the happening of an event perfectly unknown, should lie between any two named degrees of probability, antecedently to any experiments made about it . . . must be . . . the same [probability] that it should lie between any two different degrees." Price claimed to have found "a very ingenious solution" to the inverse problem based on this postulate among Bayes' papers. However, Bayes evidently had second thoughts, fearing that this postulate "might not perhaps be looked upon by all as reasonable." Section II represented Bayes' attempts to shore up his postulate of what would now be termed a uniform prior probability distribution with a physical model and accompanying scholium.

The physical model that Bayes proposed was a level, square table, $ABCD$, onto which balls were tossed without aiming. After the first toss a line $\overline{oS}$ is drawn through the point where the ball landed and parallel to $\overline{AD}$. Bayes defined the event M as the ball's coming to rest at a point on the table between $\overline{oS}$ and $\overline{AD}$ in any one of the subsequent $p + q = n$ trials. After showing that $P(M) = \frac{Ao}{AB}$, Bayes constructed the curve $y = Ex^p r^q$ on the abscissa AB, where $x = \frac{Ab}{AB}$; $r = \frac{Bb}{AB} = 1 - x$; $b = $ some arbitrary point along AB; $E = \left(\begin{array}{c} p \\ q \end{array}\right)$; $n = $ the coefficient of the $(q + 1)^{\text{th}}$ term of the binomial expansion $(a + b)^n$. (Since p and q remain fixed, E is a constant here.)

---

[52] There are a number of modern commentaries on Bayes' essay: see L. E. Maistrov, *Probability Theory: A Historical Sketch*, Samuel Kotz, trans. and ed. (New York and London: Academic Press, 1974), pp. 87–100; also E. C. Molina's commentary in W. Edwards Deming, ed., *Facsimiles of Two Papers by Bayes* (Washington, D.C.: Department of Agriculture, 1940), pp. vii-xii.

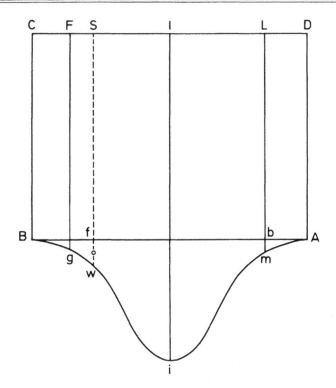

Figure 3. Bayes' table problem.

Essentially, this curve showed how the probability that a given event would happen p times out of $p + q$ trials varied as the probability of that event on a single trial took on values on the interval $[o, A]$, according to the final proposition in Section I.

Using direct probabilities inferred from the geometric properties of his model, Bayes showed that if the point o lies between points b and f along AB, the probability that event M happens p times and fails q times (call this event α) was equal to the ratio of the area of that portion of the curve *fgimb* to the area of the entire square $ABCD$. If the location of point o is totally unknown, then

$$P(\alpha) = \frac{\text{Area under the total curve } AmigB}{\text{Area } (ABCD)},$$

Since the perfectly level surface of the table dictated that any location for o along AB was as likely as any other.

Having established values for all of the prior probabilities, Bayes proceeded to apply the results of his theorem from Section I on conditional probabilities. Suppose we know nothing about the location of point o, but we guess it lies between points b and f on AB (call this event β). We are told that event M had happened p times out of $p + q$ trials. Given this information, what is the probability that our guess is correct? By Bayes' theorem,

$$P(\beta|\alpha) = \frac{P(\alpha \cap \beta)}{P(\alpha)}.$$

Substituting the prior values for $P(\alpha \cap \beta)$ and $P(\alpha)$ derived directly in the previous proposition, Bayes concluded that

$$P(\beta|\alpha) = \frac{\text{Area } (\textit{fgimb})}{\text{Area } (\textit{AmigB})}.$$

Thus far, Bayes had solved the inverse problem for a particular case, namely the one in which every value of x is assumed to be as probable as any other in the expression $y = x^p(1 - x)^q$. Bayes justified this assumption by the properties of the physical model he had chosen: on a perfectly flat table, there was good reason to believe a priori that all resting points for the ball were equiprobable, just as the physical symmetry of a coin or die rendered all outcomes equiprobable. But in order to generalize the inverse theorem to the genuinely interesting cases in which there was no such clue to a priori probability values, Bayes was obliged to add a scholium asserting that total ignorance served the same purpose as the level table. Bayes reasoned that the same assumption of equal prior probabilities applied to an event about which "we absolutely know nothing antecedently to any trials made concerning it" because of the Principle of Indifference:

. . . concerning such an event I have no reason to think that, in a certain number of trials, it should rather happen any one possible number of times than another. For, on this account, I may justly reason concerning it as if its probability had been at first unfixed, and then determined in such a manner as to give me no reason to think that, in a certain number of trials, it should rather happen any one possible number of times than another.[53]

[53] Bayes, "Essay," p. 393.

Bayes compared this general epistemological ignorance to the table example before the ball is first tossed to determine point *o* and before the results of the *n* trials of event *M* have been reported. Assuming this rule held for "any event concerning the probability of which nothing at all is known antecedently to any trials," Bayes concluded his essay by generalizing the theorem for the model to all cases of inverse probabilities, including a formula for integrating $y = x^p(1 - x)^q$ "in order to reduce the foregoing rule to practice." Postclassical commentators have reached very different judgments concerning the merits of Bayes' scholium. Those who know only the scholium itself have either praised or condemned it (depending on their own philosophical propensities) as the epitome of the subjective tendencies inherent in the classical interpretation of probability theory. Those who have examined Bayes' arguments more closely, including the physical model, have been perplexed by the ease with which he appears to shift between objective "doctrine of chances" and subjective "art of conjecture" perspectives. On the one hand, Bayes' physical model admits a straightforward frequentist interpretation; on the other hand, his scholium retreats into "epistemological isomorphisms."[54] This apparent equivocation is largely the product of the postclassical bifurcation of subjective and objective probabilities. As we have seen, the eighteenth-century classical probabilists found an epistemological *cum* psychological justification for a joint concept of probability which incorporated both aspects. According to Locke and his followers, in a well-regulated mind (subjective) belief grew in proportion to the (objective) evidence of experience. In more disorderly intellects, prejudice, custom, and passion might distort this psychological harmony of subjective and objective probabilities: hence the need for a calculus of probabilities. While mathematicians might share Price's misgivings about the precision of this internal balance of belief and evidence, they did not doubt the associationist equation that paired the two.

For all of his reluctance to accept extant expositions of probability theory, Bayes belonged to the mainstream of classical probability. His account of the elements of probability theory in Section I of his essay displayed a Huygenian preference for expectation as the least ambiguous

[54] Ernest Nagel, *Principles of the Theory of Probability*, in *International Encyclopedia of Unified Science*, Otto Neurath, Charles Morris, and Rudolf Carnap, eds. (Chicago: University of Chicago Press, 1955), vol. 1, part II, pp. 341–422, on pp. 372–373; Hacking, "Jakob Bernoulli," p. 225.

measure of uncertain outcomes. His identification of physical and epistemological symmetry was also typical of the classical theory: having "no reason to think" in our initial ignorance that one outcome was more likely than another was "exactly the case of the event M," that is, of the ball coming to rest within a given distance of the unknown point o. Finally, he exhibited, along with Jakob Bernoulli and De Moivre, a predilection for models that translated the workings of natural causes and the processes of induction into probabilistic terms. Bernoulli, De Moivre, and Bayes differed in the specific assumptions they made about causation and the overall configuration of nature, but their general strategy and objectives were the same. Just as probabilistic expectation would provide a mathematical expression for reasonableness in the moral sphere, so their theorems would quantify inductive reasoning applied to the natural world.

Price explored the applications and philosophical implications of Bayes' results in an appendix and sequel to the essay. He selected an example that yielded observations "strictly applicable to the events and appearances of nature," leaving no doubt that he had the inductive reasoning characteristic of natural philosophy in mind. Imagine a die of an unknown number of facets and unknown composition. Prior to any trials, we can form no judgment about the possible outcomes because all outcomes—an infinity of them—are all equally conceivable. Price implicitly equated psychological with probabilistic indifference: "Antecedently to all experience, it would be improbable as infinite to one, that a particular event, beforehand imagined, should follow the application of any one natural object to another; because there would be an equal chance for any one of an infinity of other events." Only after the first trial would speculation about causes by means of Bayes' theorem be permissible, and then only concerning the bare possibility of the event that actually occurred. Conclusions concerning the operation of a regular cause and the uniformity of nature followed only from the constant results of subsequent experiments.

Once again, Price slid easily between subjective belief and objective experience: a single instance "would not suggest to us any idea of uniformity of nature"; but an uninterrupted string of instances gives "reason to expect the same success in further experiments, and the calculations directed by the solution of this problem may be made." However, Price warned that no degree of experimental uniformity could render such reasonable expectations certain. In the case of the die, even a million successive identical outcomes would "not give sufficient reason for thinking that it would *never* turn any other side."

Price underscored this point by a more grandiose example that recurs frequently in the probabilistic literature of the period. Price imagined a person "just brought forth into this world," *tabula rasa* and with no other source of instruction but observation. He would, Price maintained, go to sleep his first night in total ignorance as to whether he would ever see the sun again. But sunrise the next morning, and for all subsequent mornings, would create an ever-growing expectation and, by Bayes' theorem, probability that the sun would rise on the morrow, even though no finite number of sunrises could produce "absolute or physical certainty." However, Price asserted that belief in the overall uniformity of nature, resulting from long observation of many phenomena, would supplement even meager experiential evidence that any given phenomenon would follow "stable and permanent" laws: "The consideration of this will cause one or a few experiments often to produce a much stronger expectation of success in further experiments than would otherwise have been reasonable"—that is, "reasonable" by Bayes' theorem.

Price insisted that this supplementary expectation, far from conflicting with the calculations based on the theorem, was actually "only one particular case to which they are to be applied." Citing De Moivre's comments on Bernoulli's theorem, Price made similar assertions concerning the "demonstrative evidence" that the course of nature was governed by "permanent causes" rather than by "the powers of chance" on the basis of the fact that the probability that the observed ratio of events will differ from its true probability by less than any given amount increases without bound with the number of trials. Like De Moivre's argument, Price's reasoning contained a loophole: if observed results deviated from the presumed a priori probability or "cause," this was reason not to doubt the permanence of causes but rather to infer the existence of "some unknown causes which disturb the operations of the known ones."

Instead of enlarging upon this important claim that Bayes' theorem could also test the global uniformity of nature, Price emphasized the ever-present possibility of exceptions to both particular and general regularity. He reiterated that conclusions drawn from uniform experience could never be ironclad:

In other words, where the course of nature has been the most constant, we can have only reason to reckon upon a recurrency of events proportioned to the degree of this constancy: but we can have no reason for thinking that there are no causes in nature which will ever

interfere with the operations of the causes from which this constancy
is derived, and no circumstances of the world in which it would fail.[55]

Experience alone could never provide certainty; only auxiliary "principles"
and considerations suggested by "reason, independent of experience"
could force the transition from probability to certainty. Despite the elu-
siveness of certainty in probabilistic reasoning, Price contended that calculations based on Bayes' theorem would show "what expectations are warranted by an experiment, according to the different number of times in
which they have succeeded and failed" and derive the probability of a given
cause in any single trial from past results. Although he admitted that in
application Bayes' theorem was encumbered by technical difficulties in
narrowing the probability interval by the problematic approximation of
the integral of the binomial expansion (especially when p and q were both
large), Price nonetheless defended its utility in giving "a direction to our
judgment" and in being far superior to De Moivre's solution when p and
q were both small.

It is instructive to compare Price's application of Bayes' theorem to inductive reasoning with Hume's celebrated critique of induction in Book
I, Section VI, of *Treatise of Human Nature* (1739) and Section IV of *An
Enquiry Concerning Human Understanding* (1758). Price's numerous citations to Hume in his *Review of the Principal Questions in Morals* (1758) shows
him to have been familiar with at least Hume's *Treatise* at the time he
edited Bayes' essay for the *Philosophical Transactions*. Although Price may
not have appreciated the full force of Hume's arguments and certainly did
not agree with their skeptical tendencies, he nonetheless admired Hume
as a "writer of the greatest talents."[56] Despite their divergent conclusions
on the philosophical validity of induction (neither doubted its utility),
Price and Hume held certain key principles in common. Both observed
sharp divisions between reasoning, demonstration, and certainty on the
one hand, and opinion, experience, and probability on the other. Both
agreed that matters of fact could be divined only from experience: Hume's
Adam could no more infer the danger of suffocation from the transparent
fluidity of water than Price's Adam could predict that the sun would rise
on the morrow. A priori, an infinite number of effects could conceivably

[55] Bayes, "Essay," p. 411.

[56] Richard Price, *Review of the Principal Questions in Morals* (1758), D. D. Raphael, ed.
(Oxford: Clarendon Press, 1974), p. 56n.

be conjoined with any "cause." Both philosophers asserted that constant experience entailed expectations about the future.

Regarding the grounds for such expectations, however, Hume and Price parted company. Price argued that both the belief in a necessary connection between cause and effect and in the regularity of these causes stemmed from "intuition," confirmed by an ever-increasing probability as the number of positive instances increased. To skeptics like Hume who countered that no reason could be given for either belief without risking tautology, Price retorted that such intuitions were "self-evident; and that what is meant by saying, that it is not reason that informs us there must be some account of whatever comes to pass . . . [is that] they are not subjects of *deduction*; that is, that they are so plain, that there is nothing plainer from which they can be inferred."[57] What Hume had relegated to the irrational (in the literal sense of being unsupported by reason) operations of human nature and custom, Price elevated to the status of an axiom. Similarly, Price accepted as wholly reasonable our expectation that a frequently repeated event like the sunrise will recur because it was based on an intuition "that there being some reason or cause of this *constancy of the event*, it must be derived from a cause regularly and constantly operating in given circumstances." Unlike Hume, Price saw no reason to subject such intuitions to further philosophical scrutiny. Significantly, writing at least five years before his presentation of Bayes' memoir to the Royal Society, Price appealed to an example drawn from mathematical probability, albeit qualitatively expressed: if a die were repeatedly to turn up on the same side, our expectation that this side would recur in future trials would grow "the more frequently and uninterruptedly we knew this had happened."

Hume's arguments against recourse to probable reasoning thus failed to sway Price. Hume had pointed out that the only basis for judgments of probability that the future will be like the past was past experience that this was the case, an argument that begged that question: ". . . probability is founded on the presumption of a resemblance betwixt those objects of which we have had experience, and those, of which we have had none; and therefore 'tis impossible this presumption can arise from probability."[58] The only way to break the circle, Hume contended, was to assume

[57] Price, *Review*, p. 29n.

[58] David Hume, *A Treatise of Human Nature* (1739), L. A. Selby-Bigge, ed. (Oxford: Clarendon Press, 1975), Book I, section 6, p. 90. For excellent summaries of Hume's argument, with special attention to probabilistic aspects, see A. J. Ayer, *Probability and*

the uniformity of nature as an a priori principle, to which habit alone attests. Price, as we have seen, believed that his hypothetical Adam gradually acquired the expectation that specific events would repeat past patterns, and that nature in general was uniform, a belief that rested on the far sturdier foundation of Bayes' theorem. Of course, Price's probabilistic solution to Hume's problem of induction did not fully come to terms with Hume's strictures against probability. In order to estimate the future probability that a given cause will produce effects like those in past trials, Bayes' theorem required the prior probability of the cause and the assumption that the cause remained constant. Bayes had dispatched the first qualification with the scholium hypothesizing a uniform prior distribution for the probabilities of causes; Price obliquely addressed the second with the contention that observed deviations from results certified as most probable did not undermine the permanence of posited causes, but only signaled the interference of heretofore unknown ones.

Price could no more than De Moivre conceive of an alternative world in which causes acted erratically and ephemerally: our intuition that causes were permanent and that nature is uniform was a sound one, superior to Hume's flimsy-sounding custom, and prior—but not inferior—to the conclusions of reason. Nonetheless, he took Hume's arguments against induction seriously enough to welcome Bayes' theorem as an antidote to the vagueness of inductive argument—"we seem to know little more than it does sometimes, in fact, convince us, and at other times not"—and as a demonstration of nature's uniformity and therefore design.[59] Indeed, Bayes' theorem put numbers to Hume's injunction that in all inferences from past to future experience we "must assign to each of them a particular weight and authority in proportion as we have found it to be more or less frequent."[60] Hume had, after all, advertised his *Treatise of Human Nature* as an answer to Leibniz's pleas for a system of logic that would redress the balance between demonstration and other forms of reasoning in favor "of probabilities and those other measures of evidence on which eye and action entirely depend, and which are our guides even in most of our philosophical speculation."[61] However, this reasoning from "a firm and unalterable

---

Evidence (London: Macmillan, 1972), pp. 3–26; David C. Stove, *Probability and Hume's Inductive Scepticism* (Oxford: Clarendon Press, 1973), pp. 44–45.

[59] Bayes, "Essay," pp. 372–373.

[60] Hume, *An Enquiry Concerning Human Understanding* (1758), Charles W. Hendel, ed. (Indianapolis: Library of Liberal Arts, 1955), p. 71.

[61] Hume, *An Abstract of a Treatise of Human Nature* (1740), *Enquiry*, p. 184.

experience" to "the last degree of assurance" was also the crux of Hume's argument against miracles,[62] and it is possible that this lay behind Price's insistence that the bulk of experience, no matter how uniform and extensive, could never exclude the possibility of an aberration. Hume's preponderant probability against miracles would therefore remain just that: a probability rather than a certainty.

Price's commentary thus situated Bayes' results within a philosophical and theological framework that reached back to the late seventeenth-century apologetics of Tillotson, Wilkins, and Boyle through Bishop Butler. In Price's hands, Bayes' theorem became a weapon to wield against the renovated skepticism of Hume (there is little indication, at least in the case of induction, that Price realized just how devastatingly novel Hume's variants on older themes were), in defense of the reasonableness of both natural philosophy and natural theology. Hume notwithstanding, reasoning from experience rested on more than habit; religion need not take refuge in faith alone. As Price duly noted, De Moivre had already enlisted Bernoulli's theorem in the service of the argument from design. Price was quick to add that Bayes' theorem served these same purposes admirably, and held out the additional promise of a calculus of induction.

5.4 The Calculus of Induction

This calculus of induction based on inverse probabilities became a central problem in mathematical probability in the late eighteenth and early nineteenth centuries. Laplace, Condorcet, and Poisson hoped to codify the methodological intuitions of natural philosophers into a method per se, into an explicit set of mathematical rules. Like Price, they believed such mathematical foundations to be the best defense against skeptical attacks on induction, but they did not follow him in using the probability of causes to support the argument from design. Indeed, Poisson went so far as to claim that the probability of causes *refuted* the argument from design. These later probabilists were as convinced as De Moivre and Price that the natural—and even the moral—order was stable, but not that stable causes were necessary to produce such stable effects. A new model of causation took shape in their work, making it possible to conceive of a world in which macroscopic order was the product of microscopic chaos. In proba-

[62] Hume, *Enquiry*, pp. 124–125.

bilistic terms, the underlying probabilities—the composition of the urn—could vary without disrupting the global regularity of the observed effects: Bernoulli's theorem became the law of large numbers.

Inverse probabilities launched the calculus of induction through the work of Laplace, not Bayes. Laplace's debt to Bayes remains a matter of debate among historians. Laplace was certainly not above reproducing and enlarging upon the results of others without citation: some of his earliest memoirs on probability theory had cannibalized De Moivre's work.[63] However, in this case Laplace seems to have discovered Bayes' *Philosophical Transactions* essay only after he had composed his "Mémoire sur la probabilité des causes par les événements" (1774). His most likely source of information on Bayes would have been Condorcet, who knew of Price's philosophical and political writings. But as late as 1780, Condorcet appears to have known only Laplace's work on inverse probabilities.[64]

There is also some internal evidence that Laplace reached his results independently of Bayes. Bayes' geometric model immediately yielded the general case of continuous probabilities. Laplace, however, began with the discrete case of a Bernoulli-like urn model filled with black and white balls in an unknown ratio that generalized to the continuous case only on the supposition of an infinite number of balls. Like Bernoulli, Laplace compared human uncertainty regarding the connection between causes and effects to an urn in which either the ratio of balls of various colors (the "cause") or the results of a certain number of drawings (the "effect") is to be inferred from the other. Direct probabilities derived effects from causes; inverse probabilities reasoned from effects to causes. Laplace envisioned the problem of inverse probabilities (which he stated as a principle rather than proved as a theorem) as a kind of average: if an event E can result from n different causes, the probability that in a specific case it is due to a given cause C_m is equal to the probability E given C_m divided by the sum of the n conditional probabilities of E given any C_i:

$$P(C_m|E) = \frac{P(E|Cm)}{\sum^n (E|C_i)} .$$

[63] Charles C. Gillispie, "Laplace," in *Dictionary of Scientific Biography*, Charles C. Gillispie, editor-in-chief (New York: Charles Scribner's Sons, 1978), Supplement 1, pp. 279–281.

[64] Condorcet mentioned the Bayes/Price essay in his introductory "Histoire" to Laplace's memoir "Sur les probabilités," *Histoire et Mémoires de l'Académie royale des Sciences. Année 1778* (Paris, 1781), p. 43.

For example, imagine two urns, each containing black and white balls mixed in different ratios, corresponding to two causes C_1 and C_2. Let E be results of a certain number of drawings from one of the urns; we do not know from which. Laplace's principle gives the probability of C_1 as

$$P(C_1|E) = \frac{P(E|C_1)}{P(E|C_1) + P(E|C_2)} .$$

Laplace's urn model, like Bayes' table, makes it plausible to assume that the prior probabilities of the causes are equal: we are equally likely to draw from one urn as from another; the ball is equally likely to stop at one point on the table rather than another. Laplace's ratio formulation assumes that $P(C_i)$ will be the same for all C, that is, the prior probabilities of all n causes will be equal, and can therefore be neglected. In contrast to Bayes' scholium, Laplace did not make this assumption explicit in his 1774 memoir, although his later exposition of conditional probabilities in *Théorie analytique des probabilités* did so.[65]

Positing an urn filled with an infinite number of black and white balls, from which p white and q black are drawn in $p + q$ trials, Laplace derived Bayes' result for the probability that a given value x is the correct ratio of white balls:

$$\frac{x^p (1-x)^q \, dx}{\int_0^1 x^p (1-x)^q \, dx} .$$

In contrast to Bayes, however, Laplace used inverse probabilities—the probabilities of causes derived from effects—as a steppingstone to a slightly different question: Bayes had asked what the probability that his guess concerning the "cause" x was correct, given the results of $p + q$ trials; Laplace asked what the probability of the next trial would be, given the same data. In terms of the urn example, this meant the probability of drawing a white ball on the next trial:

$$\frac{\int_0^1 x^{p+1} (1-x)^q \, dx}{\int_0^1 x^p (1-x)^q \, dx} .$$

Laplace generalized this "law of succession" to the probability of drawing m white balls and n black ones, given that there have been p white and q

[65] Pierre Simon Laplace, *Théorie analytique des probabilités* (1812; 3rd edition 1820), in *Oeuvres*, vol. 7, p. 371.

black in the past $p + q$ trials. If m and n are small compared to p and q, this probability,

$$\frac{\int_0^1 x^{p+m}\,(1-x)^{q+n}\,dx}{\int_0^1 x^p\,(1-x)^q\,dx}\;.$$

can be approximated by the expression[66]

$$\frac{p^m\,q^n}{(p+q)^{m+n}}\;.$$

In one of his earliest philosophical commentaries on probability theory, Laplace assimilated all applications of the theory to the probability of causes. Whereas Jakob Bernoulli had appropriated the lotterylike urn case as a model for unknown causes and observed effects, Laplace reversed the direction of the metaphor. Either the "cause" was known and the "event" unknown, as in the case of Bernoulli's urn of specified contents, games of chance involving fair coins or dice—in short, all cases of direct probabilities; or the "event" was known and the "causes" unknown—inverse probabilities.[67] (Cases in which neither the cause nor the event was known could be reduced to the class of known causes and unknown events by assuming a priori that all possible outcomes are equiprobable.) Bernoulli's theorem dealt with direct probabilities and Bayes' theorem addressed inverse probabilities, but both tended toward the same metaphysical conclusions, according to Laplace. He concluded from Bernoulli's theorem that "the relations of natural effects are almost constant, when these effects are considered in large numbers,"[68] where "natural" was broadly construed to encompass both the physical and moral realms.

Laplace envisioned nature as a composite of "regular" (or "constant") and "irregular" causes. Even the irregular causes exhibited a collective regularity, however: in the long run their effects were symmetric and canceled one another out, revealing the steady operation of the underlying constant cause. The inevitable triumph of constant, regular causes over ephemeral perturbations applied not only to the long-term profitability of insurance companies and lotteries, but also, Laplace maintained, to the salutary re-

[66] Laplace, "Mémoire sur la probabilité des causes par les événements" (1774), in Oeuvres, vol. 8, pp. 29–32.

[67] Laplace, "Recherches sur l'intégration des équations différentielles aux differences finies et sur leur usage dans la théorie des hasards" (1773), in Oeuvres, vol. 8, p. 152.

[68] Laplace, Essai philosophique sur les probabilités (1814), in Oeuvres, vol. 7, p. xlviii.

sults of submitting to "the eternal principles of reason, justice, and humanity upon which society is founded and maintained"—such were the moral "causes" of human happiness. There were "natural limits" to society, imposed by "reason and the natural rights of man," which governed the long-term fate of nations as inexorably as natural laws determined the tides:

> It is therefore important to both the stability and happiness of empires that they not extend themselves beyond the limits to which the incessant actions of causes confines them; thus the oceans, swelled by violent tempests, fall back to their basins by their own weight. This is also a result of the Calculus of Probabilities, confirmed by numerous and disasterous experiences.[69]

These were the views of the mature Laplace, set forth in the popular *Essai philosophique sur les probabilités* (1814), which in later editions became the introduction to his *Théorie analytique des probabilités* (1812; later eds.). But the seeds of this interpretation of causation as the interaction of steady uniform causes and symmetric perturbations can be found in Laplace's earliest expositions of the theory of astronomical error, in his 1774 memoir on inverse probabilities. Laplace posed the problem—a pressing one for observational astronomers—of determining the most accurate method of computing the mean of several disparate observations of the same celestial phenomenon. Just as he was later to assume that the irregular causes that temporarily masked the regular cause were symmetric, Laplace here assumed that the perturbing human and instrumental errors that masked the true observational value also followed a symmetric law. The function $\phi(x)$ expressed the probability that a given observation would diverge a distance x from the true value V. In order to specify $\phi(x)$, Laplace constructed a model for the relationship between error and truth: $\phi(x)$ achieved a maximum when $x = 0$, since error was equally likely to diverge from the true value V in the positive or negative direction, and V was the single most probable value. $\phi(x)$ was therefore symmetric around the maximum, and decreased as x increased: in other words, small errors were more probable than large ones.[70]

[69] Laplace, *Essai*, pp. xlviii, xlix.

[70] For Laplace's lifelong interest in the error problem, see O. B. Sheynin, "Laplace's theory of error," *Archive for History of Exact Sciences* 17 (1977): 1–61, especially pp. 9–10. See also Ivo Schneider, "Laplace and thereafter: The status of probability calculus in the

Here, Laplace's key assumption was that the probability of error conformed to a symmetric distribution. He dismissed the method of the simple arithmetic mean, which tacitly assumed that $\phi(x)$ was constant for all values of x, as "completely implausible by the very nature of the thing."[71] Errors, like the perturbational causes, eventually balance one another, leaving only the true mean, or constant cause. In mathematical terms, the probability values for the possible ratios of success to total trials m/n would distribute themselves symmetrically around a maximum $m/n = p$, where p was the a priori probability of the event, with an increasingly sharp peak at $m/n = p$ as n, the number of trials, increased. Perturbations corresponded to less probable values for m/n.

Laplace proposed investigation of the weak causes most easily overwhelmed by perturbations, such as animal magnetism or the efficacy of a medical regimen, by using inverse probabilities to measure both the extent to which existing evidence validated the hypothesis of a novel natural agent, and the amount of further information needed to reach a probable conclusion on controversial scientific issues.[72] For Laplace, Bernoulli's theorem provided a mathematical model of causation, particularly for the tangled causal nexus of animal magnetism or the tides; Bayes' theorem provided a mathematical model of scientific method itself, evaluating the likelihood of scientific hypotheses with respect to the available data.

In his later work on inverse probabilities, Laplace was careful to distinguish cases where uniform prior probabilities of unknown causes obtained from those where they did not.[73] In the former case, the probability of a given cause x was

$$\frac{y(x)}{\int_0^1 y(x)\,dx}.$$

where the function $y(x)$ expressed the probability of the observed effects as x varied. The probability that x lay within the interval $[\theta, \theta']$ was

nineteenth century," in *The Probabilistic Revolution*, vol. I, *Ideas in History*, Lorenz Krüger, Lorraine J. Daston, and Michael Heidelberger, eds. (Cambridge, Mass.: MIT Press, 1987), pp. 191–214; Stigler, *History*, pp. 11–61, on the importance of the method of least squares.

[71] Laplace, "Causes," in *Oeuvres*, vol. 8, p. 48.

[72] Laplace, *Essai*, in *Oeuvres*, vol. 7, p. lxxvii.

[73] See E. C. Molina, "The theory of probability: Some comments on Laplace's *Théorie analytique*," *Bulletin of the American Mathematical Society* 36 (1930): 369–392.

$$\frac{\int_{\theta}^{\theta'} y(x)}{\int_0^1 y(x)\,dx}.$$

But if the a priori values of x, independent of the observed effects, were not equiprobable, Laplace replaced the function $y(x)$ with the composite function $y(x)z(x)$:

$$\frac{\int_{\theta}^{\theta'} y(x)z(x)\,dx}{\int_0^1 y(x)z(x)\,dx}.$$

Thus, the prior probability distribution of causes could be compared to the case in which a uniform prior probability distribution held but where the observed effect resulted from two independent components, whose probabilities were expressed by the functions $y(x)$ and $z(x)$ respectively.[74]

Laplace's first attempts to deal explicitly with prior probability distributions in problems concerning the probability of causes dated from a memoir published in 1781, in which Laplace applied inverse probabilities to problems in games of chance, observational error, and the ratio of male-to-female births in Paris and London. In this "Mémoire sur les probabilités," Laplace invoked the Principle of Indifference in several contexts: the probability of either of two players of unknown skill winning equals 1/2 at the beginning of the game, for the probability that a given player is the stronger (or the weaker) of the pair is equal "relative to our ignorance"; the "laws of possibility" governing the birth of an infant boy being unknown, all values of $P(\text{boy}) = x$ between 0 and 1 must be assumed "equally probable."

Despite the apparently strong subjective slant to Laplace's appeal to the Principle of Indifference, particularly with respect to inverse probabilities, he clearly distinguished between "relative possibilities" derived from the assumption of equiprobability from ignorance, and the "absolute possibilities" derived from "the nature of the events themselves." On the one hand, absolute possibilities, whether computed a priori from repeated trials concerning the true shape of the coin or the relative skill of the players, corresponded to the constant causes of the events in question. On the other, relative possibilities were just that: relative to the state of our knowledge about these constant causes.[75] Laplace repeatedly warned that confusing absolute and relative possibilities would lead to mathematical

[74] Laplace, *Théorie analytique*, in *Oeuvres*, vol. 7, pp. 370–371.
[75] Laplace, "Mémoire sur les probabilités" (1781), in *Oeuvres*, vol. 9, pp. 384–385.

inconsistencies. Probabilists must continually correct their estimates of relative possibilities in light of new knowledge concerning absolute possibilities: hence the utility of inverse probabilities, which calculated the probabilities of causes in light of observed effects and the probabilities of future events given the record of past ones.

In practice, however, Laplace almost always assumed that all cases were "equally possible a priori," justifying this hypothesis on the grounds that more complex problems could be reduced to this simplest case by conceiving the observed results as composites of independent results. This alternative formulation of the problem only transformed the difficult business of finding the actual prior probability distribution into the equally difficult task of estimating the fictitious composite function $y(x)z(x)$, but the restatement of the problem in these terms apparently satisfied Laplace that his version of Bayes' scholium was legitimate.[76] His interpretation of Bayes' theorem as a mathematicized scientific method may have smoothed the way for this sweeping assumption of equal prior probabilities. Scientific hypotheses are confirmed relative to our knowledge, or lack thereof, of the natural world. Thus, epistemological indifference among alternative hypotheses at the outset of a scientific investigation justified the assumption of uniform prior probabilities in the interest of openmindedness. Experiments would soon correct the balance in favor of the better theory. Assumptions like this one have led many historians and mathematicians to describe Laplace as an unreconstructed subjectivist; indeed, as the foremost proponent of that interpretation. Many passages in Laplace's writings on probability theory could be—and have been—adduced in support of this view. Among the most often cited is Laplace's famous rendering of Jakob Bernoulli's opposition of an omniscient superintelligence for whom all events, past, present, and future, are fully determined, and the feeble human understanding that glimpses the certainties of natural law through a glass darkly. Probability theory, Laplace asserted, corresponds to nothing real; it measures human ignorance.

Yet Laplace's interpretation of probability theory was by no means so wholly psychological as the subjective label would suggest. Laplace insisted that probability theory apply itself to ratios of combinations, rather than to the states of mind occasioned by those ratios. Appropriating Daniel Bernoulli's term "moral expectation," Laplace gave it a more general meaning in this context: the distinction between mathematical and moral

---

[76] Laplace, *Théorie analytique*, in *Oeuvres*, vol. 7, p. 371.

expectation corresponded to that between the probabilistic definition and "the state of the human mind when the occurrence of some benefit depends on suppositions which are only likely." Probabilists confused the two at their peril, and Laplace applauded d'Alembert's criticisms of the theory of probability insofar as they exposed the errors of mathematicians who had blurred the distinction. Laplace exhorted probabilists to follow the general practice of "the physicomathematical sciences," which measured the objective causes of subjective states rather than the states themselves: ". . . we measure the intensity of light, the different degrees of heat in bodies, their forces, their resistances, etc. In all of these investigations, the physical causes of our sensations, and not the sensations themselves, are the object of Analysis."[77]

Although it would not be hard work to find cases, in both philosophical and mathematical contexts, in which Laplace violated his own precept, most of the counterexamples occur in situations that demanded *faute de mieux* assumptions in order to make any mathematical headway at all. Most of the cases in which Laplace invoked the Principle of Indifference to justify uniform prior distributions of probabilities fell into this category: the very fact that in most applications of inverse probabilities nothing whatsoever was known a priori about the antecedent causes made this branch of probability theory especially prone to such assumptions. For this reason, Laplace regarded inverse probabilities as the mathematical instrument par excellence of the moral sciences, where the a priori probabilities of causes were almost always totally unknown. Moral causes surpassed natural causes in obscurity and complexity: "So many unforeseen, or hidden, or inappreciable causes influence human institutions, that it is impossible to judge their results a priori." Laplace hoped to imbue the moral sciences with the same combination of "observation and calculation" that had achieved such impressive success in the natural sciences. He apparently envisioned a method that employed Bernoulli's and Bayes' theorems in tandem, supplemented by statistical data. Well-kept records would, according to Bernoulli's theorem, eventually suggest good guesses for the a priori values of causes, which could in turn be tested by Bayes' theorem against the available evidence. At present, the moral sciences lacked data, and Laplace urgently recommended government-sponsored statistical projects.[78]

[77] Laplace, "Recherches," in *Oeuvres*, vol. 8, p. 147.

[78] See Stephen M. Stigler, "Napoleonic statistics: The work of Laplace," University of Wisconsin Department of Statistics Technical Report No. 376 (June 1974).

In the meanwhile, Laplace preached caution, enlisting the calculus of probabilities in support of the status quo. However flawed the present social order, Laplace maintained that our ignorance of the possible pernicious consequences of change outweighed all disadvantages: "In this state of ignorance, the theory of probabilities prescribes against all change; it is especially necessary to avoid abrupt changes, which, in the moral as in the physical order, never operate without a great loss of *vis viva*."[79]

Laplace applied inverse probabilities to a wide spectrum of problems, ranging from the theory of observational error to demography. His treatment of the proportions of male and female births was typical of his mathematical approach, as well as of his determination to apply the calculus of probabilities to problems raised by statistical data in the moral realm. John Arbuthnot had noted the persistent discrepancy between the annual totals of male and female births in a controversial paper published in 1710.[80] Statistics compiled for Paris and other European cities confirmed Arbuthnot's London observations that (registered) male births preponderated year after year. Over the period 1745–84, the ratio of male to female births recorded in Paris, for example, was 25/24. On the basis of this data, Laplace asked: What is the probability that the a priori probability of a male birth in Paris was greater than 1/2? Let

$$x = P \text{ (male births)}$$
$$1 - x = P \text{ (female birth)}$$
$$p = \text{total number of male births observed}$$
$$q = \text{total number of female births observed.}$$

Then the probability of the observed ratio p/q for any value x was

$$P\left(\frac{p}{q}\right) = \int_0^1 \left(\frac{p+q}{q}\right) x^p (1-x)^q \, dx.$$

So by Bayes' theorem,

$$P(x > 1/2 \,|\, p/q) = 1 - \frac{\int_0^{1/2} \left(\frac{p+q}{q}\right) x^p (1-x)^q \, dx}{\int_0^1 \left(\frac{p+q}{q}\right) x^p (1-x)^q \, dx}.$$

[79] Laplace, *Essai*, in *Oeuvres*, vol. 7, p. lxxviii; compare Adolphe Quetelet, *Sur l'homme et le développement des ses facultés, ou Essai de physique sociale* (Paris, 1835), vol. 2, pp. 291–292.

[80] See n. 42, above.

Using the mathematical methods developed in Book I of the *Théorie analytique* to approximate these integrals for the very large values of p and q involved in this case, Laplace concluded that the probability that $x > 1/2$ for Paris was at least equal to that of "the most well-attested historical facts" and approached certainty when the figures for all of Europe were used to evaluate p and q.[81]

Here, Laplace believed, was a paradigm case of the utility of probability theory to the moral sciences. Observations conducted from London to St. Petersburg had suggested the hypothesis that boys were more likely to be born than girls. Bayes' theorem showed precisely how probable this hypothesis was, on the strength of the available evidence. A fundamental law concerning the propagation of the human species had been impressively confirmed by inverse probabilities. Conversely, the same method could be applied to weed out chance aberrations. If, for example, certain smaller towns reported a preponderance of female births, Laplace denied that such anomalies betrayed any variation in the natural law. Further observation would, he was confident, affirm the results gathered for the big cities. If some philosophers had been misled by these anomalies to futilely seek "the cause of phenomena which are only due to chance," their fruitless efforts only proved the need for testing all hypothetical causes by the method of inverse probabilities.

More controversial was Laplace's law of succession, first set forth in a 1774 memoir in the *Savants étrangers* of the Académie des Sciences. If an event E has occurred p times out of $p + q$ trials and its complement E' has occurred q times, what is the probability that the next event will be of the E class? Using Bayes' theorem, Laplace computed the probability to be[82]

$$\frac{\int_0^1 x^{p+1}(1-x)^q \, dx}{\int_0^1 x^p (1-x)^q \, dx} = \frac{p+1}{p+q+2},$$
$$\text{where } P(E) = x$$
$$P(E') = 1 - x.$$

In a related problem, Laplace sought the probability that if an E had happened m times in a row, it would also happen n more times in succession:[83]

[81] Laplace, *Théorie analytique*, in *Oeuvres*, vol. 7, p. 387.
[82] Laplace, "Causes," in *Oeuvres*, vol. 8, p. 31.
[83] Laplace, *Théorie analytique*, in *Oeuvres*, vol. 7, p. 402.

$$\frac{m + 1}{m + n + 1}.$$

Laplace placed no restrictions on the use of either formula, which could presumably be applied even to cases in which the number of past observations was very small or even zero. In the case of Price's Adam, the probability that the sun would rise on the morrow, after only one observation, would be 2/3, presuming that $x = P$ (sunrise) could take on any value in the interval $[0, 1]$ with equal likelihood.[84]

Condorcet immediately perceived the methodological potential of Laplace's treatment of inverse probabilities. He announced the memoir with a flourish in the introductory précis to the 1774 volume of the *Savants étrangers*—where Condorcet praised inverse probabilities as "the only useful part of this science [of probability], the only one worthy of the serious interest of philosophers"—and recommended it to correspondents such as Turgot. He himself proceeded to exploit the implications of Laplace's findings with respect to the evaluation of scientific hypotheses.

His first opportunity came in 1775 with a joint appointment, along with C. Bossut and d'Alembert, as inspector of navigation by Turgot, then Comptroller-General, with special responsibility to advise the government on technical problems concerning canal building.[85] The result of their labors was the *Nouvelles expériences sur la résistance des fluides* (1777), to which Condorcet appended a piece entitled "Essai d'une méthode pour trouver les loix des phénomènes d'après les observations." Inspired by Lagrange's 1772 memoir on the reduction of data from tables of planetary motions, Condorcet hoped to provide a practical guide to physicists at a more elementary mathematical level. Laplace's law of succession arose in the context of verifying a scientific hypothesis: suppose a function $\phi(x)$ accurately predicts the values of an observable quantity x for m trials. What is the probability that the x value observed in the next n trials will also conform to $\phi(x)$?

Condorcet returned to the urn analogy, in which the unknown number of white balls represented all observed x values that would satisfy $\phi(x)$, and the unknown number of black balls corresponded to anomalous x val-

[84] For criticisms of this assumption, see John Stuart Mill, *A System of Logic* (London, 1843), Book III, chapter 18; George Boole, *An Investigation of the Laws of Thought* (1854) (New York: Dover, 1958), pp. 367–370.

[85] Keith M. Baker, *Condorcet: From Natural Philosophy to Social Mathematics* (Chicago: University of Chicago Press, 1975), p. 68.

ues. If m white balls in a row have already been drawn, the probability the next n balls drawn will also be white was given by Laplace's law of succession: $\dfrac{m + 1}{m + n + 1}$.[86] Condorcet warned that if $n > m$, that is, if the predicted span of observations equaled or exceeded the observed run, the probability that the values of $\phi(x)$ would continue to match observations waned. Condorcet also invoked the same rule to justify the widespread practice in the natural sciences of considering laws confirmed by only a few observations as nonetheless "certain," because "phenomena of the same genre subject to similar laws" had been confirmed by numerous observations that could be added to the sum total of past confirming instances m. For Condorcet, the law of succession here seconded the results of a more intuitive use of the scientific method: "The mind links together this entire class of phenomena, and no longer judges according to the probability that each one in particular is subject to certain laws, but rather according to the probability that the entire system is so governed."[87]

Condorcet addressed the probability of causes at greater length in Part IV of his six-part memoir on probability theory in the 1783 volume (published 1786) of the *Mémoires de l'Académie royale des Sciences*: "Réflexions sur la méthode de déterminer la probabilité des événements futurs, d'après l'observation des événements passés." In this somewhat convoluted memoir, Condorcet attempted to extend the law of succession to cases in which the causes varied over time. Instead of modeling induction with drawings from a single urn containing a fixed ratio of black-to-white balls, Condorcet supposed each observed event corresponds to a drawing from a different urn, each with a different black-to-white ratio: the causes would thus vary by some unknown amount for each event. In such cases, Condorcet contended that probabilists must apply Bayes' theorem to the *average* value of the probability of the cause, which remained constant.

Condorcet's revised urn model incorporated a number of assumptions about both the nature of natural causes and the method of discovering them. Like Laplace, he was explicit in assuming all possible ratios of black-to-white—that is, all a priori probabilities of causes—to be equally probable, on the grounds that our ignorance gave no reason to prefer one hypothesis to another. Condorcet also invoked a global determinism to

---

[86] Condorcet, d'Alembert, and Bossut, *Nouvelles expériences sur la résistance des fluides* (Paris, 1777), pp. 198–199.

[87] Condorcet et al., *Expériences*, p. 201.

justify his averaging of the individual probabilities: unless the probabilities, however variable, of the causes of the individual observed events followed some general law, this average value (*valeur moyenne*) would be meaningless. Indeed, Condorcet claimed that the only difference between the case of variable and constant causes lay in our ignorance of the general law that regulated the former. Because such a law was presumed to exist, recourse to average values was a valid interim method.[88] In Condorcet's hands, the probability of causes remained a theory based on the imperfections of human knowledge—a model of the making and testing of hypotheses about causes as well as of the operation of the causes themselves.

Condorcet also raised the issue of order in long series of observations. If we know that A has happened 51,000 times and N, 49,000 times, out of the past 100,000 trials, we would be much more confident in assuming $P(A) = x$ to be constant were it known that out of every one hundred successive observations the ratio remained roughly 51:49. However, if the 100,000–event sequence began with A predominant, with N gradually gaining ground, would there not, Condorcet queried, "be reason to believe that N would surpass A in the future? . . . The results of the hypothesis that x is constant are thus here in contradiction with what reason would seem to indicate." Hence, the rules of inverse probabilities must be modified to suit the practice of the alert natural philosopher.

Condorcet interpreted this result as a mandate for modesty in hypotheses generalizing past regularities to future instances: the force of such generalization diminished as the number of past events decreased or future events increased. Once again, Condorcet noted, the probabilistic results confirmed accepted practice.[89] Condorcet did not take the radical step of doubting the existence or stability of natural laws, only the durability of human hypotheses concerning them. Unlike Hume, he did not sever all connections between past experience and future expectation: observations of past events did render future repetitions more probable. However, the probability was limited by the extent of both past experience and future predictions, as in Locke's parable of the King of Siam's view on ice. Inevitably formed on the basis of incomplete data, scientific hypotheses could at best be regarded as first approximations to more complex expressions.

---

[88] Condorcet, "Réflexions sur la méthode de déterminer la probabilité des événements futurs . . . ," *Mémoires de l'Académie royale des Sciences. Année 1783* (Paris, 1786), p. 540.
[89] Condorcet, "Réflexions," pp. 542, 548.

Through a process of extended observation and consequent adjustment of hypotheses, natural philosophers could improve—but not perfect—their approximation of the true causes. The variation in the probabilities of causes was only apparent, the result of human failure to grasp the ruling natural law in its entirety:

> We would not dare to assert that [even] the most regular law which we observe in phenomena will persist without any alternation for an indefinite time. We suppose in truth that there may exist a more complicated constant law, which for a time seems the same to our eyes as the one first posited, and which subsequently deviates perceptibly from it, but it is easy to see that this is precisely the case in which the first law having ceased to be constant, we substituted another which embraces both the phenomena encompassed by the first law and those which appeared to diverge from it.[90]

From these epistemological reflections on the restricted probability of scientific hypotheses, Condorcet derived several methodological rules. Out of the three possible hypotheses concerning the probability of causes—constant probabilities, variable probabilities independent of time and order, and variable probabilities dependent on time and order—Condorcet recommended the third as the most general. Although Condorcet retained the maxims of simplicity and comprehensiveness, advising natural philosophers to select the simplest law that also subsumes the largest proportion of observations, he implicitly discarded—at least at the epistemological level—the precept that like causes produced like effects. His revised urn model had postulated a string of indistinguishable effects A (e.g., drawing a black ball), each produced by a different cause (the particular ratio of black balls to white balls). Nonetheless, Condorcet remained faithful to Jakob Bernoulli's thoroughgoing determinism, never doubting that all phenomena were governed by constant laws.

Condorcet did not implement inverse probabilities as a description of the actual interplay of causes nor yet as a description of mental states, but rather as a description of the interaction of these objective and subjective elements. His urn model was epistemological, rather than ontological or psychological: it represented the rational method of investigating the laws of nature and society. Like Laplace, Condorcet recognized a distinction

---

[90] Condorcet, "Réflexions," p. 548.

between probabilities and beliefs (*motifs de croire*), between the physical causes of sensations and the sensations themselves. However, Condorcet located probability theory in the intermediate zone between the two: his mathematical research in this area explored the connection between the objective and subjective aspects of probability.

In his earliest writings on probability, Condorcet had emphatically denied any connection between probabilities and the "true reality."[91] In later works, he was at pains to enforce a parallel distinction between probabilities and belief. In the philosophical Preliminary Discourse to his *Essai sur l'application d'analyse à la probabilité des décisions* (1785), Condorcet elaborated both distinctions. If a ball is drawn from the canonical urn filled with black and white balls in a given ratio, and then covered with a veil, Condorcet argued that the probability that the ball is white is the same prior to the drawing and after, until the veil was removed. Even though the prior probability pertained to a future possibility and the posterior probability to a realized certainty, our knowledge—therefore the probability—of the two events is identical: there existed no "immediate connection between . . . the probability and the reality of events."[92]

Nor was this probability to be confused with subjective belief, although the two were related. Defined as the ratio of the number of combinations favorable to the given event to the total number of possible combinations (all presumed equiprobable), probability qualified as a "rigorous truth" of mathematics. Condorcet posited a proportionality between the "motive to believe" and the combinatorial probability: if the combinatorial probability of an event was known to exceed that of its complement, there was more reason to believe that the former would occur than the latter. This motive for belief grew with the combinatorial probability, and, moreover, it increased *proportionally* to this probability. It followed that equal probabilities yielded equal motives for belief, and that, by Bernoulli's theorem, even a small difference in probabilities ultimately warranted a proportionate discrimination in belief, measured by the long-term ratio of correct to total judgments based on this scale of credibility.

Condorcet mathematicized Locke's qualitative treatment of probabilities proportioned to degrees of assent. Scientific hypotheses, like all other conjectures, depended on probabilities to provide "un juste motif de

[91] Condorcet, Bibliothèque de l'Institut MS. 875, f. 93.

[92] Condorcet, *Essai sur l'application de l'analyse à la probabilité des décisions rendues à la pluralité des voix* (Paris, 1785), p. x.

croire." It would be tempting, given Hume's express interest in probabilities, to find Humean, as well Lockean, overtones in Condorcet's work. However, neither external nor internal evidence supports such a reading. Keith Baker suggests that Condorcet may have read Hume at Turgot's urging, but acknowledges that there is no extant record that he did so.[93] If Condorcet did know Hume's work, it appears to have made little impact on his philosophy of probability and induction. Condorcet is in fact far closer to Price's views on both subjects. It is perhaps more illuminating to consider all three within the postpyrrhonist tradition, with Hume representing a rejuvenated skepticism against the mathematicized moderation that Price and Condorcet adapted from Mersenne, Boyle, Wilkins, Locke, et al. Condorcet, like Price, believed past experience to be a valid basis for future expectations. Probabilities, not just habit or instinct, vouchsafed the legitimacy of inductive generalizations. All our knowledge about the future, Condorcet claimed, rests upon two principles: first, that nature follows invariable laws; and second, that we are capable of discovering these laws through observation.

At first glance, these principles appear to parallel Hume's a priori assumption of the uniformity of nature. However, Condorcet's rationale for both principles was a posteriori and hence vulnerable to Hume's charge of circularity: "Our constant experience that facts conform to these principles is for us the only reason to believe in them."[94] If all of the relevant observations could be assembled even the probability that nature is uniform could be evaluated. Moreover, the exact weight of belief could be computed from inverse probabilities. Condorcet reproduced Price's sunrise example, as well as other instances of "intuitive" judgments based on probabilities derived from experience. All knowledge, aside from the intuitive self-evidence of consciousness, may be probabilistic—and Condorcet extended this maxim to apply to even mathematical demonstrations—but it was genuine knowledge nonetheless.

It is thus Condorcet, rather than Laplace, who must be praised or blamed for a thoroughgoing epistemological interpretation of probability theory. Whereas d'Alembert's criticisms of probability theory prompted Condorcet to clarify the relationships between reality, probability, and belief, they apparently persuaded Laplace to take a more robustly objective stance, exemplified by his sharp distinction between mathematical and

[93] Baker, *Condorcet*, p. 139.　　[94] Condorcet, *Essai*, p. xi.

moral probabilities. Early in his career, Laplace voiced his preference for a more objectivist interpretation of probabilities, a preference that intensified in later works. Despite the attention they have received from critics beginning with Boole, Laplace's epistemological assumptions—almost all of which concern the a priori equiprobability of combinations or of causes—are few and explicit. Moreover, Laplace expended much mathematical energy attempting to overcome these assumptions by appeal to experience—as, for example, in his treatment of asymmetric dice and coins and the distribution of observational error.

Laplace's secular version of Jakob Bernoulli's reconciliation of probability theory to a thoroughly determined universe was so eloquent that he has sometimes been called its originator. However, it would be more accurate to say that Laplace's work, particularly his enthusiasm for statistics, marked the beginning of the end for the classical interpretation that had dominated eighteenth-century probability theory. Laplace is a figure of transition; Poisson, Laplace's avowed disciple, made the first unambiguous distinction between objective and subjective probabilities, which Cournot and Ellis expanded into a full-scale critique of the classical interpretation.

Siméon-Denis Poisson, a member of the first generation of mathematicians to graduate from the Ecole Polytechnique, was a protégé of Laplace.[95] There is little doubt that Poisson's interest in probability theory, like many of his other research interests, was sparked by Laplace's work on the subject. Although Poisson published a sprinkling of papers on the application of probability theory to demography and to artillery aim, he presented his major contributions and incorporated earlier results in his treatise *Recherches sur la probabilité des jugements en matière criminelle et en matière civile* (1837).

As the title suggests, the principal problem addressed in the treatise was the one raised by Condorcet in his *Essai sur l'application d'analyse à la probabilité des décisions* (1785) and later discussed by Laplace in Chapter 11 of the *Théorie analytique des probabilités* (1812): how to determine the probability that a jury, tribunal, or deliberative assembly would reach a correct decision. The history of probabilistic attempts to solve this problem will be treated in the next chapter. However, Poisson intended the mathematical and philosophical framework expounded in the first four chapters of

[95] Maurice Crosland, *The Society of Arcueil* (Cambridge, Mass.: Harvard University Press, 1967), pp. 127–128; 212–213.

284

the treatise to serve as the basis for a broad range of problems in both the natural and moral sciences. In particular, Poisson believed that his generalization of Bernoulli's theorem, which he christened the "general law of large numbers," stated a universal truth about both the physical and moral order that was at once a mathematical theorem and a fact of experience, "which never lies." For Poisson, the law of large numbers was both an inductive result itself and the guarantee of the validity of all other inductive results: it was the mathematical expression of the uniformity of nature.

Poisson, like Condorcet, regarded the model of causation implicit in Bernoulli's theorem as an oversimplification. Reasoning from his urn model, Bernoulli had assumed the probability of any given outcome to be fixed for the sequence of trials. In response to Leibniz's objection that some of Bernoulli's intended applications—for example, to the causes of human mortality—might well involve causes that varied in number and probability over time, Bernoulli had acknowledged that separate runs of trials should be planned to detect such variations. However, he did not explain how this precaution might be incorporated into his theorem. Poisson took up Condorcet's emphasis on average values and used it to revise Bernoulli's theorem. Bernoulli's theorem estimated the proportions of the terms of the binomial expansion by various approximation techniques, assuming that the probabilities of the causes were constant for each trial, but Poisson observed that this assumption seldom held in reality: "Yet, in the application of the calculus of probabilities either to diverse physical phenomena or to moral matters, these chances often vary from one trial to the next, and also often in a completely irregular way."[96]

Poisson amended Bernoulli's theorem by considering a large number n of consecutive trials. The probability of the given event E on the first trial, due to cause C_1, was p_1; on the second trial, due to cause C_2, p_2; etc. The average chance p' was therefore:

$$p' = 1/n(p_1 + p_2 + \ldots + p_n).$$

If E occurred m times out of the n trials, then the law of large numbers stated that for any ϵ, however small,

$$\lim_{n \to \infty} P(|p - m/n| < \epsilon) = 1.$$

[96] Siméon-Denis Poisson, *Recherches sur la probabilité des jugements en matière criminelle et en matière civil* (Paris, 1837), p. 139.

Moreover, if E happened m times out of a large number n trials, there was a high probability that it would happen m' times out of a large number of future trials n', such that

$$m/n \approx m'/n'.$$

If m/n and m'/n' are observed to differ significantly, Poisson maintained, as Bernoulli had to Leibniz, then there was good reason to believe that some of the causes C_1, C_2, C_3, . . . have been replaced by others, which altered the probabilities of producing E in a given trial. If there existed only a single constant cause C, the law of large numbers reduced to Bernoulli's theorem. Poisson interpreted the law of large numbers as "a universal law" that governed phenomena of all kinds, moral and physical alike. All phenomena, even those apparently due to "blind chance," eventually could be shown to result from the interplay of permanent and variable causes. In all cases patient observation sufficed to reveal a stable ratio m/n, which approximated the average probability p' ever more closely with repeated trials. De Moivre had drawn similar conclusions from Bernoulli's theorem in the *Doctrine of Chances*. However, Poisson's claims for the law of large numbers differed from De Moivre's interpretation of Bernoulli's theorem in two important respects: first, Poisson emphasized that the law of large numbers applied to all phenomena, not just to the natural realm; and second, Poisson denied that the law of large numbers implied providential design.

While De Moivre had praised John Arbuthnot's use of demographic data and probabilistic reasoning to support the argument from design, he did not systematically explore the potential of Bernoulli's theorem, combined with statistical compilations, to uncover regularities in the moral realm. The lack of data may have discouraged any such project. In contrast, Poisson was able to profit from a growing store of official statistical information, including the *Compte général de l'administration de la justice criminelle en France* begun by the French Minister of Justice, Charles-Ignace de Peyronnet, in 1825 and published annually thereafter. Moreover, Poisson knew Adolphe Quetelet's recently published *Sur l'homme et le développement de ses facultés, ou Essai de physique sociale* (1835), which showed that the rates of marriages, crimes, suicides, and other moral phenomena remained surprisingly constant from year to year. Quetelet, eager to implement Laplace's suggestions for a moral science modeled on the methods of natural sciences,[97] maintained that moral phenomena considered collectively obeyed laws as regular as those that governed physical phenomena. Poisson corresponded with Quetelet (they exchanged copies of French and Belgian statistical digests), and many of his examples for the universal applicability of the law of large numbers, such as the constant proportion of criminal convictions, may well have been drawn from the Belgian statistician's work.[98] Poisson's comparison of the irregular effects of human will, passions, and intelligence on these "moral" laws to the influence of variable causes on the permanent causes of physical phenomena[99] precisely paralleled Quetelet's account of how the "immutable" social laws ultimately obscured the "perturbing forces" of individual will and circumstance.[100]

The law of large numbers applied equally to the birthrates of males and females, coin tosses, shipwrecks, tax revenues, molecular intervals, and crime rates, according to Poisson. Its generality stemmed from its abstract consideration of causes. Although the calculus of probabilities could demonstrate the existence of a constant cause, it could tell us nothing of its nature. In the case of Buffon's coin-tossing experiment, the law of large numbers confirmed that a "permanent cause" favored one side, but only the science of mechanics could identify the cause.[101] Although the law of large numbers admitted mathematical demonstration, it was primarily an experiential truth: "From these examples of all kinds, it follows that the universal law of large numbers is for us already a general and incontestable fact, resulting from experience, which never lies."[102] It was also the "basis for all applications of the calculus of probabilities," and since the law of large numbers held for both physical and moral phenomena, the calculus

[97] Even the frontispiece quotation of Quetelet's *Sur l'homme* was taken from Laplace's *Essai philosophique*: "Appliquons aux sciences politiques et morales la méthode fondée sur l'observation et sur le calcul, méthode qui nous a si bien servi dans les sciences naturelles."

[98] Poisson-Quetelet correspondence, Fonds Quetelet, MS. 2030. In letters dated 27 March 1836 and 26 June 1836, Poisson requested statistical data regarding the Belgian judicial system from Quetelet. Poisson may also have been drawing on the work of the lawyer André Michel Guerry, whose *Essai sur la statistique morale* (Paris, 1833) also emphasized the regularities in judicial statistics.

[99] Poisson, *Recherches*, p. 12.

[100] Quetelet, *Sur l'homme*, vol. 1, pp. 18–19.

[101] Poisson, *Recherches*, p. 169.

[102] Poisson, *Recherches*, p. 12.

or probabilities applied equally to both, despite the reservations voiced by several "fine minds" regarding its extension to the moral sciences.[103]

For De Moivre, the stability of statistical ratios and the underlying constant causes thus discovered could only be ascribed to the orderly, providential arrangement God had imposed on the universe. Poisson, however, dissociated the law of large numbers from the argument from design. While he recognized that the remarkable tendency of all kinds of phenomena to exhibit statistical regularities tempted many to posit an "occult power, distinct from the physical and moral causes of events, and acting with an eye toward order and conservation," Poisson insisted that the law of large numbers needed no such prop. The calculus of probabilities showed that the law was "the natural state of things, which subsists by itself without the help of any foreign cause and which would, on the contrary, require such a cause in order to undergo a significant change."[104] The statistical regularity enforced by the law of large numbers was built into the universe in the same way that inertia was: a body at rest or in uniform rectilinear motion required no external sustaining cause.

Poisson's nature was orderly but not under orders. Despite its universality, the law of large numbers was a secondary cause, not a token of divine providence. Poisson agreed that the manifest "harmony" of nature could not be the product of chance alone, but interpreted statistical regularities as indicative of physical causes, not the veiled hand of God. Analogy persuaded Poisson that the as yet unknown causes of other physical phenomena would be as regular as those already discovered.[105] Neither Poisson nor De Moivre could conceive of a universe ruled by chance alone, but for De Moivre the alternative was divinely imposed order; for Poisson, the inherent order of nature itself.[106]

---

[103] Poisson, *Recherches*, p. 7. Poisson may have had Auguste Comte or Destutt de Tracy in mind here, both of whom had criticized Condorcet's program for "social mathematics."

[104] Poisson, *Recherches*, pp. 144–145.

[105] Poisson, *Recherches*, p. 118.

[106] I. J. Bienaymé criticized Poisson's law of large numbers from the standpoint of the practicing statistician, claiming that it was formally identical to Bernoulli's theorem. Bienaymé suggested that Bernoulli's theorem be modified in certain cases to take account of short runs of nonindependent trials: Bienaymé, "Communication sur un principe que M. Poisson avait cru decouvrir et qu'il avait appelé Loi des Grands Nombres," *Séances et Travaux de l'Académie des Sciences Morales et Politiques*, 3rd series, 11 (1855): 379–389; see also C. C. Heyde and E. Seneta, *I. J. Bienaymé: Statistical Theory Anticipated* (New York: Springer-Verlag, 1977), pp. 40–49.

The law of large numbers dominated Poisson's treatment of causality and inverse probabilities. Far more than any of his predecessors, including Laplace, Poisson distrusted the convenient assumption made by Bayes' theorem that the a priori probabilities of causes were equal. Poisson faulted specific aspects of Laplace's treatment of the probability of judgments on these grounds, and argued that unless inverse probabilities took stock of all relevant information, including that pertaining to prior probability distributions, its solutions would contradict observation and good sense. Bayes' theorem should be supplemented by the observational estimates of probabilities derived from the law of large numbers and by sound reasoning. Particularly with respect to the probability of judgments, Poisson urged that "if one had a priori some reason to believe more in the existence in one cause than in that of another, it would be necessary to take account of this inequality of the chances of C_1, C_2, C_3, etc. [i.e., the causes of event E], prior to observation in the evaluation of the probabilities which these various causes have acquired after the occurrence of E."[107]

In a Humean argument against the credibility of reports of miracles, Poisson invoked our anterior knowledge of the "general laws of nature" and the overall harmony of their interaction to estimate the a priori probability of such a report as nearly null, which only the testimony of an infallible witness could balance. Poisson went so far as to recommend doubting one's *own* experience in cases of apparent miracles: the infinitesimal a priori probability, derived from eons of human observations, made any violation of the laws of nature less likely than hallucination or insanity.[108]

Poisson's model of causation admitted multiple causes that issued forth in the same observed event. As a rule, each causal strand C_i created a different probability value for the event E: that is, $P(E|C_i) \neq P(E|C_j)$. Poisson distinguished between two classes of causes, according to the strength of the bond between cause and effect: those which produced an event necessarily—that is, $P(E|C) = 1$; and those which did so only contingently: $P(E|C) < 1$. The latter class of causes produced events only in combination with "hazard" or "variable" or "accidental" causes. Poisson defined "hazard" as the nexus of causes that influenced the occurrence of an event without altering its "chance." By "chance" Poisson meant the objective propensity "independent of our knowledge" of an outcome, as opposed to "probability," the subjective reason to believe. In practice, Poisson usually

---

[107] Poisson, *Recherches*, pp. 93–94.   [108] Poisson, *Recherches*, p. 99.

identified "chance" with the ratio of favorable to total combinations. The distinction between "chance" and "hazard" could be a delicate one. For example, in a dice game, the outcome of the toss is the result of the number and markings of the die faces, the irregularities in the form and density of the die, and the shaking of the tumbler. According to Poisson, only the first two factors affected the chance of the outcome; the last was only hazard. The shaking of the tumbler merely served to erase "the influence of the position of the die in the tumbler before these motions, for fear that the initial position might be known by one of the players."[109] Hazard was epistemological, canceling out the effects of any foreknowledge, while chance depended solely on physical features of the situation. To draw a ball "at hazard" from an urn filled with otherwise identical black and white balls meant "not to look at their arrangement inside the urn before drawing." Although the mechanisms of hazard were often physical—shaking dice, shuffling cards, mixing balls—its purpose was to insure that the Principle of Indifference obtained. If all players were equally ignorant concerning initial position of the die, cards, or balls, only the physical circumstances of chance entered into the evaluation of probabilities.

Variable causes, as opposed to hazard, could alter the average chance of an event p' over a long run of trials, and therefore affect the observed ratio m/n. Variously labeled "variable," accidental," or "irregular," these causes masked the constant effects of "a permanent cause." Poisson's variable causes closely resembled Laplace's irregular causes. Depending on the "amplitude" of the variations, lesser or greater numbers of trials would be required to establish the constant ratios discovered everywhere by the law of large numbers. Like the irregular causes of his mentor, Poisson's variable causes were symmetric. Because they were truly haphazard, these causes were equally likely to shift the observed ratio m/n from the average probability p' in either direction, ultimately canceling out one another's effects in the long run.

Poisson deployed his distinction between necessary and partial causes to respond to Hume's rejection of any necessary connection between cause and effect. Poisson conceded that Hume's claim that habit joined cause to effect by "a sort of association of ideas" held for most people "who do not examine the principle of their belief and its degree of probability" and who associate cause and effect by convention, as they associate word and thing.

<hr>

[109] Poisson, *Recherches*, p. 80.

However, Poisson maintained that "reason or the calculus," as well as habit, provided "a great assurance that in the future causes will always be followed by their effects."[110] Poisson argued his point with a characteristic appeal to both the practice of reasonable men—in this case, natural scientists—and the results of probabilistic calculation. True to the classical tradition in probability theory, Poisson assumed that reasonableness and the calculus of probabilities were interchangeable.

Hume had contended that the mind proportioned future expectations to the weight of past experience through the association of ideas:

As the habit, which produces the association, arises from the frequent conjunction of objects, it must arrive at its perfection by degrees, and must acquire new force from each instance, that falls under our observation. The first instance has little or no force: The second makes some addition to it: the third becomes still more sensible; and 'tis by these slow steps that our judgment arrives at a full assurance.[111]

Hume seconded Locke's view that the "wise man" carefully apportioned "his belief to the evidence," weighing experiences that confirmed and contradicted the hypothetical connection between cause and effect.[112] Poisson objected that scientists regularly achieved full conviction that a phenomenon would recur from only a few, or even a single, observation. Citing H. C. Oersted's discovery that an electric current deflected a nearby magnetic needle and J. B. Biot's experiments on the polarization of light, Poisson remarked that both scientists and their colleagues had been confident that the novel effects could be reproduced in future experiments without actually conducting these additional trials. He therefore concluded that mere constant conjunction alone, no matter how frequently reiterated, was not the sole source of the belief that the future would resemble the past: "The confidence of the mind in the recurrence of effects following their causes cannot be uniquely grounded in prior observation of this succession more or less frequently repeated."[113] Hume's associationist theory could not account for the conduct of Oersted, Biot, and other scientists.

Instead, Poisson proposed that their confident expectation sprung from the "possibility" that a cause might be of the strong sort, which necessarily

[110] Poisson, *Recherches*, p. 162.   [111] Hume, *Treatise*, p. 130.
[112] Hume, *Enquiry*, p. 118.   [113] Poisson, *Recherches*, p. 165.

produced the observed effect on each and every occasion. After all, for a given phenomenon E it was not absolutely impossible, prior to observation, that there existed a cause C which produces E necessarily. Poisson believed that one could assign an a priori probability p to C, from "particular considerations which rendered it more or less plausible." It remained to determine the a posteriori probability w of C by inverse probabilities. Poisson made use of his hierarchy of weak and strong causes and the deceptive interference of "hazard" to solve the problem. Whatever measures were taken to screen out the effects of causes other than C, it would be impossible to entirely eliminate their influence. Suppose there existed weak causes $B_1, B_2, \ldots B_n$ (known or unknown) capable of producing E by "combining with hazard," without C, such that B_1 produces E in the first trial, B_2 in the second trial, etc. Let the probability that some weak cause B_i will produce E on a given trial be

$$r_i = P(B_i) \times P(E|B_i).$$

Let the probability of producing E in all n trials by the B_i be ρ, such that

$$\rho = r_1 \times r_2 \times \ldots r_n.$$

Since the probability of not-C is $1 - p$, then the probability of E occurring in all n trials without C would be

$$(1 - p)\rho.$$

However, the probability of E happening in all n trials with C would be

$$P(C) \times P(E|C) = p,$$

since C is assumed to be a strong cause. Using Bayes' theorem, Poisson derived the probability of C, given that E had been observed in n trials:

$$w = \frac{p}{p + (1 - p)\rho}.$$

According to Poisson, this formula vindicated the conduct of Oersted and his fellow scientists. Thus, Poisson argued, even a minuscule a priori probability of C yielded "a probability very close to certainty of the occurrence of this phenomenon in future experiments" after only a few trials.[114] Poisson believed that he had used the calculus of probabilities to describe the psychology of rational belief more accurately than Hume's associa-

[114] Poisson, *Recherches*, p. 161.

tionism, since the latter's account contradicted the practice of manifestly reasonable men like Oersted and Biot.

Poisson's argument lies almost wholly on the psychological plane. He did not use his results to make the metaphysical claim that necessary causes did indeed exist in nature, or even that belief in necessary causes was rational because affirmed by calculation. Rather, he claimed that if we admitted the possibility that a necessary cause might be responsible for the observed phenomenon, then inverse probabilities showed that a very small a priori probability could be parlayed into a large a posteriori probability on the strength of very few experiments. Since reasonable men often acted as if a few experiments had created such a large probability—that is, conviction—that a necessary cause existed, then, Poisson asserted, they must believe in the existence of necessary causes: ". . . the object of the preceding calculations was to show that our belief in the future repetition [of the phenomenon], when it has been observed only a few times, can only be founded on the idea that we have of a cause capable of producing a phenomenon of this nature necessarily."[115]

Of course, Poisson's formula was subject to many of the same strictures he himself had urged against the uncritical use of Bayes' theorem. The values assigned to p and (more important) to the r_i in his example were totally arbitrary, although they could be partially rationalized by an appeal to Poisson's model of causation, which estimated both $P(B_i)$ and $P(E|B_i)$ very low. Philosophically, it did not prove that necessary causes existed in nature (although Poisson certainly assumed that they did), only that such an assumption was consistent with the behavior of natural scientists— which Hume had never doubted.

5.5 *Conclusion*

Despite its dubious validity, Poisson's rebuttal to Hume served to dramatize the most important features of the attempts made by classical probabilists to construct a mathematics of induction, and thus to summarize the major points made in this chapter. Poisson, like his predecessors from Jakob Bernoulli through Laplace, applied the probability of causes by means of what Laplace had described as "a very delicate metaphysics,"[116] a

[115] Poisson, *Recherches*, p. 168.
[116] Laplace, "Mémoire," in *Oeuvres*, vol. 9, p. 383.

model of causation. In addition to the ubiquitous urn model, Poisson erected a hierarchy of necessary and partial causes, distinguished sharply between the objective and subjective components of "chance" and "probability," added the additional, quasi-epistemological element of "hazard," and admitted the existence of causes with probabilities that varied from trial to trial. Finally, Poisson believed in the existence of a determinate and self-sufficient order of causes that regulated both the natural and moral realms. Although these causes might vary at the level of the individual event, order emerged en masse and over the long run.

This last insight guided quantitative social research for much of the nineteenth century, through the work and influence of Quetelet. The probabilities that this person might commit suicide or that letter go astray fluctuated from case to case, but the grand totals of suicides and dead letters seemed astonishingly stable from year to year. It was typical of the probability of causes that it should find its most important applications in the moral rather than the natural sciences. Although the classical probabilists hoped to solve the old problem of how to reason rigorously from effects to causes, and although they paid close attention to the actual practice of natural philosophers, their idea of a cause as an underlying probability was a strange one for scientists intent on uncovering physical mechanisms. Probabilists were in fact not so much interested in answering the question, "What caused this effect?" as they were in replicability: "How many times must this effect happen for it to be reasonable to expect it the next time?" Since the most dramatic phenomena then under investigation—electrical and chemical effects—interested natural philosophers precisely because they always recurred, it is not surprising that physicists and chemists had little use for the calculus of induction. It was in fields plagued by variability even at the level of effects—such as medicine, demography, and the moral sciences—that the classical probabilists found their best examples and most dedicated followers in the nineteenth century. They were largely ignored by the physical scientists and reviled by later probabilists, who dismissed most of the probability of causes as so much armchair algebra in lieu of hands-on empiricism.

It is tempting to connect the probability of causes with post-Galtonian statistical inference into a single, long-lived tradition. There are striking similarities in the kinds of questions posed, the enthusiasm of, above all, social scientists for these techniques, and even in the methodological ideal of a calculus of induction, a mechanized form of empirical inference. But there are equally striking contrasts: the classical probabilists looked upon their calculus of induction as descriptive, not prescriptive of scientific practice; they were genuinely interested in causes rather than correlations, despite their odd construal of the word; and they were not afraid to introduce metaphysical considerations about chance and causation when the occasion arose. Indeed, it is almost impossible to disentangle the methodological from the metaphysical assumptions in the probability of causes. How scientists worked and the workings of nature were assumed to fit together hand in glove: no classical probabilist concerned with the one could afford to ignore the other. We cannot surgically excise a respectable, modern core from the work of the classical probabilists without losing the spirit of the enterprise as well. Several of their assumptions, like those underpinning Laplace's law of succession, eventually discredited the probability of causes, but it was those same assumptions that made the application thinkable in the first place.

CHAPTER SIX

Moralizing Mathematics

6.1 Introduction

Almost from the inception of mathematical probability, the classical probabilists had hoped that their calculus would mathematize what were then
called the moral sciences: jurisprudence, political economy, and other
studies of social relationships. Part IV of Jakob Bernoulli's *Ars conjectandi*
(1713) was the first full-dress attempt to realize this hope, and it set the
pattern through Siméon-Denis Poisson's *Recherches sur la probabilité des jugements* (1837) in at least two important respects. First, the major applications still smacked of the courtroom, centering on problems of evidence
and judgment, in contrast to the nineteenth-century focus on mass social
phenomena. Second, in order to use mathematical tools like Bernoulli's
theorem and later Bayes' theorem, assumptions about the stability and
uniformity of natural causes were controversially extended to the moral
realm.

As in the cases of probabilistic expectation and the probability of causes,
the application of probability theory to the moral sciences inevitably involved the mathematicians in further, more specific assumptions about the
subject matter. Definitions of expectation required them to speculate over
the criteria for reasonableness; the probability of causes demanded a model
of nature and causation. The chief classical applications to the moral sciences, the probabilities of testimony and of judgments, rested upon assumptions about the nature of belief, historical tradition, and judicial decision, as well as the optimal balance of individual and societal rights. Just
as in the case of the controversy over expectation, these assumptions varied
according to person and circumstance. In the hands of M.J.A.N. Condorcet on the eve of the French Revolution, the probability of judgments was
an arrow in the quiver of liberal reformers; wielded by Poisson fifty years
later, it was a conservative apologetic.

296

Like every other aspect of classical probability theory, these applications to the moral sciences were subject to the dictates of sovereign "good sense." Condorcet insisted that his calculations on judicial tribunals concurred with what "the simplest reason would have dictated," and reassured his Lycée audience that the applications of probability to such problems would lead them "by a sure route, considering only the common interest and justice, to the same humane maxims and magnanimity that you find in your hearts, and teach you the first cry of nature did not lead you astray."[1] Similar passages, albeit without the Rousseauian flourishes, repeat in the works of Pierre Simon Laplace and Poisson like a refrain. Once again, the probabilists modestly claimed only to clarify and explain what reasonable men had believed all along.

Yet John Stuart Mill spoke for the majority of mid-nineteenth-century probabilists, mathematicians and philosophers alike, when he condemned attempts to apply the calculus of probabilities to the credibility of witnesses and to the accuracy of judicial verdicts as an offense to good sense. According to Mill, these abuses of probability theory had made it "the real opprobrium of mathematics."[2] The same calculations that had affirmed Condorcet's common-sensical conclusions concerning legal procedures blatantly contradicted those of Mill. Over a period of some fifty years, good sense had been transformed almost beyond recognition, and assumptions that had once seemed self-evidently true to the classical probabilists—for example, that every individual possessed a certain amount of *lumières* that determined the accuracy of his judgment on each and every occasion, or that good judgment is simply the expression of intuitive calculations—now struck their successors as suspect and even ludicrous.

My aim in this chapter is to explain the rise and fall of the probabilistic approach to the moral sciences, primarily through a close examination of the work of its main exponents: Condorcet, Laplace, and Poisson. How-

[1] Marie-Jean-Antoine-Nicholas Condorcet, *Essai sur l'application de l'analyse à la probabilité des décisions rendues à la pluralité des voix* (Paris, 1785), p. ii; "Discours sur l'astronomie et le calcul des probabilités, lu au Lycée" (1787), in *Oeuvres de Condorcet*, F. Arago and A. Condorcet-O'Connor, eds. (Paris, 1847–49), vol. 1, p. 502.

[2] John Stuart Mill, *A System of Logic* (1843), 8th edition (New York, 1881), p. 382. Mill's antipathy toward the probability of testimony and judgments may well have been influenced by Bentham's critique in his *Rationale of Judicial Evidence*, which Mill edited for publication from Bentham's manuscript drafts: see Jeremy Bentham, *Traité des preuves judiciaires* (Paris, 1823), vol. 2, pp. 53–55 (this two-volume French edition preceded the five-volume English version).

ever, the mathematical writings alone cannot reveal *why* the assumptions these probabilists made about the behavior of witnesses and judges seemed plausible at one time and absurd at another. Nor can they explain why the probabilists believed their calculus applied to the moral sciences in the first place. We must understand something of the Enlightenment program for the moral sciences, the *philosophe* campaign against official injustice and credulous superstition, and the legal background of the mathematical theory of probability to answer these questions. Hence, Section 6.2 describes how the orientation of the moral sciences meshed with that of classical probability theory; Section 6.3 situates the probability of testimony against the background of the debate over miracles; and Section 6.4 examines the probability of judgments as part of a long effort to reform the French judicial system over the course of the Old Regime, the Revolution, the Empire, the Restoration, and the July Monarchy.

6.2 The Moral Sciences

The Enlightenment moral sciences were continuous but by no means identical with the social or human sciences that emerged in the course of the nineteenth century. Students of the moral sciences took the individual rather than society or culture as their unit of analysis; when they discussed society as a whole, it was as an aggregate of such individuals. They were, in the current lingo, methodological individualists, and their explanations of human phenomena were largely psychological. In contrast, the luminaries of the nineteenth-century social sciences—Auguste Comte, Herbert Spencer, Karl Marx, Émile Durkheim—insisted upon the primacy of whole societies as their subject matter, and emphasized structural features such as "social facts" that emerged only at the group level. They shunned psychological explanations for the most part, and were particularly fond of physiological analogies to highlight the emergent properties of the "social organism." Society was the object of rational explanation, not the reasoning individuals who composed it. When the *homme éclairé* of Condorcet's "social mathematics" became the *homme moyen* of Quetelet's "social physics," moral science had become social science.

Another, deeper gulf divided the moral and social sciences. Although it would not be difficult to find many examples of values shaping the work of the nineteenth-century social theorists, their methods and goals were meant to be descriptive, not normative, and they understood the distinction well enough to use it to discredit an opponent's work. However, in Enlightenment moral sciences, as in Enlightenment probability theory, the two were regularly conflated. Moral scientists vied with one another to be the Newton of their field, and talked incessantly of the "natural laws" of the moral realm. But they also spilled much ink trying to persuade governments and citizens to obey these laws: evidently the imperative was moral, not physical. The discoveries of the moral sciences were of what ought to be; social scientists aspired to lay bare what is. Arguments in the moral sciences sounded in some ways remarkably like those of classical probability theory: both were often attempts to convince the recalcitrant—gamblers, insurers, judges, monarchs—that on the basis of reason and self-interest they should alter their practice.

However, the probabilists brought a radically new perspective to the moral sciences that was largely the result of the mathematical tools at their disposal, namely Bernoulli's and Bayes' theorems. The mathematicians conceived of both the probabilities of testimony and of judgments as simply a branch, albeit a particularly important branch, of the probability of causes. This meant assuming that moral and physical causes existed on the same ontological plane, in order to be susceptible to the same mathematical treatment. Condorcet claimed that the moral like the physical sciences were founded on the "observation of facts" and must therefore "follow the same method."[3] Laplace predicted that the same combination of observation and mathematics that had served the physical sciences so well would produce comparable successes in the moral sciences, and even suggested that the causes that regulated moral phenomena were precisely analogous to those governing physical effects: sudden changes in the moral order "as in the physical order, never operate without a great loss of *vis viva*," and are therefore to be avoided. (Laplace had himself weathered enough of these political ruptures to know whereof he spoke.)[4] Poisson believed his law of large numbers to be a universal truth, demonstrating the applicability of the probability of causes to moral as well as physical phenomena.[5]

[3] Condorcet, "Discours prononcé dans l'Académie Française" (1782), in *Oeuvres*, vol. 1, pp. 392–393.

[4] Pierre Simon Laplace, *Essai philosophique sur les probabilités* (1814), in *Oeuvres complètes*, Académie des Sciences (Paris, 1886), vol. 7, p. lxxviii. See also Quetelet's "Principes de mécanique qui sont aussi susceptible d'application à la société," Quetelet MS. 110, Bibliothèque Albert 1er, Brussels.

[5] Siméon-Denis Poisson, *Recherches sur la probabilité des jugements en matière criminelle et en matière civile* (Paris, 1837).

Such claims of homogeneity ran counter to the philosophical foundations of the moral sciences. Most eighteenth-century social theorists accepted the Cartesian distinction between brute, passive matter governed by entirely determinate, mechanistic laws, and active, immaterial mind where volition reigned with complete freedom. This was why the purported laws of the moral realm were matters of choice. Charles de Secondat de Montesquieu was typical not only in defining laws as "necessary relations arising from the nature of things," but also in qualifying that necessity in the moral realm:

> But the intelligent world is far from being so well governed as the physical. For though the former has also its own laws, which are of their own nature invariable, it does not conform to them so exactly as the physical world. This is because, on the one hand, particular intelligent beings are of a finite nature; and consequently liable to error; and on the other, their nature requires them to be free agents.[6]

Even the physiocrats, the most militant spokesmen for a law-governed moral science on the Newtonian model, believed that intelligent agents chose to obey moral laws, and that these were therefore of a weaker order than those operating on the physical universe. While man could not escape "the machine of Nature" and its "universal law," he was free to defy moral laws, albeit at his own risk. François Quesnay and his followers maintained that the natural order existed for human benefit, and that it was therefore in our best interests to take advantage of our privileged position in creation by obeying the laws discerned by human intelligence. Failure to do so, either through ignorance or perversity, resulted in economic disaster, but we were free to err. The laws of the moral order exacted obedience in the sense that mathematical demonstration coerced assent, through the exercise of reason. Natural laws and natural rights might coincide by a kind of providential harmony, but their only necessity derived from evidence acting upon reason, as this Panglossian passage from Quesnay makes clear:

> The physical laws which constitute the natural order most advantageous to the human species and which comprise exactly the natural rights of all men, are perpetual, inalterable laws, decidedly the best laws possible. Their evidence imperiously subjugates all human rea-

[6] Charles de Secondat de Montesquieu, *De l'esprit des lois* (1748), G. Truc, ed., 2 vols. (Paris: Garnier, 1961), Book I, chapter 1; vol. 1, p. 6.

son and intelligence, with a precision that demonstrates geometrically and arithmetically in each detail, and which leaves no subterfuge for error, impostiture, or illicit claims.[7]

Thus mathematicians seeking to apply the calculus of probabilities to the natural and moral realms faced opposite problems: natural causes had to be weakened and moral causes strengthened to make the probabilistic approach of Bernoulli's urn model a plausible one. The "natural history of man" pioneered by Johann Süssmilch, George Leclerc Buffon, and others investigated the statistical regularities exhibited by the tables of mortality and the records of birth—for example, the steady preponderance of male to female births—but skirted the issues that touched on man's nonmaterial aspects.[8] Although it did not deal with the vexed questions of action and free will, the growing literature on the statistics of human life may have nonetheless encouraged probabilists to extend their speculations to moral issues. Jakob Bernoulli had assumed that constant probabilities determined the mortality rates for given ages,[9] and the eighteenth-century statistical compilations bore him out. Paul-Edmé Saint-Cyran, a captain in the Corps de Génie and the author of *Calcul des rentes viagères* (1779), drew attention to the grim regularity of mortality statistics:

Those who observe with some care the workings of nature, will discover, amidst an infinity of particular irregularities, a certain general order from which nature deviates hardly at all: thus, even though each man dies as if by chance, and in such a way that one cannot determine the term of his life, yet it is certain that in considering simultaneously a large number of men of the same age, and in examining the order according to which they die, passing from one age to another, one will find almost the same sequence of mortality, if one made a similar investigation on another collection of persons.[10]

[7] François Quesnay and Victor de Riquetti Mirabeau, *Philosophie rurale* (Amsterdam, 1764), vol. 1, pp. 100–101; vol. 3, pp. 319–322; Quesnay, "Despotisme de la Chine" (1767), in *Oeuvres économiques et philosophiques de Quesnay*, Auguste Oncken, ed. (Paris, 1888), p. 645.

[8] George Leclerc Buffon, "Des probabilités de la durée de la vie," *Histoire naturelle. Supplément* (Paris, 1777), vol. 4, pp. 149–323.

[9] Jakob Bernoulli, *Ars conjectandi* (1713), in *Die Werke von Jakob Bernoulli*, Basel Naturforschende Gesellschaft (Basel: Birkhäuser, 1975), vol. 3, pp. 248–249.

[10] Paul-Edmé Saint-Cyran, *Calcul des rentes viagères* (Paris, 1779), p. 11.

These were of course just the results one would expect if the assumptions of Bernoulli's theorem held: constant probabilities and numerous independent trials. In their capacity as academicians, Laplace and Condorcet regularly served on the commissions that reported on the ongoing statistical projects of Morand and Michodière.[11] Laplace, Condorcet and Pierre-Achille Dionis Du Sejour collaborated on a number of memoirs[12] which analyzed these demographic results, and the use of probabilistic techniques to distinguish coincidence from genuine regularity therein remained a lifelong interest for Laplace.[13] Though restricted to the physical aspects of human existence, the manifest regularity of the burgeoning store of birth, death, and even marriage statistics must have impressed the mathematicians. Condorcet's praise for Buffon's *Essais d'arithmétique morale* for showing the close relationship between the calculus of probabilities and the natural history of man hinted at an even more general program: "It has contributed to the progress of a science which, by submitting to the calculation events directed by laws which we call irregular because they are unknown to us, appears to extend the empire of the human mind beyond its natural boundaries."[14] Although the efforts of Condorcet and Laplace to apply the calculus of probabilities antedated the collection of legal statistics (begun only in 1825), the example of demographic regularities may have emboldened them to extend probabilistic analysis further.

It is more difficult to assess the indirect influences of Enlightenment theories of human nature. Although the systems of David Hartley, P.H.D. Holbach, Étienne Condillac, and Claude Adrien Helvetius differed widely in their particulars, all were founded upon the assumption that at least certain features of human nature were fixed and universal. In order to apply the probability of causes to moral phenomena, mathematicians, at least prior to Poisson, were obliged to assume that the probability

---

[11] The MS. *Procès-Verbaux* of the meetings of the Académie des Sciences record many presentations of demographic statistics by Morand; see, for example, the minutes of 23 February 1774, 6 July 1776, 14 August 1778, 14 April 1779, 12 April 1780, 12 July 1780, 9 January 1782, Archives de l'Académie des Sciences.

[12] See, for example, Laplace, Condorcet, and Du Sejour, "Essai sur la population du royaume," *Mémoires de l'Académie royale des Sciences. Année 1784* (Paris, 1787), pp. 703–718.

[13] See Laplace, "Sur les naissances, les mariages et les morts à Paris, dupuis 1771 jusqu'en 1784 . . .", *Mémoires de l'Académie royale des Sciences. Année 1783* (Paris, 1786), pp. 693–702.

[14] Condorcet, "Éloge de Buffon," in *Oeuvres*, vol. 3, p. 348.

value of these "causes"—the veracity of a witness or the perspicacity of a judge—remained constant throughout a large number of trials. Contemporary emphasis on the constant features of human nature, combined with the suggestive regularity of the demographic statistics, may have lent credence to the mathematical assumption.

Condorcet's repeated emphasis on uniformity in both the physical and moral realm is suggestive in this context. Attacking the relativism of Montesquieu's *Esprit des lois*, Condorcet argued that since "truth, reason, justice, and the rights of man, interest in property, liberty, and security are the same everywhere," criminal, civil, and commercial laws should also be standardized.[15] Uniform human nature required uniform laws. Condorcet impatiently swept aside Montesquieu's erudite references to the laws of other nations and cultures, which seemed to imply that statutory laws were mere games with arbitrary rules. Even the apparently arbitrary rules of games, Condorcet retorted, were founded upon reasons (presumably of fairness) that mathematicians could divine using the calculus of probabilities. Beneath the bewildering diversity of individual customs, there existed an underlying uniformity based on the nature and rights of man.

The pervasive ambiguity in the concept of moral law, which embraced both necessity and choice, imbued the moral sciences with a double nature. On the one hand, the moral sciences, like the physical sciences, studied the immutable order of phenomena, those concerning human thought and action; on the other, they recommended changes in the existing social, political, and economic order. For eighteenth-century practitioners, the moral sciences were both descriptive and normative. It was superfluous for physicists to urge compliance with the law of gravitation; it would have been absurd to suggest that there existed any choice in the matter. Yet eighteenth-century social thinkers complained ceaselessly that current social arrangements violated one or another law of the moral realm. Even Montesquieu, who advised that legislation be used to curb certain natural tendencies among the citizenry, cautioned that such artificial correctives should be used sparingly, "for we do nothing so well as when we act with freedom, and follow the bent of our natural genius."[16] These writers as-

[15] Condorcet, "Commentaire sur le vingt-neuvième livre de *l'Esprit des lois*," in Antoine Louis Claude Destutt de Tracy, *Commentaire sur l'Esprit des lois de Montesquieu* . . . (Paris, 1828), pp. 380–381.

sumed that the natural order was best, but not necessary. The moral sciences therefore guided policy and reform by discovering the laws to which governments and individuals *should* conform. The probabilists also presented their attempt to apply mathematics to the moral sciences as aids to policy decisions both general and specific. Condorcet endorsed the moral sciences as an essential means to human improvement and recommended the abolition of the death penalty on the basis of probabilistic arguments; Laplace advocated gradual social change and condemned the French jury system for what he computed to be scandalously high rates of erroneous decisions; Poisson quarreled with Laplace's results and urged that the true criterion of judicial success was the security of the society at large, rather than the danger of an erroneous conviction in individual cases.

The probabilists made legal applications their entry into the moral sciences for several reasons, some inherent in the calculus itself and some relating to the political climate of the time. The close historical associations between legal proofs, including testimony, and mathematic probability prepared the way for the probability of testimony. Probabilistic treatments of witness credibility appeared as early as 1699 and persisted at least until 1837, with Poisson's *Recherches sur la probabilité des jugements*. As a result of the critical historiography of Pierre Bayle and his eighteenth-century successors, the focus of interest in the probability of testimony shifted from the courtroom to historical tradition, particularly historical and scriptural reports of miracles.

The probability of judgments originated much later, with Condorcet's *Essai sur l'application de d'analyse à la probabilité des decisions rendues à la pluralité des voix* (1785), and appears to have been prompted by the ardent interest of philosophes such as Voltaire, Ann-Robert-Jacques Turgot, and Condorcet in legal reform. Stimulated by Cesare Becarria's influential treatise, *Dei delitti e delle penne* (1764), translated by the Abbé André Morellet as *Des délits et des peines* in 1766, and enraged by the infamous trials of Jean Calas and the Chevalier de La Barre, the *philosophes* waged a polemical battle for a more humane legal system. Condorcet energetically summed up the roster of reforms in a letter to Turgot, dated 20 July 1774: "There are so many things to do for the public good! Proscribe fanaticism and bring the assassins of [Jean-François LeFévre, Chevalier de] LaBarre to jus-

---

[16] Montesquieu, *Lois*, Book XIX, chapter 5; vol. 1, p. 320.

tice; assign to each crime a legal punishment; abolish barbaric tortures."[17] The redesign of judicial tribunals to replace the detested *parlements* figured prominently among these reforms.[18]

The fact that jurists had already cast both problems in qualitatively probabilistic terms invited a more precise mathematical treatment. The testimony of witnesses, along with all other legal evidence, created only presumptions of guilt or innocence, never certainty; the decisions reached by tribunals or juries attained at best to moral certainty, which was, as Beccaria admitted, "only a probability."[19] The mathematicians were in the business of measuring probabilities, and historical links already existed between legal and mathematical probabilities. Moreover, here mathematics could perform an additional service for the moral sciences. Quantitative probabilities were not only clearer and more precise than the qualitative ones of the jurists; they were also backed by the authority of mathematics itself. Given the importance of persuasion in the moral sciences, particularly where policy recommendations were at issue, this was no slight advantage, at least in the eyes of the mathematicians. It was just this appropriation of mathematics' reputation for certainty in support of conclusions about moral matters that most offended later critics of the classical probabilists.

They were also shocked by the insouciance with which their predecessors had quantified the unquantifiable: veracity, credibility, enlightenment, perspicacity were all assigned numerical values. But for those primed by contemporary psychological theories to think of judgment as calculation and combination of ideas, and of belief as exquisitely proportioned to frequencies, it was a natural enough step. With the exception of their assumptions about the stability and regularity of moral causes, the mathematicians operated within the conceptual framework of the moral sciences. They posed problems about fairness and rationality; they took individual witnesses and judges as their units of analysis; they believed that an orderly society could only result from summing its rational members; they concluded what ought to be, rather than what was the case. In the service of the moral sciences, mathematics itself took on a moral tinge.

[17] Condorcet to Turgot, circa 20 July 1774, in *Correspondance inédite de Condorcet et de Turgot 1770–1779*, Charles Henry, ed. (Paris, 1883), p. 184.

[18] Condorcet, in Henry, ed., *Correspondance*, p. 201.

[19] Cesare Bonesana Beccaria, *Traité des délits et des peines* (1764), Abbé Morellet, trans. (Lausanne, 1766), p. 39.

6.3 Testimony and the Probability of Miracles

The seventeenth-century jurist Jean Domat instructed his readers in the art of assessing the degree of certainty rightly accorded to the intrinsic evidence of things and the extrinsic evidence of testimony in legal proceedings. Although the legal rules of evidence had been systematized in the hierarchy of proofs, the judicial weighting of evidence remained an art rather than a science in Domat's eyes. Ultimately it rested with the judge to apply the detailed apparatus of presumptions, conjectures, and indices to an individual case: "The use and application of all of these rules, according to the quality of the facts and circumstances, depends upon the prudence of the Judge."[20] With the advent of the calculus of probabilities, some mathematicians attempted to replace the intuitions of Domat's prudent judge with quantitative comparisons. Given the early background of mathematical probability,[21] it was natural for probabilists to attempt to convert the qualitative legal probabilities assigned to various classes of evidence to the quantitative probabilities defined by the new theory.

However, how to translate intuition into calculation was not straightforward. Jakob Bernoulli devoted a chapter of Part IV of the *Ars conjectandi* to the mathematical evaluation of "proofs by conjecture," both intrinsic and extrinsic. His cumbersome system of necessary and contingent proofs and dubious assumption of equiprobable cases did not inspire any imitators, but the problem continued to intrigue classical probabilists from Nicholas Bernoulli through Poisson. Although the problem originated in the context of legal rules of evidence, contemporary controversies over the reliability of historical evidence, which was frequently and explicitly compared to legal testimony, also attracted the attention of the mathematicians. Whether the testimony was legal, historical, or biblical, the question posed by the probabilists was the same: Under what conditions is belief rational?

Examples drawn from courtroom practice abound in the philosophical literature of the seventeenth century. Many of these legal references were used to illustrate the criteria for rational belief of testimony in any context. The model of legal evaluation of testimony served two different purposes for these writers: first, to guard against credulity by distinguishing fact

[20] Jean Domat, *Les loix civiles dans leur ordre naturel* (1689–94), M. de Hericourt, ed. (Paris, 1777), p. 285.

[21] See Chapter One above.

from fable; and second, to guard against skepticism by setting standards for warranted belief. The Port Royal *Logique* attributed both obstacles to right reasoning to intellectual sloth: "For some do not want to take the trouble to discern errors: the others do not want to bother to conceive of the truth with the care necessary to appreciate the evidence for it."[22] Francis Bacon's legal examples addressed the former problem of separating true (if extraordinary) accounts from the fabulous traditions of marvels, wonders, and prodigies that had dominated natural history since Pliny. He established a set of guidelines, borrowed directly from the legal rules of evidence concerning testimony (Bacon was trained in the law), for weighing the credibility of natural history reports, which Robert Boyle scrupulously adhered to in his own accurately titled "Strange Reports."[23] The number and social standing of the witnesses, as well as immediacy of experience (i.e., eyewitness versus hearsay) concerning the event in question, all entered into Bacon's reckoning of credibility.[24]

John Locke warned against the opposite extreme of incredulity. Locke related the anecdote of the King of Siam's refusal to believe in the Dutch ambassador's description of ice to show that skepticism based solely on one's own experience (water never froze in Siam) could lead to error: some marvels are real. In particular, Locke argued that analogical reasoning justified belief in miracles: this was "one case, wherein the strangeness of the fact lessens not the assent to a fair testimony," although such reports by definition contradict experience. However, assent should not depend on the mere opinion of others; in fact, Locke regarded such uncritical acceptance of authority as the major source of human error. To guard against unquestioning belief, Locke presented a list of criteria for evaluating the testimony of witnesses similar to Bacon's guidelines: the number, integrity, skill, personal interest of the witnesses, as well as the existence of internal inconsistencies and contradictory testimony, increased or diminished the "probability" of the report.[25]

[22] Antoine Arnauld and Pierre Nicole, *La logique, ou l'Art de penser* (1662), Pierre Clair and François Girbal, eds. (Paris: Presses Universitaires de France, 1965), p. 18.

[23] Robert Boyle, "Strange Reports," in *The Works of the Honourable Robert Boyle* (London, 1772), vol. 5, p. 604.

[24] Francis Bacon, *The Works of Francis Bacon*, J. Spedding, D. Heath, and R. L. Ellis, eds. (London, 1870), vol. 1, p. 401.

[25] John Locke, *An Essay Concerning Human Understanding* (1689; all copies dated 1690), Alexander Campbell Fraser, ed. (New York: Dover, 1959), vol. 2, pp. 367, 382, 365–368; see also Locke's *Discourse on Miracles* (1702).

Gottfried Wilhelm Leibniz upheld not only the necessity but also the trustworthiness of a graduated scale of probabilities in assessing the grounds for rational belief in lieu of complete certainty. As a trained jurist, Leibniz rebuked Locke for the latter's pessimistic description of a system of reasoning that worked so well in jurisprudence as a mere "twilight of probabilities." Mindful perhaps of Blaise Pascal's satirical attacks on the Jesuit doctrine of moral "probabilism" in the Fifth and Sixth Letters of the *Lettres provinciales* (1657), Leibniz distinguished between the probability of opinion treated by the "moralists" and the genuine probability of testimony. The latter consisted of the "likelihood" (*vraisemblance*) of both "the nature of things" and the testimony of trustworthy witnesses. Leibniz also contrasted mere opinion with the weightier legal testimony delivered under courtroom conditions and appraised by a judge.[26]

The distinction between the probabilities of mere opinion and of trustworthy testimony (or tradition) lay at the heart of the critical school of historiography, particularly Biblical historiography, which emerged from the theological and skeptical controversies of the Reformation and Counter-Reformation.[27] It also documents the shifting sense of the term "probability." In medieval theology, probability meant authoritative opinion.[28] It was this sense of the word which gave rise to the moral doctrines of probabilism lambasted by Pascal, who accused Jesuit casuists of excusing even the most outrageous transgressions on the grounds that one theologian or another had mitigated the severity of the sin. The learned Jesuit of Pascal's Fifth Letter claimed that even a single opinion could render a moral doctrine "probable."[29]

Probability could also result from tradition, a consideration of particular importance in the bitter disputes over biblical interpretation waged by seventeenth-century Catholic and Protestant theologians. Protestant writers attacked the extrascriptural sources of received Catholic dogma as hu-

[26] Gottfried Wilhelm Leibniz, *Nouveaux essais sur l'entendement humain* (composed 1703–1705), in *Sämtliche Schriften und Briefe*, Deutschen Akademie der Wissenschaften zu Berlin (Berlin: Akademie Verlag, 1962), Sechste Reihe: Philosophische Schriften, vol. 6, pp. 456–457, 372–373.

[27] Richard Popkin, *The History of Scepticism from Erasmus to Descartes* (Assen, The Netherlands: Van Gorcum, 1964), chapter 1.

[28] Edmund F. Byrne, *Probability and Opinion* (The Hague: Martinus Nijhoff, 1968), p. 139.

[29] Blaise Pascal, *Lettres provinciales* (1657), J. J. Pauvert, ed. (Holland, 1964), p. 91.

man additions to the word of God; Catholic apologists retaliated with the critical hermeneutics based on the comparison of various biblical texts, undermining Protestant claims to have returned to some pristine "original" text of the scriptures. Some reliance on tradition, oral and written, was inevitable. The author of the preface to the revised edition of Richard Simon's *Histoire critique du Vieux Testament* (1685) turned the tables on the author of a rebuttal to an earlier edition of the *Histoire*, who had challenged the credibility of a text received from the Jewish Pharisees, by affirming the necessity of tradition:

> We have no other Scripture than that which we have received from the Jewish Pharisees. If they are as wicked as the Author of l'Examen depicts them to be, on what will our religion be founded if we reject the Traditions? The Author of the Critique [Simon] would doubtless ask him if he could produce Originals of the Scriptures? and if in default of these Originals, he can in good conscience depend on Copies, which have no other authority than that which they receive from the testimony of people more wicked than the Pagans?[30]

Despite the fact that scriptural interpretations rested upon tradition, tradition, like the probability of opinion, also came under attack during the latter half of the seventeenth century. Pierre Bayle, in his *Pensées diverses sur la comète* (1683), ridiculed the widely held belief, passed down from classical sources, that comets presaged disaster, and made cometary portents the occasion for a more general attack on the authority of tradition. Sheer bulk of testimony was insufficient to warrant belief, Bayle claimed, for the vast majority of people credulously accept what they are told without careful examination of sources and circumstances. In fact, the very force of collective opinion militated against discriminating belief: "One is finally reduced to the necessity of believing what everyone believes." Bayle pointed to other "fabulous opinions" recently discredited which had been supported by the testimony of innumerable persons, and concluded that "six million" such witnesses, insofar as one merely aped the other, should count as "only one." The probability of testimony derived not from the total number of witnesses, but only from the handful who had diligently examined the issue:

[30] Richard Simon, *Histoire critique du Vieux Testament*, Nouvelle Édition (Rotterdam, 1685), Preface (n.p.).

. . . an opinion cannot become probable by the multitude of those who endorse it except insofar as it appears true to some [of these] independently of all prejudice and by the sole force of a judicious examination and of a wide knowledge of the things [in question] . . . one may rest assured that an intelligent man who pronounces only upon that which he has long pondered, and which he has found proof against all his doubts, gives greater weight to his belief, than one hundred thousand vulgar minds who only follow like sheep.[31]

Bayle maintained that not only the integrity and intelligence of the witnesses but also the content of the testimony should be weighed before assenting. Reports of events that violated the laws of nature were particularly suspect. Although Bayle did not deny that miracles were possible, he nonetheless urged that a principle of moral utility be applied to uncertain reports. God availed himself of miracles to inspire or confirm faith. Since prodigies and marvels manifestly encouraged idolatry, they could hardly be divine handiwork. If purported miracles could be explained by "general laws," it was the greater part of piety to attribute phenomena to God's ordinary, as opposed to extraordinary, providence, "because there was nothing less worthy of a general cause which set in action all the others by a simple and uniform law, than to violate this law at any moment, to anticipate the babbling and superstitions to which feeble, ignorant men are prey."[32] Similar views on the primacy of natural causes in explanation were expressed by Nicolas Malebranche in his *Traité sur la nature et la grâce* (1680) and by Thomas Burnet in his *Sacred History of the Earth* (Latin edition, 1681; 1st English edition, 1684).

Leibniz's attempts to redefine probability in the *Nouveaux essais sur l'entendement humain* reflected the critique both of the probability of opinion and of the probability of tradition. Just as in the courtroom, the probability of testimony depended on the character of the witness and his competence concerning the matter at hand, and on the intrinsic content of his testimony. Thus, the opinions of Copernicus were more probable in the true sense than the consensus of all the rest of humanity because his painstaking examination of the evidence for the heliocentric hypothesis showed his intellectual integrity and his insight into "the nature of things." Coper-

[31] Pierre Bayle, *Pensées diverses sur la comète* (1681) A. Prat, ed. (Paris: Societé Nouvelle de Librairie et d'Edition, 1911), vol. 1, pp. 37, 134–135.

[32] Bayle, *Pensées*, vol. 2, p. 232.

nicus belonged to Bayle's elite of "seven or eight men" whose considered opinion counted for more than the lazy assent of "many nations and many centuries."[33] These two criteria of the credibility of the individual witness and the plausibility of the fact in question became the parameters that shaped the mathematical treatment of the probability of testimony.

Jakob Bernoulli's discussion of the probability of evidence in Part IV of the *Ars conjectandi* turned upon this joint assessment of the "intrinsic" probability of a "thing's cause, effect, subject, accessory, or sign" and the "extrinsic" probability "derived from the authority and testimony of men."[34] Bernoulli's intricate computations of the probabilities of "pure" and "mixed" proofs appears to have been stillborn, mathematically speaking. Even his nephew and disciple Nicholas Bernoulli attacked the problem from fresh premises in his *De usu artis conjectandi in jure* (1709).

Like his uncle, Nicholas presented his theory of the probability of testimony as part of a general attempt to mathematize the legal theory of proofs. Compared to Jakob's convoluted system of assessing "pure" and "mixed" proofs, Nicholas's approach was remarkably straightforward, if no less curious. Assuming that all of the evidence is circumstantial and that the accused is twice as likely a priori to be innocent, Nicholas computed that if no evidence were submitted against the accused, then the probability of innocence would be one or "beyond doubt"; if one piece of incriminating evidence were presented by the plaintiff, the probability of innocence would drop to 2/3; after two pieces of evidence, to 4/9; etc. In the case of a single piece of evidence, Nicholas reasoned that out of a total of three chances (two for innocence; one for guilt), two favored a probability of complete innocence and one indicated guilt (in which latter case $P(\text{innocence}) = 0$), so the probability of evidence would be

$$\frac{2(1) + 1(0)}{3} = 2/3.$$

Reasoning regressively for the two articles of evidence, Nicholas derived

$$\frac{2(2/3) + 1(0)}{3} = 4/9.$$

<hr>

[33] Leibniz, *Essais*, p. 373; Bayle, *Pensées*, vol. 1, p. 38.

[34] See Section 1.3, above; also Glenn Shafer, "Non-additive probabilities in the work of Bernoulli and Lambert," *Archive for History of Exact Sciences* 19 (1978): 309–370.

Thus, for n pieces of evidence the probability of innocence would be $(2/3)^n$.[35]

Nicholas's method of evaluating evidence harkened back to his belief in the central importance of averages in mathematical probability. Just as the fundamental rule of expectation was nothing else but a way of determining "the center of gravity of all the probabilities," so the computation of evidence multiplied the number of chances of guilt and innocence by the respective "values" of the outcome. However, at this point Nicholas's legal training betrayed his mathematics to confound outcome value with probability. In this exceptional case, the value was indeed a "probability" in the older legal sense of a proof of guilt or innocence. Nicholas quantified these proofs as one and zero according to legal, rather than mathematical, practice. While Nicholas's assumptions concerning the a priori probabilities of guilt or innocence appear to have been as arbitrary as Jakob's, his treatment of the probability of testimony was strikingly empirical, if impractical. Nicholas proposed that the "degree of trustworthiness" of each witness be measured by dividing the number of times the witness's testimony had been confirmed as truthful by the total number of his testimonies. Nicholas's rough-and-ready a posteriori approach was no doubt inspired by his uncle Jakob's theorem and belief that the theorem would provide a way of accurately estimating probabilities from experience. Nicholas's striking assumption that there existed a "true proportion" of individual veracity for a particular witness, to be discerned by numerous observations, was shared by almost every other mathematician who addressed the subject of the probability of testimony. Uniform causes prevailed in the moral as well as the natural world.

The models underlying this assumption of uniform veracity were occasionally made explicit, as in the case of John Craig's treatment of the probability of testimony in his *Theologiae christianae principia mathematica* (1699). Craig was a Cambridge-educated Newtonian whose research on the integral calculus was known to Nicholas Bernoulli, Pierre Montmort, and other Continental mathematicians, and was among the first British mathematicians to use the Leibnizian dx and $\int$ notations in print. Craig took religious orders and in 1708 became prebendary at Salisbury, under

[35] Nicholas Bernoulli, *De usu artis conjectandi in jure* (1709), in *Die Werke von Jakob Bernoulli*, vol. 3, p. 324.

Bishop Burnet, to whom his *Principia* is dedicated.[36] A student and protégé of Newton, Craig attempted to harness his mathematical skills to
theological ends; the title and format of his work showed the influence and
inspiration of his master. In the treatise Craig drew upon the fledgling
calculus of probabilities and analogies from Newtonian mechanics to construct a mathematical treatment of historical tradition, both oral and written, in order to defend the evidence of scripture and to predict the date of
the millennium.

Craig took the familiar distinction between "natural" (stemming from
"our own observation or experience") and "historical" (derived from the
testimony of others regarding their experience) probabilities as his departure point, making the latter type the special topic of his treatise. Craig's
interpretation of probability was psychological: "suspicion" arose concerning historical probabilities through "the application of the mind to the
contradictory sides of an historical event." The mind was a "moving
thing" impelled by the "motive forces" of argument at a particular "velocity of suspicion" through a "space" which represented the "degree of assent
which the mind gives the opposed arguments of history." Adding the "hypothesis" that all men have "an equal right to be believed" on the basis of
roughly similar natural endowments and the "common practice of men,"
Craig proposed to compute the probability of historical testimony.

According to Craig, the probability of historical testimony could be
diminished in three ways: by the number of witnesses through whom the
account was transmitted; by the distance through which the report trav-
eled; and by the amount of time that had elapsed since the original event.
In practice, Craig considered only the effects of the third, temporal factor
as important, dismissing the effects of the first as "very insignificant" and
the second as not so easily "reduced to the calculus." Craig took as his
minimal unit of computation the degree of probability produced by the
testimony of one primary witness, which he regarded as a kind of infinitesimal of belief:

> It can indeed happen that such a degree of probability may be so
> small that our mind can scarcely perceive its force, just as in the
> movement of bodies the degree of velocity is sometimes so little that

---

[36] See J. F. Scott, "John Craig," in *Dictionary of Scientific Biography*, Charles C. Gillispie,
editor-in-chief (New York: Charles Scribner's Sons, 1971), vol. 2, pp. 458–459.

we are unable to discern it with our own eyes. But (even as the latter degree of velocity, so too) this degree of probability is of a determined magnitude and when repeated many times produces a perceptible probability.[37]

Craig's treatment thus hinged upon at least two standardizing assumptions: every witness possessed a constant "amount" of credibility; and each witness possessed the *same* unitary amount of credibility, which produced a corresponding unit of probability or belief in the mind of each and every auditor or reader. These assumptions of uniformity had a radical import, for they implied that psychological aspects of mind, such as belief and veracity, could be idealized to a state of homogeneity comparable to the brute, unindividualized matter described by Newtonian mechanics. Craig's comparison of the mind to a moving body passively impelled by the force of argument must have struck contemporaries catechized in the absolute distinction between brute, passive matter and active, self-determining mind as even more heterodox, almost a contradiction in terms. Although none of the subsequent treatments of the probability of testimony adopted Craig's elaborate mechanical conceit, some variant upon his standardizing assumptions entered into every such discussion. The probabilists' models for belief thus challenged the orthodox metaphysics of their time at a fundamental level, and prepared the way for social theories that viewed society as an aggregate of undifferentiated individual units.

Craig calculated that the credibility of testimony conveyed only by word of mouth would sink below the perceptible level of probability in 800 years. The more permanent written record would expire in perceptible credibility in 3150 years, which Craig, citing Luke 18:8, fixed as the year of the second coming.[38] He mixed his chiliastic speculations with attempts to incorporate hypotheses concerning the critical evaluation of historical evidence into his mathematical treatment. "First" historians—that is, those who derived their account from firsthand experience—and exemplars—that is, those written or printed by the first historian—commanded greater probability than secondary sources which introduced copying errors. Although the probability of an account might sink below the psy-

---

[37] John Craig, "Craig's rules of historical evidence," excerpts from *Theologiae christianae principia mathematica* (1699), *History and Theory*, Beiheft 4 (The Hague: Mouton, 1964), pp. 3, 5.

[38] Craig, "Rules," p. 27.

chological threshold of belief, it never entirely vanished; also, the unit of suspicion varied according to whether the source of the account was permanent or ephemeral. All of these precepts might be found in the critical historiography of the time, and it is likely that Craig hoped to turn the historical phyrrhonists' own weapons against them.

In the same year that Craig's *Principia* appeared, an anonymous paper (probably by George Hooper) entitled "A Calculation of the Credibility of Human Testimony" was published in the *Philosophical Transactions* of the Royal Society, which approached the subject of the probability of testimony from an entirely different perspective, using a method based on Huygenian expectation. In place of expectation, the author posited degrees of "Moral Certitude Incompleat," equal to the product of the probability of testimony and the amount hinging thereupon. For example, if someone whose testimony you "would not give above One in Six to be ensur'd of the Truth of what he says" reports that you have just acquired £1200 by, say, the arrival of a ship, you may count on having £1000, "or five sixths of Absolute Certainty for the whole Sum." Within this expectation model, the author distinguished among individual witnesses according to their "integrity" and "ability": gone is the uniformity hypothesis of the *Principia*. If each of n successive witnesses has an individual credibility of p, then for n concurrent witnesses, the total probability would be

$$P(\text{True Account}) = (1 - p)\,(p) + (1 - p)^2(p) + \ldots (1 - p)^{n-1}(p).$$

As in the *Principia*, the probability or oral testimony vanished approximately five times more quickly than that of written testimony, for in "the Chances against the Truth or Conservation of a single Writing are far less; and several Copies may also be easily suppos'd to concur; and those since the invention of Printing are exactly the same."[39]

The anonymous *Philosophical Transactions* treatment was far more typical of the probability of testimony of later mathematicians than the mechanical model advanced in Craig's *Principia*. Although religious issues were not broached explicitly in the *Philosophical Transactions* paper, allusions to texts "taken by different Hands, and preserv'd in different Places or

[39] [George Hooper], "A Calculation of the credibility of human testimony," *Philosophical Transactions of the Royal Society* 21 (1699): 359–365, on pp. 359, 363. See also Glenn Shafer, "The combination of evidence" (typescript, February 1984), pp. 1–8.

Languages" probably referred to the Bible.[40] The *Philosophical Transactions* paper included the traditional legal considerations of the credibility of individual witnesses and the content of the narrative: in order to be a priori equally probable, parts of a narrative had to be assumed "equally remarkable."

Pierre de Montmort voiced reservations regarding the approaches of both Craig and the anonymous author of the *Philosophical Transactions* piece, and the viability of applications of the calculus of probabilities to problems of civil life in general. Montmort had not seen Jakob Bernoulli's *Ars conjectandi* when he published his own *Essai d'analyse sur les jeux de hazard* (1713, 1708), but he knew of the work and of its contents, at least in outline, from an account in the *Journal des Savants* (1706) and from Bernard de Fontenelle's *éloge* of Bernoulli. Intrigued by Fontenelle's announcement that the unfinished treatise had extended mathematical probability "even to Moral and Political matters,"[41] Montmort reflected on the possibilities for such moral applications in the introduction to his own work on mathematical probability. Although he could envision how the calculus of probabilities could be used to compute expectations in order to maximize personal advantage prudently, to discriminate between "the true and the probable," and to apportion assent to evidence, Montmort abandoned his own attempts to carry the "glory of mathematics" further into the moral realm when he was unable to discover "hypotheses which were supported by certain facts to guide me in my research."

Montmort believed that the excesses of Craig's *Principia* vindicated his own hesitation to formulate speculative hypotheses. Although Montmort professed the highest personal and mathematical regard for Craig and praised his pious motives, he regarded the mathematical treatment of the probability of testimony as fanciful and misconceived. Montmort criticized Craig's hypotheses as "arbitrary," and maintained that he could have posited "equally plausible" ones and derived different results. Craig's assumed values for the units of time, distance, and suspicion could easily have missed the mark by "a half or a third or a fourth," rendering the precision with which he purported to predict the date of the second coming ludicrous in Montmort's eyes. Concluding that the *Principia* was better suited to "exercise mathematicians than to convert Jews and infidels,"

[40] Hooper, "Calculation," p. 364.

[41] Bernard de Fontenelle, "Éloge de M. Bernoulli," *Histoire de l'Académie royale des Sciences. Année 1705* (Amsterdam, 1746), p. 187.

Montmort hinted that all such attempts to mix mathematics and religion were predestined to failure: "The clarity of mathematics and the sacred obscurity of faith are very much opposed to one another."[42]

Mathematicians who attempted to penetrate the "infinity of obscurities" that surrounded economic, political, moral, and religious issues soon lost their way in a labyrinth of hypotheses of their own making. Montmort judged that the most valuable contribution that mathematics could make to civil life was "that force and accuracy of Mind" acquired from mathematical study, and claimed that the best works on metaphysics, physics, and even medicine and ethics were for this reason written by mathematicians. Otherwise, mathematicians risked reproving the obvious or defending absurd or impractical conclusions deduced from dubious hypotheses. In Montmort's opinion, the anonymous author of "A Calculation of the Credibility of Human Testimony" erred in both directions. On the one hand, the claim that a report conveyed by n witnesses, each with credibility p, had probability p^n struck Montmort as "true and even self-evident"; on the other hand, he castigated any attempt to apply such theories as "impractical and perhaps impossible."

According to Montmort, two formidable obstacles blocked broader application of the calculus of probabilities to the moral realm. First, the "fixed and invariable" laws that governed the physical universe did not operate in the sphere of human action. Even self-interest failed to regulate conduct completely: relative advantages are only dimly perceived; caprice battles reason; a swarm of competing motives confuses decision. Second, the limited human intellect cannot comprehend the vast number of relationships involved in daily transactions: we are continually surprised by an unexpected turn of events. Despite his pessimism, Montmort nonetheless suggested two guidelines for mathematicians who persisted in these applications to "economic, political and moral" problems. Prompted by Fontenelle's tantalizing preview of Part IV of the *Ars conjectandi* and by the example of Edmund Halley's tables of mortality, Montmort advised mathematicians to restrict their problems "to a small number of suppositions, founded upon certain facts"; and to "abstract all circumstances pertaining to human freedom, that shoal upon which our understanding perpetually comes to grief."[43]

---

[42] Pierre Raymond de Montmort, *Essai d'analyse sur les jeux de hazard*, 2nd edition (Paris, 1713), pp. xv, xviii, xxxix.

[43] Montmort, *Essai*, pp. xix, xxxvi–xxxviii.

Montmort thus made both the method and the dangers of a mathematical approach to the moral sciences explicit. Mathematicians who undertook such a treatment would have to ignore, if not actually deny, the ontological dualism of mind and matter. They would also be obliged to ceaselessly amend and winnow their hypotheses in light of "certain facts." Just what those "certain facts" might be concerning the probability of testimony varied with circumstances and with the mathematician.

Montmort was practically alone among eighteenth-century probabilists in his skepticism concerning applications of mathematical probability to moral matters. The project, if not the results, of Part IV of the *Ars conjectandi* won converts among philosophers as well as mathematicians, perhaps because legal and mathematical probabilities of evidence were seldom distinguished. The "Probabilité" article of the *Encyclopédie* illustrates to what extent the mathematical and legal senses of probability were still intertwined. The article is unsigned, normally indicating Denis Diderot as author. Although the author was thoroughly conversant with Jakob Bernoulli's *Ars conjectandi*, the article is more philosophical than technical.[44] The article borrowed heavily from Part IV of the *Ars conjectandi*, reproducing Bernoulli's general rules for weighing evidence and conjecturing, and interpreting Bernoulli's theorem as a demonstration that "past experience is a principle of probability for the future; we have reason to expect events similar to those we have seen happen." While the article touched upon the issues of moral certainty, human life expectancy, statistical tables (recommended as a way of painlessly imparting to youth "all the experience of age"), and the probability of causes, the single longest section was devoted to the probability of testimony: What was "the degree of assent that we can give to testimony, and what is its *probability* for us?"

Following the familiar legal distinction between the "intrinsic" probability of "the nature of the thing" and the "extrinsic" probability of the witness, the author addressed each category in turn. Under intrinsic probability, he considered the cases of metaphysical, physical, and moral impossibility. No amount of witness credibility sufficed to make an account of a metaphysical impossibility, such as the squaring of the circle, believ-

[44] In the *Dictionnaire encyclopédique des mathématiques*, a compilation of mathematical articles from the *Encyclopédie*, the editors supplemented Diderot's original "Probabilité" article with one by Condorcet on the more technical aspects of the theory. See "Probabilité," in *Dictionnaire encyclopédique des mathématiques*, d'Alembert, Bossut, Lalande, Condorcet, et al., eds. (Paris, 1789), vol. 2, pp. 649–663.

able. However, reports of physical impossibilities sometimes warranted belief, if buttressed by unimpeachable testimony, because God was empowered to suspend his own laws. Reports of moral impossibilities—for example, that a man of sober character should commit an indecent act in public without motive—required substantial, but not ironclad, testimony to be plausible. The extrinsic credibility of witnesses depended upon "competence" and "integrity," which the author expressed as the product of two fractional ratings. Like Nicholas Bernoulli, the author hoped to derive these ratings, conceived as fractions of "certitude," from past experience concerning individual witnesses. However, he acknowledged that such a posteriori determinations were not always feasible, and fell back upon the traditional legal criteria: education, personal interest, ulterior motives, familiarity with the question at hand, and so forth.

Once the credibility of a given witness had been fixed, the probability of testimony varied with the circumstances of testimony, which once again followed legal usage. Numerous concurrent witnesses should increase the probability of testimony; two witnesses with individual credibilities of 9/10 gave a joint corroborative probability of 99/100. That is, the probability that each witness gave false testimony was 1/10, and the probability that both would do so was

$$(1/10)(1/10) = 1/100.$$

Therefore, the probability of true testimony was

$$1 - 1/100 = 99/100.$$

Here, a mixture of available mathematical techniques and hypotheses dictated the model. The author of the article was evidently unaware of conditional probabilities: Bayes' theorem had been published two years before in 1763, but was unknown in France, even among mathematicians, until about 1780. Laplace's version of the theorem did not appear until 1774. Therefore, the probabilities of the two witnesses had to be assumed to be independent. However, independence did not preclude an alternative approach that would have computed the joint probability of testimony as $(9/10)^2$, but hypothesis did. Because $81/100 < 9/10$, this method would imply that the testimony of one witness was more trustworthy than the concurrent testimony of two witnesses with identical credibility—a blatant contradiction of the evidentiary rule that insisted on the corroborative testimony of at least two unimpeachable witnesses for a full proof. It is important to note that in the actual reckoning of the probability of testimony, the *Encyclopédie* author took into account only extrinsic probability, that is, the credibility of witnesses, without regard to the intrinsic probability of the event in question.

However, the probability of hearsay evidence did diminish with the number of links in the testimonial chain, so the multiplications of credibilities produced the desired results in this case. In a computation reminiscent of Craig's *Principia* (and using the same generational unit of twenty years), the *Encyclopédie* author calculated the waning probability of historical accounts transmitted by oral tradition from generation to generation. Yet he denied that his results strengthened the arguments of the "historical pyrrhonists": although the probability of any one such testimonial chain inevitably decreased, collateral chains affirming the same account raised the overall probability significantly. Again, the author assumed these chains and the testimony of each witness within them to be independent—a problematical assumption in many historical cases. Moreover, if the medium of historical transmission was written rather than oral, "the *probability* increases infinitely." Not only did written texts endure many times longer than ephemeral human witnesses (the author suggested one hundred years); each exemplar launched the equivalent of an independent collateral chain, so that the total probability of testimony "will differ infinitely little from total certainty and will greatly surpass the oral assurances given by one or even several eyewitnesses."[45]

The belief that written testimony, even of an event centuries remote, counted as more reliable than the oral testimony of an eyewitness reflected changing canons of both legal and, especially, historical evidence. Medieval jurisprudence preferred oral to written testimony ("Témoins passent lettres"), but by the sixteenth and seventeenth centuries jurists had reversed this adage. In France, the Ordinance of 1667 further weakened the force of oral testimony vis-à-vis written evidence.[46] A parallel shift occurred in seventeenth-century historiography. Historians came to rely increasingly on textual comparison and criticism and to discount the credibility of popular tradition. Since popular tradition was often synonymous

[45] [Diderot], "Probabilité," in *Encyclopédie, ou Dictionnaire raisonné des sciences, des arts et des métiers*, d'Alembert and Diderot, eds. (Neuchâtel, 1765), vol. 13, pp. 395–399.

[46] John Gilissen, "La preuve en Europe du XVIe au debut du XIXe siècle. Rapport de synthèse," in *La Preuve. Deuxième partie: Moyen Age et temps modernes. Recueils de la Société Jean Bodin pour l'histoire comparative des institutions*, vol. 17 (1965), pp. 821 et passim.

with oral tradition, the evidentiary force of the latter also diminished in the eyes of historians. Enlightenment historians shared Bayle's contempt of the "vulgar" marvel-mongering tradition of antiquity and folklore, and associated this credulous appetite for wonders with illiterate peoples. Hume's argument in the "Essay on Miracles" was typical:

> It forms a strong persumption against all supernatural and miraculous relations, that they are observed chiefly to abound among ignorant and barbarous nations; or if a civilized people has ever given admission to any of them, that people will be found to have received them from ignorant and barbarous ancestors, who transmitted them with that inviolable sanction and authority which always attends received opinions.[47]

In a similar vein, Voltaire railed against the "tales" and "legends" retailed as fact by historians from Herodotus to Jacques Bénigne Bossuet. In an age that had disabused itself of so many illusions about the physical world, why, Voltaire queried, were children still reared on the folktales of "the fabulous infancy of Cyrus?"[48]

Enlightenment histories enshrined the invention of printing as the true starting point of modern civilization. In contrast to oral tradition, the printed word brought permanence and comparative accuracy. The transmission of texts in manuscript had been a painstaking, unreliable process: it took a great deal of time to make even a single copy, and copying errors proliferated with each transcription. Printing multiplied identical exemplars of a single edition with speed and fidelity. Although printing did not eliminate copying error, it greatly reduced it. Moreover, a manuscript that existed in only a few copies might be easily lost to the vicissitudes of war and neglect (the fate of the library of Alexandria served as a cautionary tale), while of the numerous copies of any single printed edition, at least one stood a good chance of surviving even a cataclysm of worldwide proportions to instruct future generations. Diderot, for example, envisioned the *Encyclopédie* as a kind of time capsule that would preserve all of human knowledge in both the liberal and mechanical arts for posterity, in the event of a second Dark Ages: "Let the *Encyclopédie* be a sanctuary where

---

[47] David Hume, *Essays Moral, Political and Literary* (1741–42) (London: Oxford University Press, 1963), p. 529.

[48] Voltaire, *Le pyrrhonisme de l'histoire* (1769), in *Oeuvres complètes* (Paris, 1785), vol. 31, p. 13.

human knowledge will be sheltered from Time and Revolutions."[49] In his inaugural address to the Académie Française, Condorcet hailed printing as "that preserving art of human reason," which would make each new discovery instantly and permanently part of the "patrimony of all nations," thus insuring human progress: "It is vain to resist a new truth, once it is deposed in books."[50]

In keeping with the spirit of this paean to printing, the mathematical treatments of the probability of testimony presented assumptions and results that affirmed the superior credibility of written testimony. These treatments also placed increasing emphasis on historical, rather than legal, testimony, although the legal model for evaluating testimonial evidence never vanished wholly from view in these discussions. The critical reexamination of ancient and religious sources prompted by the critiques of historical pyrrhonists such as François de la Mothe le Vayer,[51] and the hermeneutic polemics of the Reformation and Counter-Reformation inspired a historical variant of the "mitigated skepticism" described in Chapter Two. Moral certainty pertained to "both that kind of Evidence arising from the Nature of things, and likewise to that which doth arise from Testimony, or from Experience," according to John Wilkins: testimony certifies the existence of Queen Elizabeth I; experience convinces us that night follows day in endless succession.[52] Joseph Glanvill contended that "indubitable" (as opposed to "infallible") certainty could be founded upon testimony if it were "general . . . full, plain, and constant," as it was for the existence of Rome or Constantinople.[53] Just as religious arguments rested upon "such evidence as a prudent considering man, who is not credulous on the one hand, and on the other is not prejudiced by any interest against it, would rest satisfied,"[54] so historical evidence should also be supported by probability great enough to persuade a reasonable man. In order to separate truth from legend, Enlightenment historians

[49] Denis Diderot, "Prospectus," in *Encyclopédie* (Paris, 1751), pp. 4–5.

[50] Condorcet, "Académie française," in *Oeuvres*, vol. 1, p. 393.

[51] See François de la Mothe le Vayer, *Des anciens et principavs historiens grecs et latins* (Paris, 1646); *Discours de l'histoire* (Paris, 1638).

[52] John Wilkins, *Of the Principles and Duties of Natural Religion* (1675), 4th edition (London, 1699), p. 9.

[53] Joseph Glanvill, *Essays on Several Important Subjects in Philosophy and Religion* (London, 1676), p. 49.

[54] John Tillotson, *The Works of the Most Reverend Dr. John Tillotson* (Edinburgh and Glasgow, 1748), vol. 8, p. 25.

submitted the testimony of historical witnesses to the same criteria which jurists applied to legal testimony. Historical evidence, like legal evidence, became a matter of graduated certainty, that is, of probabilities.

Voltaire, one of the leading historians of his time, briefly summarized this historiographical perspective in his article "Histoire" in the *Dictionnaire philosophique*, reprinted under the same heading in the *Encyclopédie*: "All certainty which is not a mathematical demonstration is only an extreme probability: there is no other kind of historical certainty."[55] Pierre Bayle had discounted the testimony of all past historians, claiming to find in their writings only the "spirit, the prejudices, the interests, and the taste of the party embraced by the historian."[56] Voltaire's avowed "pyrrhonism" hardly merited so radical a label by comparison, and Voltaire himself admitted that he wanted "neither an outrageous pyrrhonism, nor a ridiculous credulousness" in the critical reexaminations of Bossuet, Livy, Thucydides, and other historians which comprised his *Le Pyrrhonisme de l'histoire*.[57] Eighteenth-century historians sought neither to categorically deny nor indiscriminately endorse the accounts of their predecessors, but rather to obey Locke's instructions to "apportion assent to probability."

Mathematicians followed historians in extending the probability of testimony from legal to historical examples. As Craig's *Principia* and the Port Royal discussion of Saint Sylvester would suggest, the earliest examples concerned what the *Encyclopédie* termed "sacred" history, which included both scriptural and ecclesiastic traditions. By the mid-eighteenth century, however, the focus of probabilistic interest in religious history had narrowed to the problem of miracles, which taxed to its limits the opposition of intrinsic and extrinsic categories that constituted the probability of testimonies: what degree of extrinsic probability would counterbalance or outweigh the great intrinsic improbability of a violation of the laws of nature? Not only philosophers such as Hume and Richard Price, but mathematicians such as Condorcet, Laplace, and Poisson deliberated over this question; indeed, in the latter half of the eighteenth century it became the primary problem in the probability of testimony. Mathematicians, philosophers, and historians, all inspired by the older legal probabilities of testimony, converged upon this extreme test of rational belief.

[55] Voltaire, "Histoire," in *Dictionnaire philosophique* (1764), in *Oeuvres*, vol. 52, p. 266.

[56] Bayle, *Critique de l'histoire du Calvinisme* (1682), in *Oeuvres diverses* (1727), Elisabeth Labrousse, ed. (Hildesheim: Georg Olms, 1965), p. 11.

[57] Voltaire, *Pyrrhonisme*, p. 11.

Locke had posed the problem of testimony concerning miracles in his chapter "Of the Degrees of Assent" in the *Essay Concerning Human Understanding*. By framing the question in terms of probability, evidence, and assent, Locke established the ground rules for future discussions of the problem. Locke, it will be recalled, exempted miracles from the ordinary standards of experiential evidence: in such cases, we must rely on testimony alone, provided we can be sure of its divine source. That is, the intrinsic probability of the miraculous event being almost nil, extrinsic probability must bear the entire burden of proof. Eighteenth-century writers adopted Locke's framework of the probabilities of testimony and experience, but rejected both the exceptional status he had granted to miracles and his pessimistic conclusion that because of the "great variety of contrary observations, circumstances, report, different qualifications, tempers, designs, oversights & c. of the reporters," it would be "impossible to reduce to precise rules the various degrees wherein men given their assent" to testimony.[58] Locke was content to settle for a simple preponderance of evidence, but Condorcet, Laplace, and Poisson sought the "precise rules" in the calculus of probabilities.

Miracles excited much discussion in the late seventeenth and eighteenth centuries as the most dramatic point of conflict between belief in the articles of Christian faith and belief in an inviolable natural order. This theme spawned a voluminous literature and a spectrum of subtly graduated views on the relationship of divine will to natural law.[59] Only those arguments that directly touch upon questions of evidence and probability fall within the purview of this discussion, however. The most famous of these, at least to post-eighteenth-century readers, was David Hume's *Essay on Miracles*, first sketched as Section X of the *Philosophical Essays Concerning Human Understanding* (1748), revised and republished as *An Enquiry Concerning Human Understanding* (1758). As seen in Chapter Four, Hume's critique of induction, including induction based on probability, did not prevent him from asserting that probability, in the sense of rational belief, accrued from past experience. Depending on the uniformity of that experience, there existed "all imaginable degrees of assurance, from the highest certainty to the lowest species of moral evidence." The "wise man" was he who accordingly "proportions his belief to the evidence." In cases of com-

[58] Locke, *Essay*, vol. 2, pp. 377, 382–383.

[59] See R. M. Burns, *The Great Debate on Miracles* (Lewisburg, Pa.: Bucknell University Press, 1981).

pletely uniform experience, he would have "the last degree of assurance and regards his past experience as a full *proof* of the future existence of that event."

Such was the case of miracles, defined as "a violation of the laws of nature." Hume argued that because "a firm and unalterable experience has established these laws, the proof against a miracle, from the very nature of the fact, is as entire as any argument for experience can possibly be imagined." Proceeding in legal fashion from probabilities to "full proofs," Hume argued that the intrinsic probability of a miracle—from "the nature of the fact"—was zero:

> There must, therefore, be a uniform experience against every miraculous event, otherwise the event would not merit that apellation. And as a uniform experience amounts to a proof, there is here a direct and full *proof*, from the nature of the fact against the existence of any miracle, nor can such a proof be destroyed or the miracle rendered credible but by an opposite proof which is superior.[60]

The only available type of proof that could counter the intrinsic probability of experience was the extrinsic probability of witnesses. Hume actually regarded the credibility of witnesses to be a subdivision of the probability of causes generally conceived, although he maintained the traditional distinction between intrinsic and extrinsic probabilities for the purposes of discussion. Since we are unable to discover necessary connections between *any* phenomena, all "constant and regular conjunctions" of events must qualify as a causal relation if any does. Our faith in human testimony stemmed from past experience that the tenacity of memory and integrity of others vouched for their testimony. Hence, the credibility of a given witness counted "either as a *proof* or a *probability*, according as the conjunction between any particular kind of report and any kind of object has been found to be constant or variable." The character, number, comportment, and interest of witnesses all entered into an assessment of individual credibility. In the case of miracles, what Hume claimed to be the universal human appetite for the marvelous, coupled with the partisan religious motives that whetted this appetite, considerably detracted from the credibility of any witness reporting on a purported miracle. Hume

[60] Hume, *An Enquiry Concerning Human Understanding* (1758), Charles W. Hendel, ed. (Indianapolis: Library of Liberal Arts, 1955), pp. 118, 121, 122–123.

concluded that no testimony had ever carried sufficient force to convince a rational auditor that "its falsehood would be more miraculous than the fact which it endeavors to establish." Theologians must therefore admit that "our most holy religion is founded on *faith*, not on reason."[61]

Of the several theologians who replied to Hume's arguments of the unreasonableness of believing in miracles, the most important from the standpoint of the probabilistic content of the counterattack and possible influence on later mathematical treatments of the probability of testimony was Richard Price's "On the Importance of Christianity, the Nature of Historical Evidence, and Miracles" (1767). Price was a mathematician in his own right, as well as a leading liberal theologian and moral philosopher, who was particularly interested in the mathematical theory of probabilities.[62] As in his rebuttal to Hume's critique of induction, Price enlisted the aid of the calculus of probabilities, especially Bayes' theorem, to argue for the reasonableness of belief in miracles. Price's terms of analysis were novel to the subject of the probability of testimony, and Condorcet's 1783 memoir on the topic appears to have been indebted to Price's dissection of the problem, if not to his ultimate conclusions.

Price's attack on Hume centered upon three pivotal assumptions: that the uniformity of past experience could provide a "proof"; that the credit given to testimony derives solely from the frequency of past experience; and that reasonable assent consisted of comparing the probability of the testimony, $P(W)$, with $1 - P(E)$, where $P(E)$ is the probability of the event. At stake, as Price was well aware, was the meaning of "reasonableness." Price attempted to destroy Hume's argument with a combination of arguments drawn from the calculus of probabilities, common practice, and Hume's own conflicting positions on induction and on miracles.

Price accepted Hume's contention that future expectations were founded on the "proportion to the greater of less constancy and uniformity of our experience." Indeed, Price considered this principle, suitably strengthened by Bayes' theorem, to be the best response to Hume's own critique of induction. A reasonable "intuition" predicated on the constant operation of natural causes rather than Hume's blind "instinct" was responsible for our belief that the future would resemble the past. Yet while Price was willing to countenance the question-begging assumption of con-

---

[61] Hume, *Enquiry*, pp. 120, 123, 140.

[62] See Sections 3.4.2 and 5.3, above, for Price's other probabilistic interests.

stant natural causes, he did not believe that these completely precluded miracles. Although the complete uniformity of past experience created an ever-increasing probability of future recurrence, no amount of past experience could ever produce a certainty. This was the point Price had hammered away at in his commentary on Bayes' theorem, and in his dissertation on miracles he returned to Bayes' theorem to support his claim:

> Upon observing, that any natural event has happened often or invariably, we have only reason to expect that it will happen again, with an assurance proportioned to the frequency of our observations. But, we have no *absolute proof* that it will happen again, in any particular future trial; nor the least reason to believe that it will always happen.[63]

Contrary to Hume's claim, therefore, uniform past experience did not produce a "proof" but only a "probability" that miracles would not occur in the future, and, Price noted, "between *impossibilities* and *improbabilities*, however apt we are to confound them, there is an infinite difference." Turning the full force of Hume's own skeptical weapons against him, Price argued that certain knowledge that any event was truly impossible was itself impossible, "for it can be no part of any one's experience, that the course of nature will always continue to be the same."

Having responded to Hume's arguments regarding the intrinsic probability of miracles, Price next turned to the extrinsic probability of witnesses. Price flatly denied Hume's contention that the credibility of witnesses reduced to another probability based on past experience, in this case, the individual witness's past record of veracity. No reasonable man, Price protested, actually weighs testimony "in proportion to the number of instances, in which we have found, that it has given us right information, compared with those in which it has deceived us." Instead, he would have recourse to more traditional legal criteria, such as the comportment of the witness, the circumstances of the narrative, and corroborative testimony. Price drew a sharp distinction between the grounds for belief in (future) experience, which derived from the frequency of past experience, and the grounds for belief in testimony, which often stemmed from a single conversation. Reasonable men were able, Price insisted, to gauge a witness's integrity without any prior experience concerning his veracity. The prob-

---

[63] Richard Price, *Four Dissertations* (1767), 5th edition (London, 1811), pp. 226–227, 229.

ability of testimony was emphatically not simply a special case of the probability of causes; Bernoulli's theorem did not apply, Hume and Nicholas Bernoulli notwithstanding.

Not only were the types of conviction created by experience and testimony of two qualitatively different sorts; the conviction of testimony actually surpassed that of experience: ". . . we are not so certain that the tide will go on to ebb and flow, and the sun to rise and set in the manner they have hitherto done, a year longer, as we are that there has been such a man as Alexander, or such an empire as the Roman."[64] Moreover, much of what Hume termed "experience" actually owed to the testimony of others concerning *their* experience; in this sense, "TESTIMONY is truly no more than SENSE at second hand."[65] In sum, extrinsic probability was less, and intrinsic probability more certain than Hume would have us believe.

Finally, Price examined the crux of Hume's argument, namely, that reasonable belief reduced to the comparison of the prior probabilities of the event reported and of the witness, both computed from past experiential frequencies. Price objected that the daily practice of reasonable men flew in the face of this criterion, for they regularly accepted even dubious testimonial reports on matters with a prior probability "of almost infinity to one." Price summoned the familiar dice and lottery examples of the calculus of probabilities to substantiate his argument. In a lottery of fifty numbered tickets, all drawn in succession, the probability of any one sequence of numbers is the same as that of any other sequence: $1/50!$—a tiny probability indeed. Yet even the testimony of a witness of slight credibility, with $P(W) = 1/2$, concerning the results of the drawing suffices to persuade us, unless we have reason to suspect some ulterior motive to deceive. Price contended that many reports of mundane events possessed equally tiny prior probabilities: "What is similar to this is true of almost everything that can be offered to our assent, independently of any evidence for it; and particularly, of the numberless facts which are the objects of testimony, and which are continually believed, without the least hesitation."[66] If we are to follow the actual practice of reasonable men, rather than Hume's criterion, Price asserted that we would evaluate the plausibility of testimony from the credibility of the witness alone.

Price's use of mathematical probabilities against Hume's criterion in-

---

[64] Price, *Dissertations*, pp. 242, 232, 229–230.

[65] Price, *Dissertations*, p. 238. [66] Price, *Dissertations*, p. 292n.

volved at least three tacit assumptions: first, that the criterion of reasonable belief should be extracted from the practice of reasonable men, and that these paragons acted as he claimed they did; second, that there was no distinction between the probability of a particular outcome from the set of all possible alternatives and the probability of *any* outcome; and third, that "independently of any evidence," all possible outcomes were equally probable. Price indirectly gave ground on the first and second assumptions, conceding that reasonable men did in fact sometimes weigh intrinsic against extrinsic probability when the narrative pandered to the human hankering after marvels or was otherwise suspected of serving ulterior ends. He also acknowledged that particular outcomes, as in the lottery example, deserved a skeptical hearing if the reporter was known to have some stake in the matter. In both cases, suspicion of intent to deceive waived normal standards of evidence.

But Price stood firm on the Bayesian principle that in the absence of any prior information, all alternative outcomes were equally probable, or—given the innumerable conceivable alternative outcomes for even ordinary events—equally improbable. This assumption that all events had a lotterylike equipartition of probabilities downplayed the improbability of miracles by heightening the improbability of every other event. After all, Price observed, to the "common man" the discoveries of natural philosophy were as strange and wondrous as any biblical miracle. Once again turning the tables on Hume, Price suggested that true prejudice was not vulgar credulity but narrowminded refusal, like that of Locke's Siamese potentate, to believe anything not contained within the limited compass of one's own experience: there were more things in heaven and earth than were contained in the skeptical philosophy.[67]

Condorcet dimly perceived that this Bayesian assumption was the linchpin of the rebuttal to the skeptical argument against miracles, but attempted to combat one fallacy with an even greater one of the same variety. It should be noted that there is no direct evidence that Condorcet had read either Hume or Price's writings on miracles, although he was certainly familiar with both authors by reputation.[68] However, Condorcet's discussion of what he termed the "probability of extraordinary facts" fol-

---

[67] Price, *Dissertations*, pp. 243–245, 250–251.

[68] For a brief account of Condorcet's relations with Hume, see Keith M. Baker, *Condorcet: From Natural Philosophy to Social Mathematics* (Chicago and London: University of Chicago Press, 1975), pp. 138–140.

lows the line of the argument set out in Hume and countered by Price so closely that it seems likely that he had at least secondhand knowledge of the British debate, particularly since there seems to have been no alternative French source for this probabilistic perspective on miracles.[69]

Buffon had briefly discussed how credibility decreased as the extraordinary nature of the fact reported increased, using the example of a report of a monstrous birth, but had related the problem to the more general issues of judgments based upon analogy.[70] The *Encyclopédie* article "Certitude," written by the Abbé Jean-Martin De Prades and introduced by Diderot, touched upon some of the same issues, but in a nonmathematical—indeed, antimathematical—fashion. De Prades rebuked the "géomètre anglais" who had attempted to calculate "the different degrees of *certitude*" assigned to testimony for misunderstanding the nature of certainty, and the relationship between certainty and probability. He understood probabilities in a wholly legal, qualitative fashion, as a personal judgment of a witness's integrity from the traditional legal indicators. The "probability" supplied by one witness grew into "certainty" when confirmed by others because the auditor passed from the individual particulars entailed in the "study of the heart" of a single witness to the "general laws followed by all men" embodied by several concurrent witnesses. This transition from particular cases to general conclusion had nothing to do with calculation, according to De Prades. De Prades, like Price, maintained that there was no inherent difference in the way in which human beings apprehended ordinary and extraordinary facts, and that testimony concerning the two should be submitted to the same critical criteria. However, the closest De Prades got to a probabilistic argument was to suggest a sort of metaphysical struggle for existence among all possible facts which rendered all equipossible:

> Descend into the abyss of not-being [*néant*]; you will see there natural and supernatural facts mingled together, one no closer to coming-into-being than any of the others. Their degree of possibility of emerging from this chasm and becoming realized is precisely the

[69] For example, both Voltaire's article "Miracle" in the *Dictionnaire Philosophique* and Diderot's article "Miracle" in the *Encyclopédie* are primarily concerned with the theological problem of reconciling divine volition with divine order.

[70] Buffon, "Essais d'arithmétique morale," in *Histoire naturelle. Supplément* (Paris, 1777), vol. 4, pp. 46–148, on p. 54.

same; because it is as easy for God to bring a dead person to life, as it is to preserve the life of a living person.[71]

Neither Buffon nor De Prades came to grips with the mathematical implications of the testimonial arguments for or against miracles which Hume had hinted at and Price had made explicit. It was Condorcet who attempted a mathematical treatment of the probability of testimony that balanced both extrinsic and intrinsic probabilities to affirm the Humean conclusion that the extraordinary content of a report weakened the force of the testimony that conveyed it.

Condorcet began his memoir "Sur la probabilité des faits extraordinaires," the fifth section of his six-part series on probability published in the *Mémoires de l'Académie royale des Sciences*, with the distinction between the intrinsic probability of an event and the extrinsic probability of a witness's testimony. The overall probability of testimony concerning an extraordinary event E would be

$$\frac{uu'}{uu' + ee'} \qquad \text{where } \begin{aligned} u &= P\ (E \text{ occurred}) \\ u' &= P\ (\text{testimony is true}) \\ e &= P\ (E \text{ did not occur}) \\ e' &= P\ (\text{testimony is false}). \end{aligned}$$

Condorcet cautioned that the value of u must refer to the probability of a "determinate" event as compared to that of another determinate event, *not* of an indeterminate event compared to the sum of all possible events. Condorcet's lottery example made it clear that he hoped to correct assumptions like that of Price which rendered all events equally improbable. However, Condorcet's solution depended on the same Bayesian assumption of equiprobability.

By distinguishing (albeit on dubious grounds) between the probabilities of determinate and indeterminate events, Condorcet had driven a wedge between the two critical parts of Price's argument. In the case of ordinary events, only indeterminate probabilities were involved, and Condorcet's method of evaluating these vitiated Price's key contention that ordinary events were as improbable as extraordinary ones. Computed according to Condorcet's method and probability values, the overall probability of tes-

---

[71] Abbé De Prades, "Certitude," in *Encyclopédie, ou Dictionnaire raisonné des sciences, des arts et des métiers*, d'Alembert and Diderot, eds. (Paris, 1751), vol. 2, pp. 845–862, on pp. 848–850.

timony in the case of ordinary events reduced to the probability of the witness, just as Price had claimed. This was consonant with the practice of reasonable men, which Condorcet, like Price, took as his guide in all of his research on mathematical probability. However, in the case of a single determinate probability, the overall probability of testimony plummeted. Condorcet's treatment implied that the value dropped because of the low intrinsic probability of the event rather than suspicion of motives for deceit—that is, for Hume's rather than Price's reasons.

Condorcet also attempted to analyze and quantify the difference between "vulgar" and "enlightened" witnesses, assuming both sorts to be of good faith. According to Condorcet, the probability of a witness varied with the difficulty of the observation, the sources of error that might color testimony, and the complexity of the event itself. Condorcet seized upon this last decrement in credibility in order to make a Lockean distinction between simple and complex facts the basis for a parallel distinction between orders of intellect. A "simple fact" was one which "a man of ordinary capacity can grasp both the whole and the details in a single glance without a great effort of attention." Complex facts were combinations of such simple facts. If u' was the probability that the testimony of a given witness regarding a simple fact was accurate, then $(u')^2$ would be the probability that his testimony regarding a complex fact composed of two such simple facts was reliable, and so on. Therefore, a witness capable of comprehending three simple facts at a glance while another's attention would have been fully absorbed by only one simple fact would boast a credibility of u', as opposed to his duller comrade's considerably smaller $(u')^3$, since $u' < 1$.

Thus "unenlightened" witnesses who fail to understand that the global truth of the fact in question presupposes the truth of several constituent simple facts could easily be sincerely but erroneously convinced of the truth of the fact. Since most extraordinary facts were complex, being the sum of many ordinary facts, the force of testimony for ordinary witnesses affirming such facts waned accordingly. Condorcet concluded his memoir with the claim that he had shown "how one can explain by calculation the weakening of testimony which concerns extraordinary facts."[72] In this round, the calculus of probabilities sided with Hume over Price.

[72] Condorcet, "Sur la probabilité des faits extraordinaires," *Mémoires de l'Académie royale des Sciences. Année 1783* (Paris, 1786), pp. 553–559, on pp. 555–557, 559.

In the sixth and final part of his series of memoirs on probability, Condorcet began by reviewing his distinction between the probabilities of determinate and indeterminate events made in the previous memoir. Although he still insisted on the importance of the distinction and on its value in explaining when and how the extraordinary nature of the fact reported influenced the overall probability of testimony, Condorcet confessed that he now found his original method of estimating indeterminate probabilities to be "too hypothetical"; "arbitrary" when applied in practice; and "too remote" from the results ordained by "common reason." In its place, Condorcet proposed the "probability proper" of an event, defined as the ratio of the conventional probability of the event to the average probability of all other possible outcomes. Condorcet pointed out that when the probability proper of an event was greater than $1/2$, the overall probability of testimony of an event increased significantly, even if the credibility of the witness was scanty:

> So it is, for example, that an astronomical fact consonant with the theory of gravitation would be easily believed on the assertion of a single Savant, even by those who had not verified his calculations, but if the same Savant had announced a fact contrary to that theory, he would need a very great authority, even for anyone to think it reasonable just to check his calculations.[73]

Condorcet attempted to apply his new concept of probability proper to the question of determining the reigns of seven Roman rulers, prefacing his calculations with a brief history of earlier accounts to apply mathematics to historical chronology. With a deferential nod in the direction of Newton's attempts to revise the accepted chronology of ancient history from astronomical considerations, Condorcet examined the academician Nicolas Fréret's excoriation of Newton's effort in a 1724 address to the Académie des Inscriptions et Belles-Lettres.[74] Fréret had also had harsh words for John Craig's *Principia*, advising probabilists to restrict their efforts to games of chance. Condorcet admitted that Fréret's distinction be-

---

[73] Condorcet, "Application des principes de l'article précédent à quelques questions de critique," *Mémoires de l'Académie royale des Sciences. Année 1784* (Paris, 1787), pp. 454–468, on pp. 456, 460.

[74] For an account of Fréret's criticisms and of Newton's chronology, see Frank Manuel, *Isaac Newton, Historian* (Cambridge, Mass.: Harvard University Press, 1963), pp. 26–30 et passim.

tween the finite number of combinations in a game of chance and the indefinite number of combinations in the occurrence of a natural event had been a trenchant criticism of the probability of testimony when Fréret had aired it in 1724, several decades prior to the discovery of inverse probabilities. However, Condorcet reproached Fréret, otherwise well versed in mathematics and the exact sciences, for not accepting the a posteriori approximations sanctioned by Bernoulli's theorem for large numbers of observations. Although Condorcet acknowledged that Fréret's skepticism concerning the mathematical treatment of history owed to applications which were "too hypothetical, founded upon principles which were false or quite bizarre," he maintained that the erudite academician had never denied "the general principle that one must take account of the probability proper, be it physical or moral, of events."

Condorcet applied his own mathematical method to the problem of estimating the reigns of seven elected Roman rulers, a problem to which Voltaire had already submitted to quasi-probabilistic considerations with what Condorcet described as "zeal and success." Supplementing his method of evaluating the probability proper of an event with several assumptions concerning age at election and uniform mortality rate between the ages of thirty and ninety, Condorcet assessed the probability of the 257-year figure given by the chronicles at approximately 1/4. In order to boost the overall probability of testimony to .9999, the figure designated by Condorcet as providing a secure basis for further inferences, the reporter would have to be a historian who erred only once in 30,000 trials. Condorcet, like Nicholas Bernoulli, understood the credibility of a witness to be a problem in the probability of causes.

Condorcet's second historical example recalled his earlier discussion of extraordinary facts: What was the probability of the report transmitted by Roman historians, that razors cut flintstone? In order to compute the probability proper to this event, Condorcet supposed that a million past attempts to cut flintstone contradicted this report. Using the law of succession derived from Bayes' theorem, he computed the probability proper of this extraordinary event as so astronomically small that even "the most excessive credulity could not lend the historians the necessary authority to give a reason for belief."[75] Condorcet had incorporated several of Hume's

[75] Condorcet, "Application," pp. 462, 463, 466. Jakob Bernoulli's and Buffon's figure for moral certainty was also .9999.

assumptions into his mathematical treatment of the probability of testimony, notably the hypotheses that the intrinsic improbability of an event mitigated the force of the extrinsic credibility of the witness, and that the latter was simply a special case of the probability of causes, to be gauged from past experience. It was therefore not surprising that his results bore out Hume's criterion of reasonable belief.

Laplace also affirmed Hume's criterion, albeit by means of a different and mathematically simpler model. Laplace brushed aside Condorcet's attempts to impose such subtle distinctions as those between simple and complex facts in favor of a streamlined treatment that abstracted as many details as possible from the actual circumstances of testimony. Laplace also restricted his example to the simplest case, reporting upon numbered balls from an urn, giving specific historical applications a wide berth. Yet Laplace retained the categories of intrinsic and extrinsic probability, and declared that his results endorsed the Hume criterion (again, Hume was not mentioned by name). His treatment followed Condorcet's in outline if not in particulars.

Laplace devoted the last chapter of the *Théorie analytique des probabilités* (1812; later eds.) to the probability of testimony, indicating that the problem still counted as one of the standard applications of the calculus of probabilities. Positing that the probability of testimony depended on the intrinsic probability of the fact reported and the extrinsic probability of the witness (which in turn depended on the witness's integrity and competence), Laplace distinguished among four possibilities in the case where the reported event E was drawing a particular ball i from an urn containing n balls, $P(E) = 1/n$; $p = P$ (witness tells the truth); and $r = P$(witness is accurate):

1. the witness doesn't lie and doesn't err: $P_1 = pr/n$
2. the witness doesn't lie but does err: $P_2 = p(1 - r)/n$
3. the witness lies and errs: $P_3 = (1-p)(1 - r)/n$
4. the witness lies and doesn't err: $P_4 = (1 - p)(r)/n$.

Therefore the probability of the testimony that the ith ball had been drawn would be

$$P_i = \frac{P_1 + \dfrac{(1 - p)(1 - r)}{n(n - 1)}}{P_1 + P_2 + P_3 + P_4} = pr + \frac{(1 - p)(1 - r)}{n - 1} .$$

335

In essence, this was an expanded version of Condorcet's formula $\dfrac{uu'}{uu' + ee'}$, with the extra terms arising from the double criterion of integrity and competence. The second term of the numerator emerged from the following line of reasoning: What if the witness is prone to both deception and error? A certain number i' has been drawn instead. However, he may have unintentionally confused himself as well: this is case (3): $P_3 = (1 - p)(1 - r)/n$. There is a possibility, $1/n - 1$, that the deceptive and befuddled witness might trick himself into actually announcing the correct number i, rather than i'. The product of the two probabilities P_3 and $1/n - 1$ would give the probability that this double deception and self-deception would accidentally result in true testimony.[76]

Laplace generally set $r = 1$ (i.e., the witness is presumed never to make mistakes), eliminating cases P_2 and P_3 and collapsing his formula to that of Condorcet. Laplace also reproduced Condorcet's distinction between the probabilities of determinate and indeterminate events, although without Condorcet's convoluted computations and redefinitions. Imagine a witness with a probability p of $9/10$, who announces that number i has been drawn from an urn containing $n = 1,000$ balls. In this case,

$$P_1 = 9/10,000;$$
$$P_4 = P\,(\text{not-}i) \cdot P\,(\text{not any other particular number } i) \cdot (1 - p)$$
$$= (999/1000)(1/999)(1/10) = 1/10,000.$$

Here, Laplace analyzed the intrinsic probability of the complementary event into two distinct probabilities: that of any other outcome except i, and that of any *particular* outcome besides i. The overall probability of testimony (omitting the terms that disappear when $r \neq 1$), would be $9/10$, or simply the probability p. This was Condorcet's comparison of two determinate cases, and coincided with Price's conclusion that the probability of such testimony was identical to that of the witness's veracity.

However, Laplace proposed an alternative example with more Humean overtones. Imagine the urn to be filled with $n = 1,000$ balls, 999 black and 1 white; the same witness announces that a white ball has been drawn. As usual $P_1 = (1/1,000)(9/10) = 9/10,000$; but $P_4 = (999/1,000)(1/10) = 999/10,000$. Here, the extrinsic probability of the complementary event compared a determinate to an indeterminate outcome (all of the

---

[76] Laplace, *Théorie analytique des probabilités* (1812), 3rd edition (1820), in *Oeuvres*, vol. 7, p. 457.

black balls were assumed to be indistinguishable). As a result, the overall probability of testimony was considerably diminished. Admitting the possibility of error ($r \neq 1$) did not alter this conclusion. Moreover, as n grew larger, and the intrinsic probability of the reported extraordinary fact in the indeterminate case decreased accordingly, so did the overall probability of testimony. Laplace called these results to the attention of those who, like the Abbé De Prades in the *Encyclopédie* article "Certitude," had claimed that there existed no inherent difference between an ordinary and extraordinary fact and the probability of testimony reduced to witness credibility in both cases: "Simple good sense rejects such a strange assertion; but the Calculus of Probabilities, by confirming the indication of common sense also gauges the unlikelihood of testimony concerning extraordinary facts."[77]

Laplace was under no illusions regarding the highly idealized nature of the assumptions upon which his treatment of the probability of testimony was based. He repeatedly confessed to the limitations of his results when applied to specific cases: for example, it was obviously more likely for a witness deliberately to report the wrong number in the first urn example if he had a stake in another outcome, thereby vitiating the assumption that all $n - 1$ other alternatives were equiprobable choices. However, Laplace waved aside such caveats and refinements—perhaps with Condorcet's tortuous exposition in mind—as too difficult to formulate mathematically.[78] Yet this admission did not dissuade Laplace from pursuing such computations and defending their utility. Although somewhat oversimplified, the probabilistic model of reasonable belief in testimony was "highly analogous" to many real situations; that is, it seized the essentials and provided reliable guidelines for assent: ". . . [these] solutions can be regarded as approximations suited to guide us and to preserve us from the errors and dangers to which bad reasoning exposes us. An approximation of this sort, when it is done well, is always preferable to the most plausible reasoning." Laplace was particularly firm in his claim that the calculus of probabilities as applied to the probability of testimony demonstrated the fallacy of belief in miracles:

We must conclude from this that the probability of the constancy of the laws of nature is for us superior to that of the event in question; a probability in itself superior to the majority of historical facts which

[77] Laplace, *Essai*, p. lxxxii. [78] Laplace, *Théorie analytique*, p. 458.

we regard as incontestable. One can judge from this what immense weight testimony must carry in order to admit the suspension of natural laws; and how great an abuse it would be to apply the ordinary rules of criticism to such cases.[79]

Just as the probability of causes had affirmed the uniformity of the natural order, so the probability of testimony could be used to invalidate all reports to the contrary.

Poisson's discussion of the probability of testimony in his *Recherches sur la probabilité des jugements* (1837) followed Laplace's method and conclusion closely. Like that of Condorcet and Laplace, Poisson's basic formula was based on the intrinsic probability of the event reported and the extrinsic probability of the witness's veracity. However, Poisson interpreted this formula wholly in the context of Bayesian probabilities of causes. In this case, the intrinsic probability of the event, q, corresponded to the a priori probability of the hypothesis that the controversial event did indeed happen, independent of any report. The extrinsic probability of the witness, p, represented further evidence. Poisson conceived of the overall probability of testimony as equivalent to the conditional probability, r, that, given the evidence of the witness's report, the event had actually occurred:

$$r = \frac{pq}{pq + (1 - p)(1 - q)}.$$

Although Poisson admitted that neither p nor q could ever assume the extreme values of 1 or 0 in this example, he envisioned the case in which past experience would assign an extremely small value to q and a very high value to p. The resulting value of r would be extremely small. Poisson applied this observation to the case of miracles:

This is the case of a fact contrary to the general laws of nature, yet affirmed by a witness to whom one would accord, without this opposition, a high degree of confidence. These general laws are for us the result of a long series of experiences which gives them, if not an absolute certainty, at least a very high probability, augmented still more by their appearance of harmony, and which no testimony can counterbalance.[80]

Only an infallible witness could render the probability of testimony r sufficiently great to warrant belief. Poisson argued that even the testimony of

[79] Laplace, *Essai*, pp. lxxix, lxxxv–lxxxvi.   [80] Poisson, *Recherches*, pp. 98–99.

one's own senses should be doubted in cases of apparent abrogations of natural laws: the overwhelming value of $1 - q$ amassed from past experience made it more reasonable to confess to a hallucination.

Poisson applied a similar Bayesian approach to the probabilities of concurrent testimony and chains of transmitted testimony. By viewing the probability of the witnesses' veracity as simply another type of evidence, comparable to the confirming instance in the probability of causes, Poisson was able to assert the sovereignty of the natural order. In the case of a long, traditional chain of witnesses, Poisson's results indicated that the impact of the testimony itself was minimal, and that the overall probability of testimony differed little from the intrinsic probability of the event. Although Poisson and Laplace had used nearly identical methods, they chose to emphasize different aspects of their results. Both pointed to the steady diminution of probability as the chain lengthened, but Laplace called attention to the case in which the intrinsic probability was also very small because the number of possibilities had increased, while Poisson noted the case in which the impact of the testimony became attenuated while the probability of the event remained constant.[81]

Condorcet, Laplace, and Poisson were the foremost exponents of the probability of testimony during the latter half of the eighteenth and the first half of the nineteenth centuries, but they were by no means the only ones. Johann Heinrich Lambert, Simon L'Huillier, Silvestre-François Lacroix, and less well-known figures such as Charles-François Bicquelley also wrote on the subject. By the turn of the nineteenth century, the probability of testimony was featured in texts and treatises on the calculus of probabilities as a standard application, along with the probabilities of games of chance, annuities, and the probability of causes.[82] However, in contrast to the other standard topics, the probability of testimony did not receive a single, standard treatment.

Lacroix's text, *Traité élémentaire du calcul des probabilités* (1816), is particularly revealing on this point, for Lacroix was at once a consummate textbook writer, a philosopher of mathematics, and extremely well versed in the mathematical literature of his time. Lacroix's own treatment of the probability of testimony owed much to that of his mentor Condorcet, although Lacroix tended to minimize the contributions of the intrinsic probability

[81] Poisson, *Recherches*, p. 112.

[82] See, for example, Bicquelley, *De calcul des probabilités* (Toulouse, 1783), or Condorcet, *Elémens du calcul des probabilités* (Paris, An XIII/1805).

of the event by considering only the witness's past record of veracity. For example, in Condorcet's 1783 formulation, Lacroix interpreted the "probability proper" of a fact as if it were the testimony of a previous witness. If v = number of times the first witness had been truthful; m = the number of times he had not, and with v', m' similarly defined for the second witness, the probability of testimony would be, according to Lacroix,

$$\frac{\left(\frac{v}{v + m}\right)\left(\frac{v'}{v' + m'}\right)}{\left(\frac{v}{v + m}\right)\left(\frac{v'}{v' + m'}\right)+\left(\frac{m}{v + m}\right)\left(\frac{m'}{v' + m'}\right)} = \frac{vv'}{vv' + mm'}.$$

This was Condorcet's formula, but hardly his interpretation of the probability of testimony.

More interesting than Lacroix's divergences from Condorcet and other authors on the topic was his awareness and commentary upon this divergence. Lacroix underscored the critical, if implicit, role of hypothesis in the probability of testimony. He recognized that his own treatment hinged upon an analogy between testimony and the throw of a die: each witness could be compared to a die with a certain number of sides marked "truth" and a certain number marked "false." Each side had an equal possibility of turning up on any given throw. By Bernoulli's theorem, the witness's actual ratio of truth to falsehood would reflect his inherent veracity, just as the observed ratio of T to F outcomes would with a high probability closely approximate the actual proportion of T to F sides in a large number of trials.

In addition to begging the practical question of how to find the witness's record of veracity from observation, this method assumed the consistency of the witness. Yet Lacroix, who cited Hume frequently, realized that the credibility of the witness varied with the nature of the facts, particularly for extraordinary facts: "The form of the die must change with the nature of the fact."[83] However, this was easier said than done from the standpoint of the mathematical model. As Laplace had acknowledged, an exact probability of testimony "soon becomes impossible, because of the difficulty of estimating the veracity of the witnesses, and because of the large number of circumstances which accompany the reported facts."[84]

[83] Silvestre-François Lacroix, *Traité élémentaire du calcul des probabilités* (Paris, 1816), pp. 219, 227.

[84] Laplace, *Essai*, p. lxxix.

Mathematicians must rest content with admittedly oversimplified hypotheses, as long as these lead to acceptable approximations.

"Acceptable approximations" were those that confirmed good sense. Lacroix compared the delicacy of choosing among hypotheses regarding the probability of testimony to the same difficulty in defining probabilistic expectation. In both cases, the mathematician sought to "give the calculus a form which approached the results of the insights of good sense." In this spirit, Lacroix criticized the results of the *Philosophical Transactions* author for contradicting good sense by suggesting that the probability of testimony always increased with the number of eyewitnesses. Lacroix accused the author of having arranged his calculations so as to get a particular result repugnant to good sense: ". . . it takes only brief reflections to be persuaded of the difficulty or near impossibility that numerous ignorant spectators will not err concerning the appearance of extraordinary facts."[85] Lacroix quoted Hume's arguments on miracles with approval, and confidently predicted that future experience would only strengthen conviction on the inviolability of the natural order. He defended the probability of testimony, even though occasionally "this subject cannot be readily submitted to the calculus due to the fact that the veracity and sagacity of violently agitated men undergo abrupt changes, and chiefly in the case of marvelous facts."[86]

Lacroix was able to recognize the artificial character of the hypotheses underlying rival treatments and even the inherent limitations of all such hypotheses that are due to the complexity of the subject matter. Yet he stopped short of suggesting that "good sense" was also a fluid, hypothetical notion, or of abandoning the probability of testimony, as Montmort might well have, as a doomed enterprise. Lacroix's own brand of associationism reduced good sense to the intuitive reckoning of probabilities: "The processes of reasoning take the form of a sort of calculus whose result acquires command of our belief, precisely by the effect of the repetition of these judgments or observation."[87] As long as good sense remained a kind of internal calculus, it seemed reasonable to expect unique answers.

The particular conclusions dictated by "good sense" concerning the probability of testimony varied. Initially, the mathematical probability of testimony had mirrored the legal procedures for evaluating testimony. Critical historiography concerning the reliability of oral and written tra-

[85] Lacroix, *Traité*, pp. 227, 234.　　　[86] Lacroix, *Traité*, p. 236.
[87] Lacroix, *Traité*, p. 7.

ditions also contributed problems and hypotheses. Probabilists supplemented and altered these hypotheses with assumptions of their own which derived from more physicalist premises. John Craig's extended mechanical analogy was perhaps the most striking, but all of the probabilists assimilated the probability of testimony to the probability of natural causes to a greater or lesser degree. In doing so, they effectively followed Montmort's counsel to eliminate all considerations of free will, that crucial dividing line between the natural and moral realms. The use of the mathematical techniques—in particular, Bayes' and Bernoulli's theorems—of the probability of causes presupposed constant causes that would emerge from prolonged observation. Consequently, the probability of testimony assumed constant witness veracity, to be in principle assessed from observed frequencies. Judgments about testimony and about natural events were founded on the same psychological processes and were subject to the same mathematical methods.

However, quantifying the probability of testimony was not simply a question of contorting the meaning of credibility to fit ready-made mathematical boxes. I have gone into some detail in order to show how permeable the mathematical treatment could be to philosophical assumptions: in the proper hands, the probability of testimony could be made to serve both sides of the debate over miracles. As usual, quantification required accommodations from both the mathematics and from the subject matter. The mathematicians were well aware that their assumptions did not do full justice to the complexity of all cases, but this did not discredit the enterprise in their eyes. All mathematical models abstracted and idealized the phenomena; the key question was whether the results nonetheless tallied with observation. For the classical probabilists, the test was the good sense of reasonable men, which they persisted in treating as monolithic. The probability of judgments, even more than the philosophy of testimony, strained this faith to the breaking point.

6.4 The Probability of Judgments

While the probability of testimony had had clear legal antecedents in the judicial criteria applied to courtroom evidence, the probability of judgments was a mathematical application with no preexisting analogue in jurisprudence. Devised by Condorcet on the eve of the French Revolution, it sprang from the philosophes' campaign for judicial reform in France.

The probability of judgments addressed the problem of the optimal design of a tribunal in order to minimize the probability of a wrong decision by manipulating three variables: the number of judges; the minimum majority required; and the probability that the individual judges would decide correctly. Condorcet and Laplace also applied this model to the decisions of legislative and electoral assemblies; however, the difficulties of defining the meaning of a "correct" decision in this context, in contrast to the straightforward judicial alternatives, significantly altered the context, although not the method, of this kind of application.[88] This section will concentrate on the probability of judgments per se as treated by Condorcet, Laplace, and Poisson, the context in which it arose, and the criticisms which eventually discredited it as a valid application of mathematical probability.

Judicial reform was arguably the *philosophes'* single most ardently waged campaign prior to the French Revolution. Two literary events rallied enlightened opinion to the cause: Voltaire's scathing tracts on several notorious miscarriages of justice in the cases of the Calas family, the Chevalier de La Barre, Elizabeth Canning, and others, culminating in the *Traité de la tolérance*; and the Abbé Morellet's translation of Cesare Beccaria's *Dei delitti e delle pene* (1764) in 1766. Montesquieu's emphasis on security as the mainstay of liberty, particularly in criminal proceedings, also fueled the movement for reform. Criticism focused on the abolition of torture, the openness of the trial process, the right of the accused to legal counsel, and the need for procedural reforms in the selection of judges and the rules for evaluating evidence.[89]

While Voltaire's outraged 1763 appeal to "all of the party of the *Encyclopédie*" to protest against the atrocities of the judicial system in the Calas case galvanized literary opinion, it was Beccaria's treatise which laid the theoretical foundations for the reform movement and probably inspired Condorcet's mathematical efforts. The *Traité des délits et des peines* took the French intellectual world by storm, and enjoyed greater success in France than in Beccaria's native Italy. Most of the philosophes would have seconded Condorcet's opinion that Beccaria had "rejected in Italy the barbaric maxims of French jurisprudence."[90] J. P. Brissot de Warville, in one of the numerous commentaries inspired by Beccaria's treatise, described the

---

[88] See Duncan Black, *The Theory of Committees and Elections* (Cambridge, Eng.: Cambridge University Press, 1958), chapter 8.

[89] See A. Esmein, *Histoire de la procédure criminelle en France* (Paris, 1882), pp. 357 ff.

[90] Quoted in Esmein, *Histoire*, p. 364.

"crowd of discourses, memoirs, and dissertations" which Beccaria's work had unleashed in France.[91] Voltaire himself wrote such a commentary, which attacked the French criminal procedures codified in the Ordinance of 1670, including torture, pretrial detention, closed trials, secret testimony, and the system of legal proofs, and expressed the hope that Beccaria's work "would soften the vestiges of barbarism in the jurisprudence of all nations."[92] Morellet sent Beccaria the praise of Diderot, Buffon, Helvetius, d'Holbach, d'Alembert, and Hume (then in Paris). Between February and September of 1766, Morellet's translation went through seven printings of 1,000 copies each, prompting Morellet to write enthusiastically to the author that no obstacle could halt "the universal action . . . of light, reason, and public opinion."[93]

Beccaria addressed the issues of proportioning crimes to punishment; the justification of torture; the validity of the death penalty; the contractual exchange of individual liberty for societal security; and the nature of legal proof. Beccaria's views on the last may well have caught Condorcet's attention, for they interpreted the accuracy of legal decisions probabilistically:

> One may be surprised to hear me use the word probability in speaking of crimes which in order to warrant punishment, must be certain.
> . . . But it must be noted that, rigorously speaking, moral certainty is only a probability, which is called a certainty because all men of good sense are compelled to give it their assent, which is necessarily determined by a habit following from the necessity of acting, which is anterior to all speculation.[94]

There was certainly nothing novel in this view, even in the sphere of jurisprudence: Jean Domat and other natural law jurists had already urged the probabilistic nature of judgments. Beccaria's discussion of the weighting of testimony and other sorts of evidence remained traditional and qualitative, rather than mathematical. However, Beccaria's suggestively worded explanation of how society acquired the right to punish individu-

---

[91] Brissot de Warville, "Note," in Cesare Bonesana Beccaria, *Des délits et des peines*, J.A.S. Collin de Plancy, trans. and ed. (Paris, 1823), p. 440.

[92] Voltaire, "Commentaire," in Beccaria/Collin de Plancy, *Délits*, p. xl.

[93] Morellet to Beccaria, February, 1766, and September, 1766, in Beccaria/Collin de Plancy, *Délits*, pp. 443, 458.

[94] Beccaria/Morellet, *Délits*, pp. 39–40.

als may have triggered Condorcet's mathematical reflections on the subject. Beccaria outlined a contractual model of society, in which individuals banded together for mutual sustenance and defense. Each individual member of society ceded the minimum fraction of his liberty necessary to guarantee societal order to the "common depository" of authority. According to Beccaria, the sum total of these fractions of liberty constituted the basis of the societal right to inflict punishment. Any exercise of power beyond this base was "abuse," not "justice."[95]

The balance between summed fractions of individual liberty and societal authority represented by the criminal justice system formed the framework of Condorcet's *Essai sur l'application de l'analyse à la probabilité des décisions rendues à la pluralité des voix* (1785). Condorcet translated the fraction of individual liberty into the risk that an innocent citizen might be wrongly convicted and punished for a crime. Some degree of risk was the inescapable price of social order and the fallible human institutions that maintained that order. However, Condorcet asserted that this margin of inevitable error varied according to the judicial system. Both social justice and social order required that the risk be minimized; otherwise, the citizenry would lose faith in the judiciary and refuse to cooperate with its edicts.[96] Judicial verdicts could never exceed the probability of the evidence on which they were based, but Condorcet argued that the form and composition of judicial tribunals could fix the average probability of error. The *Essai* purported to quantify the maximum acceptable error level in judicial proceedings, and to calculate the number of members, their individual degree of "enlightenment" (*lumières*), and the minimum required plurality needed to guarantee this probability.

Prior to the publication of the *Essai*, Condorcet had taken an active interest in judicial reform. Condorcet's mentor Turgot was one of a group of *philosophes*, including Chrétien-Guillaume de Lamoignon de Malesherbes and d'Alembert, who invited Beccaria to Paris after Morellet's spectacularly successful translation had made Beccaria an overnight celebrity in Paris salons.[97] Condorcet's letters to Turgot are filled with comments on related matters. Condorcet complained that "parliamentary pretensions, prejudices, conduct and the laws they follow are the principal

---

[95] Beccaria/Morellet, *Délits*, pp. 15, 45–52.

[96] Condorcet, *Essai*, p. xxi.

[97] Ann-Robert-Jacques Turgot, *Oeuvres de Turgot et documents le concernant*, Gustav Schelle, ed. (Tauners: Detlev Auvermann, 1972; reprint of 1913–23 edition), vol. 2, p. 67.

cause of the ills of France" apropos of the conviction and brutal execution of the Chevalier de La Barre;[98] chafed at the "spirit of intolerance, ignorance, pedantry, and barbarity" which ruled the old judiciary; speculated on the advantages of the jury system;[99] and argued for the abolition of torture and the need to assign specific punishments to particular crimes.[100]

At the time, no rigid code of criminal procedure existed in France. Judicial officials were empowered to modify punishments if circumstances dictated.[101] Abuses of this discretionary power were notorious: the Chevalier de la Barre was tortured to elicit a confession of impiety, then beheaded, and both body and head were thrown to the flames.[102] Condorcet intended his *Essai* as a manifesto for reform, not simply as a mathematical treatise. He summarized his method and results in a lengthy Preliminary Discourse to spare the lay reader the labor of following the calculations and demonstrations, recognizing that the work would have only a "very limited utility" if restricted to an audience of mathematicians.[103] In a letter presenting a copy of his *Essai* to Frederick II of Prussia, Condorcet underscored the results with policy implications, such as the abolition of the death penalty and the paramount need for enlightened judges (rather than political constitutions).[104]

Condorcet had also mused over the more general connections between the calculus of probabilities and jurisprudence. In a manuscript fragment, Condorcet enlarged upon the claim of the Parisian jurist and amateur astronomer Payen[105] that "the most celebrated jurists have also studied mathematics." Condorcet argued that not only did mathematics discipline the intellect in ordering "a series of subtle ideas and complicated reason-

---

[98] Condorcet to Turgot, 29 June 1770, in Henry, ed., *Correspondance*, p. 16.

[99] Condorcet to Turgot, ca. 1772, in Henry, ed., *Correspondance*, p. 80.

[100] Condorcet to Turgot, ca. July 1774, in Henry, ed., *Correspondance*, p. 184.

[101] See Carl Ludwig Von Bar, *A History of Continental Criminal Law*, Thomas S. Bell, trans. (Boston: Little, Brown, & Co., 1916), pp. 260 ff.

[102] Dominique Holleaux, "Le procès du Chevalier de la Barre," in *Quelques procès criminels des XVIIe et XVIIIe siècles*, Jean Imbert, ed. (Paris: Presses Universitaires de Paris, 1964), p. 177.

[103] Condorcet, *Essai*, p. ii.

[104] Condorcet to Frederick II of Prussia, 2 May 1785, in Condorcet, *Oeuvres*, vol. 1, p. 305.

[105] Antoine-François Payen, *Extrait d'une lettre . . . contenant l'observation de l'éclipse de soleil . . .* (n.p., n.d.); and *Senelion, ou Apparition luni-solaire en l'Isle de Gergonne* (Paris, 1666).

ings" but that many legal questions "depend absolutely on the calculus of probabilities, whether one seeks the principles upon which the laws should be founded or to apply the law as it exists."[106]

In this spirit, Condorcet began an annotated French translation of Nicholas Bernoulli's *De usu artis conjectandi in jure* (1709),[107] wrote a memoir applying the calculus of probabilities to evaluating the value of feudal land rights,[108] and cosponsored a prize for the reduction of all legal contracts to "general formulas." The prize was to be offered jointly under the auspices of the Académie des Sciences and two other learned academies, with the purse of 1,000 ducats to be paid by an anonymous (and apparently foreign) donor.[109] Apparently Condorcet had difficulty persuading two sister academies to join the Paris Académie, for the prize was never officially announced. However, Condorcet lobbied energetically for the proposal, urging the editors of journals to publicize the project. He described the problem as one which had more to do with "the calculus of combinations and the art of classifying objects" than with actual existing law. Indeed, the prize conditions stipulated that the solution be "applicable to all nations [and] embrace all contracts which could occur by the nature of things among men edowed with reason" by considering "the nature of men and that of property in an abstract and general manner." Condorcet invoked the past success of the natural sciences in solving apparently insoluble problems and queried "why politics, if cultivated by the same method as the other sciences, should not exhibit the same history."[110]

Condorcet viewed the *Essai* as another opportunity to harness the calculus of probabilities to the ends of political and social reform. Starting with the premise, which he described as "rigorously true" to Frederick II of Prussia, that "all possibility of an error in a judgment is a true injustice,"[111] Condorcet mathematically examined the means of minimizing

---

[106] Condorcet, Bibliothèque de l'Institut MS. 884, f. 1r (on the back of a letter dated Paris, 3 September 1785).

[107] Condorcet, Bibliothèque de l'Institut MS. 875, ff. 200–213.

[108] Condorcet, "Sur l'évaluation des droits éventuels," in *Mémoires de l'Académie royale des Sciences. Année 1782* (Paris, 1785), pp. 674–691.

[109] Archives de l'Académie des Sciences, MS. *Procès-Verbaux*, vol. 104, pp. 72–73. The original of this report (30 April 1785), in Condorcet's hand, is preserved in his Dossier at the Archives.

[110] Condorcet, Bibliothèque de l'Institut MS. 873, f. 181 r. and v.

[111] Condorcet to Frederick II of Prussia, 2 May 1785, in Condorcet, *Oeuvres*, vol. 1, p. 305.

that injustice. In doing so, he relied on a number of assumptions and techniques borrowed from earlier applications of the calculus of probabilities: Buffon's attempt to quantify the psychological impact of moral certainty; the constant causes and independent trials presumed by Bernoulli's theorem; the probabilists' maxim that their results simply confirmed the conclusions of good sense. Condorcet combined these assumptions with more political precepts such as the sanctity of natural rights, the social need for expediting legal procedures, and the inherent tension between individual and social interests. Condorcet's treatment was based on the tension between the inevitable injustice enacted in the long run and "sufficient assurance" that justice would be done in any individual case: "Thus, it is permissible to act upon an opinion, even though it is probable that in a large number of such actions, determined upon the same principle, an injustice will be done, if one has for each particular action a sufficient assurance that it is consonant with justice."[112]

In order to evaluate this critical level of "sufficient assurance," Condorcet turned to Buffon's technique of estimating moral certainty as the probability that a healthy man in the prime of life would die within the next twenty-four hours, as reckoned from the mortality tables. Although Condorcet criticized Buffon's method because it chose as a standard risks that were ignored not necessarily because they were negligible but because they were unavoidable, he accepted Buffon's notion that probabilities must be appreciated subjectively in order to be meaningful. Condorcet proposed that citizens in a just society should run no greater risk of being wrongly convicted and punished for a crime than that to which they would "voluntarily expose themselves without any preformed habit, for an interest so slight that it could not be compared to one's life, and without requiring any courage." He suggested the risk of sailing with the Dover-Calais packetboat, or the risk of investing in an annuity, or the *difference* between the probabilities of dying suddenly within a week for two different age groups. Only the last could be computed readily from the available tables, so Condorcet based his measure of sufficient assurance on that incremental difference, deriving 144,767/144,768 as the probability that an innocent citizen would not be unjustly punished. In essence, 1/144,768 was the fraction of liberty the individual traded for the benefits of community.[113]

---

[112] Condorcet, *Essai*, p. xxxix.
[113] Condorcet, *Essai*, pp. cix, cxiii–cxiv.

Condorcet's contractual model for both social bonds and probabilistic expectation converged in the probability of judgments. Justice was a kind of contract in which the risks to social order and individual liberty were balanced in the same way that risks and potential gains were balanced in a fair game. Condorcet's modified rule of expectation held for both types of contracts. In the judicial case, Condorcet maintained that for a long series of judgments, the probability that the convicted defendant is guilty should be in the same ratio to the probability that the acquitted defendant is innocent, as the disadvantage of condemning an innocent person is to that of freeing a guilty one. However, Condorcet recognized that this analogy was gravely flawed: "Society, if you will, would play a fair game, because it plays an indefinite number of times; but it would not be the same for an individual who, relative to the small risk he runs from freed criminals, can only play a number of rounds too small for equality to obtain for him."[114]

Hence an extremely high level of sufficient assurance was needed to safeguard individual liberties in the societal game of justice. With Voltaire's fiery tracts on the miscarriage of justice in the cases of the unfortunate Calas family or Chevalier de La Barre still vivid memories, Condorcet was sensitive to the possibilities of wrongful conviction. Because of the inevitable, if minuscule, possibility of judicial error, he argued against the death penalty as the only irreversible punishment. In response to Frederick II's contention that capital punishment should be retained for certain atrocious crimes, Condorcet replied that it was in just those cases, when popular opinion was inflamed, that judges were most likely to convict innocent suspects.[115]

The probability of a correct judicial decision depended on three variables in Condorcet's exposition: the number of judges; the required plurality; and the probability that each individual judge would decide correctly. In order to assimilate the probability of judgments to Bernoulli's and Bayes' theorems, Condorcet assumed that the individual probabilities for each judge were equal—that is, each judge possessed "equal enlightenment" (*lumières égales*)—that these probabilities were constant, and that the decisions of the individual judges were independent of one another.

---

[114] Condorcet, *Essai*, pp. lxxviii–lxxix.

[115] Frederick II of Prussia to Condorcet, 14 May 1785; Condorcet to Frederick II, 19 September 1785, in Condorcet, *Oeuvres*, vol. 1, pp. 311, 315.

Condorcet admitted that these assumptions did not always hold (his critics declared that they never did), and in the fourth section of the *Essai* attempted to introduce more complex considerations, although with little computational success. Like his attempts to refine the probability of causes, these amendments quickly encumbered Condorcet in formulas of insuperable complexity and questionable applicability. However, the major results of the *Essai*, and of subsequent treatment of the probability of judgments, derived from the assumption that each judicial verdict rendered by a tribunal of n judges, each with probability v, was mathematically equivalent to n drawings from an urn containing white and black balls in the proportion of v to e.

Condorcet's approach was followed by Laplace and Poisson in its essentials, although they introduced new hypotheses and suppressed others. If a tribunal is composed of $r + s = n, r > s$, judges, each with a probability of v of rendering a correct decision in any given case and probability e of erring, the probability that r members will give a correct decision in any given case would be the $(s + 1)$th term of the binomial expansion of $(v + e)^{r+s}$:

$$\binom{r + s}{s} v^r e^s,$$

and the probability that the r concurring judges are wrong would be

$$\binom{r + s}{r} v^s e^r.$$

Therefore, the probability of a correct decision rendered by a majority of r would be[116]

$$\frac{v^r e^s}{v^r e^s + v^s e^r}.$$

The probability that there would be a majority of any size for the correct decision would be, if $n = 2q + 1$, the sum of the first $q + 1$ terms of the expansion:

$$\binom{2q+1}{0} v^{2q+1} + \binom{2q+1}{1} v^{2q} e + \ldots + \binom{2q+1}{q} v^{q+1} e^q.$$

[116] Condorcet, *Essai*, pp. 10–11.

By means of diverse assumptions, such as having a tribunal of nearly "infallible" judges whose decisions would serve as a standard, or believing that all judges had a better than even chance of being correct ($v > \frac{1}{2}$), Condorcet was able in principle to use inverse probabilities (in particular, the law of succession) to find the probability of future judgments from past records.[117]

From these expressions, Condorcet concluded that the probability that the tribunal would render a correct decision in any given case could be increased by increasing either n, r, or v. Much of the *Essai* is devoted to examining the consequences of allowing one of these to vary while the others remained constant. For example, in order to form a tribunal of twelve members with a required majority of eight to claim the requisite 144,767/144,768 probability that Condorcet had stipulated for criminal cases, v must be at least 9/10. In order to guarantee the justice of criminal proceedings, Condorcet recommended a tribunal composed of thirty judges, with a majority of twenty-three required for conviction, and an individual probability of 9/10.[118] The single most important variable was v: if v was greater than 1/2, the probability of a correct tribunal decision increased with n, approaching certainty as n approached infinity. Condorcet therefore concluded that large democratic assemblies, convoked for either legislative or judicial purposes, were appropriate only in primitive societies, where all men were equally ignorant. In more advanced societies, an oligarchy of enlightened men was preferable, even if the general level of *lumières* was relatively high. Even an incremental decrease in the value of v would shift the probabilities in favor of a correct decision decisively to error: ". . . a pure democracy would not be suited even to a far more enlightened people, far more exempt from prejudice than any known to us from History."[119]

Condorcet summarized his results in his eulogistic *Vie de Turgot* (1786). Tribunals should be composed of "enlightened men, chosen from the classes that do not share popular prejudices, in order that neither the nature of the crime, nor the impression it produces on the mind, will expose an innocent to the risk of conviction."[120] Moreover, the number of judges

[117] Condorcet, *Essai*, pp. 213–224.
[118] Condorcet, *Essai*, pp. 285–287.
[119] Condorcet, *Essai*, p. xxv.
[120] Condorcet, *Vie de Turgot* (1786), in Condorcet, *Oeuvres*, vol. 5, p. 192.

should be sufficiently large to mount challenges to dubious testimony and to protect the defendant from "secret influences." Large majorities should also be required from criminal convictions. (However, Condorcet opposed the English system of unanimous accord on the grounds that sequestering twelve jurors until they arrived at a consensus was "a type of torture" ill-suited to careful decision.) Condorcet interpreted his calculation as a mathematical mandate for the rule of an enlightened elite: "Thus the form of assemblies which decide the lot of men is far less important for their happiness than the enlightenment (*lumières*) of those who compose it: and the progress of reason will contribute more to the good of the People than the form of political constitutions."[121]

Condorcet's mathematical forays into jurisprudence drew mixed reviews. Although the commission of academicians composed of Charles Bossut and Charles-Auguste Coulomb praised their Perpetual Secretary's treatise as "full of precious applications, luminous and profound reflections . . . [and] new views both for changing and improving the methods upon which depend the security and happiness of the people,"[122] it seems to have been eclipsed by the events of the Revolution in the minds both of its author and its readers. Condorcet's repeated assurance that his results coincided with "good sense" on all counts may have robbed the work of its novelty for the general audience, and few were prepared to appreciate its mathematical ingenuity. Frederick II thanked Condorcet for sending a copy of the *Essai*, but observed that Condorcet's chief conclusions (which Condorcet had carefully summarized in a covering letter) were anticipated in Beccaria and in Condorcet's own earlier writings (presumably *Réflexions sur la jurisprudence criminelle* [1775]).[123] The upheaval of the Revolution soon rendered all of these proposals obsolete.

The Revolution, Directory, Consulate, and Empire witnessed a succession of judiciary reforms and counterreforms that kept pace with the shifting locus of power. The Code of 1791 codified the first wave of reforms, including a guarantee of equality before the law and the abolition of discretionary punishments. Drafted by the prominent jurist Phillipe-Antoine Merlin de Douai, the Code of Brumaire, An IV (25 October 1795) set forth the new code of criminal procedure. Juries were instituted, and the

---

[121] Condorcet, *Essai*, pp. cxl, lxx.

[122] Archives de l'Académie des Sciences, MS. *Procès-Verbaux*, vol. 103, pp. 186–188.

[123] Frederick II of Prussia to Condorcet, 29 June 1785, in Condorcet, *Oeuvres*, vol. 1, pp. 309, 311.

entire system of legal proofs underwent a striking transformation. Whereas the hierarchy of proofs had spelled out the degree to which every type of evidence counted toward a full judicial proof, effectively minimizing the personal discretion of the judge, the Revolutionary codes instructed juries to decide according to their "intimate conviction":

> The law does not ask for juries to account for the means by which they become persuaded; it prescribes no rules according to which the completeness and sufficiency of a proof must depend; it directs them to ask themselves in silence and in seclusion, and to seek, in the sincerity of their conscience, what impresssion the proofs mustered against the accused and the means of his defense have made upon their reason . . . it puts but one queston to them: Do you have an intimate conviction?[124]

The formal, quasi-quantitative arithmetic of proof devised by the Roman and canon jurists had been the subject of one of Voltaire's sharpest attacks on the judiciary of the Old Regime. In the Calas case, the Toulouse prosecution tallied eighth and quarterproofs to convict Jean Calas and his son with a nominally "full" proof. In his commentary on Beccaria's *Traité des délits et des peines*, Voltaire ridiculed the Toulouse system in which "eight rumors, which are only an echo of a doubtful noise, can become a complete proof."[125] Voltaire returned to the same theme in his *Essai sur les probabilités en fait de justice*, emphasizing the tentative, potentially misleading character of the formal system of proofs and drawing sardonic contrasts between the "demonstrative proofs of justice" and those of "morals." Like Beccaria, Voltaire suspected the abstruse calculations of the jurists of obscuring the truth more often than they illuminated it. Beccaria had exclaimed that "simple good sense suffices to guide one more reliably than all of the learning of a Judge accustomed to find the guilty parties and fit everything into a factitious system. . . . Happy the nation where knowledge of the law is not a science!"[126] Voltaire declared that "the enlightened public . . . is the sole judge which prefers substance to form."[127]

---

[124] From the Instruction of 21 October 1791, quoted in Gilissen, "Preuve," pp. 831–832.

[125] Voltaire, "Commentaire," in Beccaria/Collin de Plancy, *Délits*, p. 59.

[126] Beccaria/Morellet, *Délits*, p. 42.

[127] Voltaire, *Essai sur les probabilités en fait de justice* (1772), in Condorcet, *Oeuvres*, vol. 30, p. 462.

Although Voltaire's criticisms and the Revolutionary reforms were aimed at the legal system of probabilities, the mathematical applications to jurisprudence may have been tainted by association. The probabilities of testimony and judgments smacked of the desiccated arithmetic of proof and demiproof decried by the reformers as morally bankrupt. In his preface to the *Essai sur les probabilités en fait de justice* published with the 1785 edition of Voltaire's collected works, Condorcet delicately informed the reader that the great *philosophe* was "a stranger to the brand of calculus" applicable to such problems, and that the essay should be read only as an indicator of "the path to be followed."[128] However, Voltaire's curious hodgepodge of mathematical, legal, and philosophical meanings of "probability" served to strengthen the association that already linked the calculus of probabilities to the discredited system of legal proofs. In this spirit, the literary historian Jean-François La Harpe looked back in 1797 upon Condorcet's probability of judgments as an epitome of the abuses of late eighteenth-century philosophy:

> Does one need anything more than good sense in order to find supremely ridiculous the application to which Condorcet, a modern scientist, put mathematical calculations to moral probabilities, calculations which he substituted, with a gravity as incomprehensible as it was indefatigable, and in a quarto volume thick with algebra, for judicial proofs, written and testimonial, the only ones admitted by all the tribunals of the World, by the good sense of all nations?[129]

The Ideologue philosopher and psychologist Antoine Louis Claude Destutt de Tracy directed more far-reaching criticisms against the calculus of probabilities as applied to the moral sciences. Destutt's emphatic distinction between the mathematical calculus of probabilities and its applications separated the monolithic theory of the classical probabilists into two parts. Ideology (i.e., the science of ideas), not probability, was the true science of man, according to Destutt. In fact, the "science" of probability did not exist, properly speaking, for it lacked its own subject matter. Anticipating Comte's dual classification of mathematics as both a method and a science, Destutt classed the calculus of probabilities per se as a method, and a limited one at that. In order for one to apply mathematics success-

---

[128] Condorcet, editor's preface to Voltaire, *Essai*, p. 416.

[129] Jean-François La Harpe, *Lycée, ou cours de littérature ancienne et moderne* (Paris, 1813), vol. 14, p. 9.

fully, the subject matter must be susceptible to the "clear, precise, and invariable divisions" characteristic of ideas of quantity. Such was not the case with most, if not all, moral phenomena, which Destutt labeled "refractory": ". . . there are a multitude of subjects in which it is impossible to calculate upon the data. . . . Certainly the degrees of competence, of probability among men; those of the energy and power of their passions, their prejudices, and their habits are impossible to evaluate in numbers."[130]

Although Destutt did not entirely dismiss Condorcet's dream of a "social mathematics," he warned that attempts to submit "numerous, fine, and complex" data to computation could only create "learned nonsense" like the probability of judgments. To venture beyond those simple cases well suited to mathematical treatment was to abuse both the moral sciences and mathematics. Far from identifying the calculus of probabilities with good sense, Destutt opposed the two, claiming that neither the syllogism nor the calculus of probabilities guided the intellect as well as "the simple light of good sense aided by sufficient attentiveness."

Although jurists had for the most part pointedly ignored the mathematical incursions of the probabilists into their domain, a few rallied to the defense of traditional, nonmathematical methods. Perhaps the most interesting of these "conservatives" was the philosophical radical Jeremy Bentham, whose multivolume treatise on evidence was originally published in a shorter French edition, *Traité des preuves judiciaires* (1823). Bentham objected primarily to the false aura of certainty imparted by mathematical treatments of legal matters. For example, by making the probative force of testimony inversely proportional to the witness's remoteness from the event in question, probabilists contradicted the legal maxim that witnesses must be evaluated on an individual basis. Should the judge place more confidence in the first-degree hearsay evidence of an indigent witness or the second-degree hearsay evidence of a more solid citizen? According to Bentham, only mathematicians would favor the former, confirming his argument that the oversimplified hypotheses of the probabilities of testimony and judgments yielded results that were the "inverse of common sense."[131]

[130] Antoine Louis Claude Destutt de Tracy, *Élémens d'idéologie*, 2nd edition (Paris, 1818), vol. 4, pp. 37–38.

[131] Jeremy Bentham, *Traité des preuves judiciaires*, Étienne Dumont, ed. (Paris, 1823), vol. 2, pp. 53–55.

Despite this increasingly hostile intellectual climate, Laplace continued Condorcet's investigations on the probability of judgments as well as of testimony in the final chapter and first supplement to the *Théorie analytique des probabilités*. However, Laplace diluted Condorcet's optimistic proposals with repeated disclaimers that the mathematical treatment could not take account of all the relevant factors: "The probability of the decisions of an assembly depends on the plurality of votes, [and] the enlightenment and the impartiality of its members. So many passions and special interests mingle their influence so often, that it is impossible to submit this probability to calculation."[132]

Nonetheless, Laplace added that certain "general results dictated by simple good sense" did fall within the compass of the calculus of probabilities. The prospect of practical applications may also have attracted Laplace to the probability of judgments. Under Napoleon, a new criminal code had been enacted at the end of 1808, and put into effect as of January 1811. Although many of the liberal reforms of the Revolution gave way to regulations revived from the Old Regime, the comparatively new system of juries was retained over the emperor's objections.[133] In his discussion of the probability of judgments, Laplace undertook to evaluate the probable accuracy of the jury system. Laplace pronounced the probability of a wrong decision under the extant system of special tribunals to be a "terrifyingly" high 65/256, and suggested that a majority of nine out of twelve would reduce the probability of error to 1/8192. Laplace warned that overwhelming evidence was not the only cause of complete accord. The decisions of unanimous juries depended on "the temperament, character, and habits of the jurors, and are often contrary to those which the majority of the jury would have arrived at if it had only heard the evidence, which strikes me as a great flaw in the manner of judging."[134]

With the exception of one key assumption, Laplace followed and expanded Condorcet's method. Whereas Condorcet had assumed that the individual probabilities v for each judge or juror were equal, independent, and constant for all cases, Laplace permitted v to vary with each case, while retaining the assumptions of independence and equality. In practice, the added versimilitude of Laplace's amendment made little difference to the

[132] Laplace, *Essai*, p. xc.
[133] Von Bar, *History*, pp. 337 ff.
[134] Laplace, *Essai*, p. xcix.

actual computations. For a single case, the probability that all n judges would concur in a correct decision would be

$$\frac{v^n}{v^n + (1-v)^n}.$$

In order to find v, Laplace supposed that the number of cases i in which the n judges reach a unanimous decision in m trials is known. If m is very large, then $v^n + (1-v)^n = i/m$, and it is possible to solve for v. When the equation had both positive and negative roots, good sense opted for the positive value of v, "for it is natural to suppose that each judge has a greater probability for truth than for error." In general, the probability that a unanimous decision would be correct increased as v, n, and i/m did.[135] Once again, good sense was vindicated by the calculation: "Analysis confirms what good sense tells us, namely, that the correctness of judgments is as probable as the judges are more numerous and enlightened."[136]

However, the prescriptions of good sense seemed to require a more subtle exegesis, for Laplace felt obliged to amend his treatment of the probability of judgments in the First Supplement of his *Théorie analytique*. In this discussion, Laplace broached the difficult problem of how the circumstances surrounding a legal case might legitimately influence the decision of the judges or jurors. In Laplace's opinion, this problem bore directly on the probability of a correct decision, or rather, on the definition of what it meant for a decision to be "correct":

> If I am not mistaken, this judgment reduces to the solution of the following question: does the proof that the accused committed the crime possess the necessary high degree of probability so that citizens have less to fear from the errors of a tribunal if the accused is innocent and convicted, than from his new attempts at crime, and those of the unfortunates emboldened by his impunity, if he is guilty and absolved?[137]

In contrast to Condorcet's attempt to balance the contractual obligations of the individual and society in order to safeguard the rights of the individual, Laplace's treatment considered all factors from the perspective of

[135] Laplace, *Théorie analytique*, pp. 469–470.
[136] Laplace, *Essai*, p. xciv.
[137] Laplace, *Théorie analytique*, p. 521.

dangers to society. Each judge must assess not only the immediate probability that the accused was guilty or innocent, but also the probability that
transcendent social interests would be served by a particular verdict in a
given case. Both the gravity of the crime and the severity of the punishment also must be taken into account in order to render a "just" decision.
In calculating this revised probability, Laplace contended that a well-exe-
cuted approximation "founded on the data indicated by a good sense"
would be superior to "specious reasoning."

Although Laplace admitted that to set any minimum value for the probability of a wrong decision necessarily entailed arbitrary assumptions, he
maintained that the solution of his problem required some threshold probability below which convictions would give rise to greater fears from judicial error than from freed criminals. Certainly the French case of 65/256
was intolerably high, while the 1/8192 value for the British system of a
unanimous twelve-person jury involved other procedural difficulties, in
Laplace's opinion. Laplace settled on the intermediate value of 1/1024 as
sufficient guarantee against both judicial error and released criminals: the
mean probability of a tribunal must only rarely dip below this value. For
$p + q$ judges, each with a probability x, and of whom p had voted one
way and q the other, the probability of a correct decision would be

$$\frac{x^p(1-x)^q}{x^p(1-x)^q + (1-x)^p x^q}.$$

Using inverse probabilities and assuming that all values of x between 1/2
and 1 were equally probable, Laplace computed the probability of a correct
judgment for all possible values of x to be

$$\frac{\int_{1/2}^1 x^p(1-x)^q dx}{\int_0^1 x^p(1-x)^q dx}.$$

Thus, using Bayes' theorem, and assuming that all values of x on the interval
[1/2, 1] were equally probable, it was in principle possible to evaluate the
probability that a judgment was correct from the observed majority.[138]

This formula was the basis of Laplace's computations on the probability
of error for the French jury system. Laplace evidently intended his results
to be taken seriously as policy recommendations for judicial reform, for he
appended a passage (Article 351) from the most recent Code of Criminal

[138] Laplace, *Théorie analytique*, p. 522–526.

Instruction to the section of the First Supplement dealing with the probability of judgments in order to criticize its provisions from the standpoint of his mathematical results.[139] The Code article struck Laplace as offering an insufficient safeguard of a correct decision.

Despite his reservations concerning the unmanageable complexity of the problems addressed by the probability of judgments, Laplace reiterated his view that the calculus of probabilities, when wielded with skill and perspicacity, would confirm and guide good sense. Indeed, it was the duty of the probabilist to expose the sources of potential injustice within the judicial system, as Laplace himself had done in calling attention to the low probability for a correct decision. Consensus was central to Laplace's treatment: the more competent the jurors or judges, the more likely they were to agree with one another, for reasonable men agreed among themselves. Condorcet's elite of enlightened judges, each independently exercising equal (if not constant) sagacity in every decision, remained the bulwark of the probability of judgments.

Poisson was the last major classical probabilist to investigate these problems. Cournot, influenced by Poisson's work, included sections on the probability of judgments in his expositions of probability theory, but the controversy sparked by Poisson's *Recherches sur la probabilité des jugements en matière criminelle et en matière civile* (1837), and the increasingly vigorous criticisms leveled against both the probabilities of testimony and judgment, made most mathematicians wary of the topic. Poisson benefited from the annual compilation of legal statistics begun by the French Ministry of Justice in 1825. Several observers, including André Guerry, Quetelet, and the Ministry itself, noted the striking constancy of certain figures from year to year: the national total of murders, thefts, and so on, and the proportion of convictions to acquittals varied surprisingly little. Primed to interpret such statistical regularities in terms of the probability of causes from Laplace's demographic research, Poisson viewed the guilt or innocence of defendants as an "unknown cause" of the judgment rendered. As in the case of the steady preponderance of male to female births, the regularity of the observed effects—that is, the proportion of convictions and acquittals—betokened the operation of a constant cause.

The Comte de Peyronnet, who served as Minister of Justice from 1821 to 1828, launched the annual *Compte général de l'administration de la justice*

---

[139] Laplace, *Théorie analytique*, pp. 529–530 (dated 15 November 1816).

criminelle en France[140] as a means of evaluating the efficacy of both legislation and judicial procedures, and as a diagnostic tool that might also suggest remedies by "determining the circumstances which tend to increase or decrease the number of crimes." Peyronnet expressed his conviction that precise statistics were a prime desideratum in modern government, replacing vague theories with "positive, certain knowledge of experience."[141] The original report presented statistics by *département* of the number and nature of crimes, the ratio of indictments to convictions, the frequency with which various types of punishments were handed down—all in tabular form. Later editions steadily expanded the number of tables and headings to take account of the age and sex of both the accused and the convicted, the number of jurors unable to perform their duties, apparent motives for capital crimes, recidivism, crime rate by season of the year, and, beginning in 1827, the number of decisions rendered by the Cour d'Assises by the minimum seven-to-five majority.

These statistics concerning the actual majorities, combined with several altered assumptions concerning the prior guilt of the accused and competence of the jurors, formed the basis of Poisson's approach to the probability of judgments. Modestly deferring to Laplace's magisterial authority on all matters mathematical, Poisson nonetheless took issue with his mentor's treatment of the probability of judgments. Poisson, like Laplace, marched under the banner of "good sense"; in fact, Poisson presumed to correct Laplace only in the spirit of the master's own maxim, that "the theory of probabilities is, at bottom, only good sense reduced to a calculus." Laplace's computation of the "terrifying" probability of an erroneous conviction by the French judiciary "appeared exorbitant, and counter to ideas generally held." Therefore, Poisson reexamined Laplace's hypotheses to find where mathematics had deviated from good sense. In doing so, he hoped to redeem this branch of probability theory in the eyes of those who condemned its application to moral questions.

Poisson quarreled with three of Laplace's assumptions on the grounds

[140] According to the statistician Benoiston de Chateauneuf, André Michel Guerry originated the idea of the survey. Guerry analyzed these statistics from the standpoint of legal reform in his *Essai sur la statistique morale* (1833); see Benoiston de Chateauneuf, "Sur les résultats des *Comptes de l'administration de la justice criminelle en France*, de 1825 à 1839," *Séances et travaux de l'Académie des Sciences Morales et Politiques* 1 (1842), p. 324.

[141] Garde des Sceaux, Ministère de la Justice, *Compte général de l'administration de la justice criminelle en France* (Paris, 1827), pp. v, x.

that they contradicted good sense in the matter of criminal trials. First, Poisson rejected Laplace's assumption that all values between 1/2 and 1 were equally likely to express the probability that an individual juror or judge would decide correctly as untenable. Although he defended the use of inverse probabilities, Poisson complained that there existed "an infinity of different laws of probability" that would satisfy this condition. In order to use Bayes' theorem properly, all information concerning the a priori probability distribution must be taken into account. In particular, the probability of judgments must assess information concerning the competence and experience of the jurors or judges in specific instances.

Second, Laplace had assumed that there existed no prior presumption of guilt against the accused, so the only evidence bearing upon the "unknown cause" of the probability of a correct decision was the majority by which the judgment was actually rendered. Poisson objected that the criminal investigation procedures preliminary to the trial created a prior probability of *at least* 1/2 for the guilt of the accused: "Certainly no one would hesitate to wager in an equal game on his guilt over his innocence." Once again, the proper use of Bayes' theorem demanded that all "presumptions anterior to observation" be included in the reckoning: if these considerations were omitted, it would be impossible to reconcile the results of calculation to the constant results of observation.

Finally, Poisson complained that Laplace's method failed to distinguish between the probability of an erroneous judgment rendered by a given majority and the probability of that majority. For example, Laplace had computed the probability that a seven-to-five majority would decide incorrectly to be approximately 2/7. Poisson, armed with the statistics supplied by the Ministry of Justice, pointed out that only 7/100 of all jury decisions were made on the basis of this minimal majority. Therefore, the true probability of being erroneously convicted by a seven-to-five majority would be the *product* of these two probabilities.

Like Condorcet and Laplace, Poisson reflected upon the criteria for a tolerable margin of error in criminal proceedings. Condorcet had appraised the risk in terms meaningful to the individual citizen, setting the level as low as possible so as to allow maximum play for individual liberty. Poisson dismissed Condorcet's dictionary of voluntary risks encountered in daily life as "much too subtle for such a serious issue." Although he regarded Laplace's balance of societal risks as more suitable, Poisson asserted the prerogatives of societal security over individual liberty even more emphatically. In Poisson's opinion, the jury did not decide upon the actual guilt or innocence of the accused, but rather on whether public security was better served by a conviction or an acquittal. With this criterion in mind, Poisson substituted the word "convictable" (*condamnable*) for "guilty" (*coupable*). Hence, the probability of an incorrect judgment measured "the proportion of convictions which had too low a probability, not to establish guilt over innocence, but rather to establish that conviction was necessary for public security."[142] The acceptable probability for conviction varied with circumstances, depending on the judges, the nature of the crime, and the circumstances of the trial. For example, a military tribunal trying an espionage charge in the presence of the enemy would require a far lower probability than was customary in civilian trials.[143]

Poisson admitted that it would be impossible to determine this probability for any individual case, but insisted that such particular values were irrelevant in any case. Because the probability of error related only to societal risks, only long-term results mattered. Condorcet had rejected the use of average probabilities and expectations reckoned thereupon in the probability of judgments because the individual would not have the opportunity to "break even" by playing many rounds of the "game" of justice. By eliminating any but societal concerns from his "convictable" probability, Poisson cleared the way for a statistical treatment of the probability of judgments. Judicial proceedings, like mortality rates, followed the universal law of large numbers if considered en masse. The perturbations caused by the effects of individual caprice, passion, or ignorance were no more erratic than those affecting human lifespan. As long as the specific details of individual cases could be safely ignored, the law of large numbers applied equally to all situations, both moral and physical. Poisson's redefinition of the probability of judicial error wholly in terms of public security and the judiciary statistics compiled by the Ministry of Justice enabled him to extend the law of large numbers to the probability of judgments: "In order to safeguard society, and also the accused, it is not important to know the chance of a particular judgment, but rather that of a collection of trials submitted to the Court of Assises in one or several years, derived from observation and calculation."[144]

[142] Poisson, *Recherches*, pp. 4, 6, 7.

[143] Poisson reassured his readers that although his calculations did not concern the number of innocent people wrongly convicted, he did not believe that there were many of these except in political trials; see *Recherches*, p. 6.

[144] Poisson, *Recherches*, pp. 12, 18.

Poisson posed two problems: to determine the probabilities that (1) any given juror will render a correct decision, as previously defined; and (2) the accused was guilty, prior to the trial. In order to solve these problems, Poisson consulted the statistics for convictions handed down by a seven-to-five majority from 1825 until 1831, when the minimum majority was raised to eight-to-four. Let

k = probability that the accused is guilty, prior to trial;
x = probability that any given juror will decide correctly;
X = probability that the accused will be convicted.

The probability X_i that the defendant will be convicted by exactly i out of n votes would be

$$\binom{n}{i}[kx^i(1-x)^{n-i} + (1-k)x^{n-i}(1-x)^i],$$

and the probability that the defendant will be convicted by at least i out of n votes is the sum

$$X_i + X_{i+1} + \ldots + X_{n-1} + X_n.$$

Referring to the *Compte général*, Poisson observed that between the years 1825 and 1830 inclusive, the ratio of indictments to convictions for crimes against persons totaled 5,286 to 11,016. By Bernoulli's theorem, this ratio gave a good approximation of the probability of conviction for a majority of at least seven-to-five, so

$$X_7 + X_8 + \ldots + X_{12} = 5286/11016 = .4782.$$

After the change in the majority required by law to eight-to-four in 1831, the ratio of indictments to convictions for that year was

$$743/2046 = .3631, \text{ so } X_8 + X_9 + \ldots + X_{12} = .3631.$$

Poisson assumed that k and x remained the same for both 1825–30 and 1831, and solved these two equations for the two unknown x and k, deriving $k = .5354$ and $x = .6786$.[145] Poisson developed the approximation to the binomial distribution for large n and small probabilities, and to the distribution that bears his name in the context of these calculations.

In both cases, Poisson had to choose between two possible roots of the equation and selected the larger value each time, on the grounds that it was more consonant with good sense to assume both $k > 1/2$ and $x > 1/2$.

---

[145] Poisson, *Recherches*, pp. 386–387.

However, in times of political unrest (Poisson cited the trials conducted during the Terror) when judgments bent with the reigning passions, "one must use the other solution, which gives these convictions so great a probability of injustice."[146] Under normal conditions, the "reasonable" values of k and x could be used to compute the probability of a correct decision rendered by at least a given minimum majority. Together, these parameters constituted "an important document on the moral state of our nation," one which carried the imprimatur of mathematical certainty. Despite his numerous assumptions concerning the meaning of the probability of a correct decision and the importance and value of x and k, Poisson declared that his formulas were derived "without any hypothesis, from the known general laws of the calculus of probabilities," and that the results were as trustworthy as the astronomical constants determined from empirical observation. If its conclusions could not be checked empirically, the probability of judgments was no different in this regard than many other branches of applied mathematics, "whose certainty depends solely, as it does here, upon the rigor of the demonstrations, and upon the precision of the observations."[147]

Although jurists paid little attention to Poisson's results, Poisson's colleagues at the Académie des Sciences reacted violently to his claims to have demonstrated his judicial conclusions with all the rigor of mathematics. Prior to the publication of the *Recherches*, Poisson presented his principal results and philosophical views to the Académie des Sciences in papers read at the sessions of 14 December 1835 and 11 April 1836.[148] The first memoir became the "Préambule" of the *Recherches* and outlined Poisson's criticisms of Laplace's approach to the probability of judgments, the universal applicability of the law of large numbers, and some of the results based on the Ministry of Justice's statistics. Poisson emphasized the importance of his analysis for judicial policy, then in a state of flux:

> In this important question of humanity and public order, nothing can replace the analytic formulas which express these various probabilities. Without their help, whether it is a question of changing the number of jurors, or of comparing two countries with different num-

---

[146] Poisson, *Recherches*, p. 26.   [147] Poisson, *Recherches*, pp. 19, 21, 25.

[148] Poisson, "Recherches sur la probabilité des jugements," *Comptes rendus hebdomadaires des séances de l'Académie des Sciences* 1 (1835): 473–494; and "Note sur la loi des grands nombres," ibid., 2 (1836): 377–380.

bers [of jurors], how will we know that a jury composed of twelve people, and judging by at least a majority of eight votes to four, offers a greater guarantee to both defendants and society, than another jury composed of nine people, for example, taken from the same list as before, and judging at such and such a majority.[149]

Poisson's second memoir discussed his "Law of Large Numbers," with special attention paid to how it differed from Bernoulli's theorem and how it was particularly well suited for applications to the moral sciences. Poisson noted that while Bernoulli's theorem supposed constant probabilities of causes, his own law of large numbers permitted the probabilities of causes to vary continuously, a hypothesis better adapted to complex natural and moral phenomena. In order to clarify the distinction, Poisson contrasted the case in which one coin was tossed 2,000 times with that in which 2,000 coins are tossed once. The first cast corresponded to Bernoulli's theorem, in which the constant probability of each side resulted from the physical construction of that particular coin. In the second case, the "unknown cause" varied over 2,000 exemplars. However, the law of large numbers ordained that even this average would stabilize if the experiment was repeated often enough. Poisson held up the second case as the prototype for moral causes:

This material example is an image of what occurs in moral matters, considered independently of the nature of their cause, and only with respect to their effects. In the case of criminal judgments, for example, the chances of conviction and acquittal vary from one trial to another, just as the chances for the two sides of the 5-franc piece vary from one coin to another. Yet this does not prevent the number of acquittals and convictions from being almost invariable for a large number of trials, like the ratio for the occurrence of the two sides of different coins.[150]

Poisson's memoir to the Académie des Sciences generated a lively discussion on the propriety of such applications of the calculus of probabilities to the "moral world," with Claude Navier siding with Poisson against Louis Poinsot and Charles Dupin. Navier, Poinsot, and Dupin were all graduates of the École Polytechnique, like Poisson; all three worked primarily in theoretical mechanics, with secondary interest in mathematics

[149] Poisson, "Recherches," p. 485.   [150] Poisson, "Note," p. 379.

and engineering. The open vehemence and occasional acerbity that marked the exchange, both rare in discussions among academic colleagues, especially of such similar backgrounds, suggest that the issue was already controversial. Literary figures such as La Harpe and Pierre Paul Royer-Collard had ridiculed the efforts of Condorcet and Laplace to apply the calculus of probabilities to testimony and judgments as prime examples of mathematical arrogance. Philosophers Destutt de Tracy and Auguste Comte had rejected the probabilistic model for the moral sciences.[151] Moreover, the conservative political overtones of Poisson's redefinition of the probability of judicial error in terms of "convictability" rather than guilt may have piqued the more liberal sensibilities of Poinsot and Dupin.

Poinsot attacked the probability of judgments as a "false application of mathematical science," quoting Laplace's warning on the extreme delicacy of such problems. Indeed, the potential for a mathematical misstep was so great that he ironically suggested that two probabilities were required: one of the error in a decision; and the other of the error in mathematical treatment. At bottom, Poinsot suspected any application of mathematics to situations involving human "passions and ignorance": reason had no truck with unreason, and the pretensions of a calculus of such irrationality could give rise to a "dangerous illusion" of accuracy in such inherently irrational matters.[152]

While Poinsot had opposed the probability of judgments as a slur upon mathematics, Dupin attacked the probabilistic model of moral phenomena on grounds of accuracy. A member of the recently resurrected Académie des Sciences Politiques et Morales as well as of the Académie des Sciences, Dupin charged that Poisson's treatment simply did not match the facts of the situation. Concentrating on Poisson's discussion of the Revolutionary tribunals, Dupin criticized the discontinuity of Poisson's double-root solution, which selected one value for stable periods and the complementary one for extraordinary circumstances, for ignoring "all of the intermediate degrees of insecurity and intimidation" which lay between the two extremes. Dupin argued that the causes that bore on judicial decisions were so complex and variable that any attempt to assign fixed probabilities was destined to fail. Nor was Dupin won over by Poisson's law of large numbers. Dupin doubted that data subject to such fluctuating influences

[151] Destutt de Tracy, *Élémens*, pp. 20–45; Auguste Comte, *The Positive Philosophy of Auguste Comte*, Harriet Martineau, trans., 2nd edition (London, 1875), vol. 2, chapter 2.
[152] Poinsot, *Comptes rendus*, 2 (1836), p. 380.

would approach stable limits defined by the average terms, and further-more questioned the utility of averages which differed significantly from the individual elements of the series.[153]

Navier came to Poisson's defense, decrying the schism between the natural realm, governed by invariable laws, and the moral realm, where all was assumed to be "fortuitous and accidental," and championed the homogeneity of "facts," "even political or judicial facts in which human passions and interests interfered."[154] He had expressed similar views at an academic session several months before, on 19 October 1835, concerning a report on medical statistics by Jean Civiale.[155] *Pace* Dupin's objections that both complexity and incorrigible individuality of the phenomena precluded mathematical treatment, Navier acknowledged that the mathematical results could not encompass every detail, but argued that this was true in the physical as well as in the biological and moral sciences. Navier contended that "the art of the mathematician consists above all in distinguishing the principal elements, and in formulating an abstract problem which resembles the natural problem as much as possible, and to which analytic methods may be applied." Despite the physician's claim to treat each patient on an individual basis, Navier argued that he must inevitably draw upon past experience. Medical statistics joined to the calculus of probabilities would simply render these intuitive generalizations more precise, without promising the absolute rigor of a mathematical demonstration.

Poisson replied to his critics at the next academic session, on 18 April 1836. He accused Poinsot of misconstruing Laplace's mild reservations concerning the probability of judgments, and quoted other passages from the *Essai philosophique* to show the master's support of such applications to the moral sciences. Calling attention to the striking regularities in crime and conviction rates revealed by the Ministry of Justice statistics, Poisson reaffirmed his confidence in the universal applicability of the law of large numbers. Any sufficiently large collection of phenomena—moral or physical—would eventually exhibit regularities if studied without regard to the nature of the causes.[156]

Poinsot persisted in his objections, carefully exempting the calculus of

---

[153] Dupin, *Comptes rendus*, 2 (1836), p. 381.
[154] Navier, *Comptes rendus*, 2 (1836), p. 382.
[155] Navier, *Comptes rendus*, 1 (1835), pp. 248–249.
[156] Poisson, *Comptes rendus*, 2 (1836), pp. 395–399.

probabilities per se, which he deemed to be fully as accurate as arithmetic, and its legitimate applications to games of chance, annuities, insurance, and all other cases in which it was possible to enumerate a finite group of equiprobable outcomes. None of this offended "good sense." Yet the probabilities of testimony and judgments did, Poinsot insisted, commit such an offense to reason. Despite Poisson's claims to have employed the calculus of probabilities simply to confirm, clarify and guide the "primary insights of good sense," Poinsot protested that such applications of the calculus of probabilities actually betrayed good sense in a particularly ludicrous manner. Poinsot's comments are worth quoting at length, for they reveal how far the notion of good sense had diverged from the mental calculus of the associationists, and how the new "free" system of proofs had severed the older connection between legal and mathematical probabilities:

> It is the applicaton of this calculus [of probabilities] to things of the moral order which offends the intellect. It is, for example, to represent by a *number* the *truth* of a witness; to thus assimilate men to so many dice, each of which has several sides, some for error, others for truth; to treat other moral qualities in the same way, and to convert them into so many *numerical fractions* . . . to dare, at the end of such calculations, in which the numbers derive only from such hypotheses, to draw conclusions which purport to guide a sensible man in his judgment of a criminal case, or simply to make a decision, or to give his opinion on a matter of importance—this is what seems to me a sort of aberration of the intellect, a false application of science, which it is only proper to discredit. [157]

6.5 Conclusion

Poinsot was no mathematical luddite, nor even a critic in the ranks of probability theory like d'Alembert. [158] His criticism of the probabilities of

[157] Poinsot, *Comptes rendus*, 2 (1836), p. 399. Both Poisson and Poinsot apparently prepared remarks for this exchange, judging from the records in the Dossier for 18 April 1836, Archives de l'Académie des Sciences. I am indebted to Dr. Ivor Grattan-Guinness for the information that the manuscript draft of Poinsot's remarks is preserved in the D. E. Smith Collection at Columbia University.

[158] Poinsot even lectured on the subject: his notes are partially preserved in Poinsot, Bibliothèque de l'Institut MS. 955, ff. 13–18.

testimony and judgment reflected a profound shift in assumptions concerning the nature of the moral phenomena to which the probabilists sought to apply their calculus rather than any distrust of mathematics per se. Indeed, he believed himself to be defending the good name of mathematics in firmly separating the theory of probability from its dubious applications. To its critics, the probability of judgments was not only wrong; it was dangerous, for it traded illegitimately upon the intellectual authority of mathematics, promising rigor and certainty in an area where none obtained. Poinsot's attempt to demoralize mathematics led him to distinguish clearly between the theory of probability and its applications, thereby granting the theory a new autonomy. Probability theory was no longer a branch of mixed mathematics.

By the middle of the nineteenth century, the probabilities of testimony and judgment had largely disappeared from texts and treatises as standard applications of the mathematical theory of probability. Quantification is not irreversible: weighing historical and courtroom evidence and designing tribunals once again became qualitative matters. The analogy between mathematical model and subject matter collapsed with changes in psychology and jurisprudence. Good sense or reasonableness was no longer so clearly identified with computation; the deliberately antiformal, antianalytic system of "free" proofs had replaced that of legal proofs, substituting an intuitive appeal to the "intimate conviction" of juror or judge for any formal reckoning.

Moreover, the new breed of social scientists rejected both the psychological framework of the moral sciences and the reductionism of Laplace and Poisson. Even those who remained faithful to the probabilistic program in the study of society, such as Quetelet, turned from the typical problems of individual belief and decision making to large-scale statistical phenomena described in terms of distributions. The calculus of probabilities no longer described the reasonableness of individuals but rather the hidden regularities of whole societies.

The Decline of
the Classical Theory

The demise of the probablilities of testimony and judgments foreshadowed the decline of the classical interpretation of probability in general. Neither died a sudden death: A. A. Cournot wrote on the probability of testimony and of judgments,[1] and John Herschel described Siméon-Denis Poisson's results in neutral, if not enthusiastic terms in his 1850 review of Adolphe Quetelet's *Lettres sur la théorie des probabilités* . . . (1846);[2] Augustus De Morgan continued to espouse a subjectivist interpretation of probability.[3] However, the same emphasis on statistical frequencies and distributions that characterized Quetelet's approach to the moral sciences also cast doubt on the glib assumption of uniform prior probability distributions responsible for the law of succession. Once the psychological bonds dissolved between objective and subjective probabilities, and between the calculus of probabilities and good sense, the classical theory seemed both dangerously subjective and distinctly unreasonable.

David Hartley's associationism had apportioned belief to experience on a probabilistic scale; but Étienne Condillac's associationism enslaved belief to fantasy and self-interest. Subjective belief and objective experience began as equivalents and ended as diametric opposites. The reasonableness

[1] Antoine Augustin Cournot, "Mémoire sur les applications du calcul des chances à la statistique judiciaire," *Journal de mathématiques pures et appliquées* 3 (1838): 257–334; also, his *Exposition de la théorie des chances et des probabilités* (Paris, 1843), chapters 15, 16.

[2] John Herschel, "Review of the *Lettres à S.A.R. le Duc regnant de Saxe-Coburg et Gotha sur la théorie des probabilités appliquée aux sciences morales et politiques, par M.A. Quetelet* . . . ," *Edinburgh Review* 96 (1850): 1–57, especially p. 46.

[3] See Adolphe Quetelet, *Lettres à S.A.R. le Duc regnant de Saxe-Coburg* . . . (Brussels, 1846); Augustus De Morgan, *An Essay on Probabilities*, in *The Cabinet Cyclopaedia*, Dionysius Lardner, ed. (London, 1838).

of the eighteenth-century probabilists derived from broad experience accurately tabulated by an unprejudiced mind: good sense was nothing more or less than a crude, intuitive approximation of Bernoulli's, and later Bayes', theorems. The belief that subjective mental states were formed in the image of objective experience permitted probabilists to equate epistemological and physical indifference, as Bayes had done in his table model and scholium. This confidence waned as philosophical and psychological (the disciplines were as yet poorly distinguished) interest in mental distortions of experience intensified. Critics of the classical theory like Cournot used "subjective" as an epithet: chance and probability were no longer interchangeable. Even if reckoning by Bernoulli's theorem had still epitomized good sense, early nineteenth-century psychologists no longer believed that such good sense came naturally.

But by this time good sense had ceased to be a matter of estimating probabilities and comparing expectations. Moral philosophers such as Victor Cousin suggested that good sense was more intuitive than rational, more synthetic than analytic, more complex than abstract.[4] Poisson's careful distinction between objective chance and subjective probability and scrupulous attention to prior probability distributions showed that even adherents to the older notion of good sense no longer trusted the subjective implications of the classical interpretation. Poinsot's critique of the probabilities of testimony and judgment showed that even some mathematicians no longer accepted the old definition of good sense.

The break between the classical and frequentist interpretations, and between the old and new meanings of good sense, was neither sudden nor clear. Mathematicians often straddled one or the other divide. Quetelet, Cournot, Poisson, and even Pierre Simon Laplace distinguished subjective degrees of certainty from objective frequencies and allied themselves with the latter interpretation; yet all had to some extent subscribed to a probabilistic model for the moral sciences. However, Quetelet, inspired by Laplace's own interest in statistics, had shifted the focus of that model from individual belief and decision making to statistical compilations and distributions for society as a whole. Poisson's and Cournot's treatment of the probability of judgments fell midway between the two extremes: they retained the format, assumptions, and orientation of the classical treatments

---

[4] See Cousin's criticisms of the utilitarian moral calculus of Frances Hutchenson, for example: Victor Cousin, *Cours d'histoire de la philosophie moderne*, 1st series (Paris, 1846), vol. 4, pp. 169–174.

but added statistical information to their analyses. In Quetelet, who discussed judicial statistics but not the probabilities of testimony and judgments, the transition was complete.[5]

The applications that survived this transition were those that squared with a purely objective interpretation of probability: gambling problems based on the physical symmetry of coins and dice; and, more importantly, actuarial computations based on mortality data. The attempts of the eighteenth-century demographers to collect numerical data and to ferret out large-scale regularities contained therein became the model for social scientists, bureaucrats, and insurance companies in the nineteenth century. But other applications of the classical theory did not fare so well.

After the mid-1840s, critics of what by then appeared to be the unbridled subjective tendencies of the classical interpretation were also usually critics of its applications to the probability of causes (at least those involving the law of succession), and to the probabilities of testimony and judgments. John Stuart Mill railed against the indiscriminate use of the Principle of Indifference in his *System of Logic* (1843), upbraiding Laplace for confusing "equally undecided" with "equally possible." Advocating a strict frequentist interpretation of probability, Mill criticized Laplace for encouraging mathematicians to apply probability theory "to things of which we are completely ignorant" in a vain, alchemical attempt to convert the dross of ignorance into the gold of inverse probabilities. The law of succession was not a mathematical philosopher's stone.[6]

Although in later editions of the *Logic* Mill moderated his strict injunction to suspend all judgment in the absence of data to allow recourse to the Principle of Indifference as a last resort,[7] he did not waiver in his opinion that the probabilities of testimony and judgment had made the calculus of probabilities "the real opprobrium of mathematics." To Mill, common sense rightly rejected any attempt to assess average veracity as an abridgement of the rights of the individual and an absurdity to boot:

[5] Quetelet's first text on mathematical probabilities, *Instructions populaires sur le calcul des probabilités* (Brussels: 1828), was closely modeled on Lacroix's *Traité élémentaire du calcul des probabilités* and Laplace's *Essai philosophique*, and included chapters on the probabilities of testimony and judgments. However, neither the *Lettres . . . sur la théorie des probabilités . . .* (1846) nor the *Théorie des probabilités* (1853) contained any treatment of either topic, although the latter text reproduced much of the *Instructions populaires* verbatim.

[6] John Stuart Mill, *A System of Logic* (London, 1843), pp. 72–75.

[7] See, for example, Mill, *Logic*, 8th edition (New York, 1881), p. 381.

. . . common sense would dictate that it is impossible to strike a general average of the veracity and other qualifications for true testimony of mankind, or of any class of them; and even if it were possible, the employment of it for such a purpose implies a misapprehension of the use of averages, which serve, indeed, to protect those whose interest is at stake, against mistaking the general result of large numbers of instances, but are of extremely small value as grounds of expectation in any one individual instance, unless the case be one of those in which the great majority of individual instances do not differ much from the average.[8]

A libertarian of Mill's persuasion would have found Poisson's "convictability" solution to this quandary untenable. Mill upheld the traditional criteria for evaluating testimony—the witness's comportment, consistency, ulterior motives, and so on—as far more just and reliable than the "rude standard" of the ratio of true to total statements uttered.

Mill made similar criticisms of the probability of judgments. Once again, he attacked "the fallacy of reasoning from a wide average" to individual cases, and argued further that it was in the most difficult and important cases that the judges' individual probabilities strayed furthest from the average values. Mill also rejected the premise that each judge arrived at his decision independently, and warned that increasing the number of judges would only amplify the undesirable effects of this responsibility and propagate the influence of "some common prejudice or mental infirmity" to all of the judges. As a sober British empiricist surveying the folly of the French rationalists, Mill ascribed the errors of both the law of succession and judicial probabilities to those "who, having made themselves familiar with the difficult formulae which algebra affords for the estimation of chances under suppositions of a complex character, like better to employ those formulae in computing what are the probabilities to a person half informed than to look out for means of being better informed." In Mill's view, the aberration of classical probability theory resulted from a combination of mathematical hubris and a paucity of observational data.

George Boole concurred in Mill's views on the classical interpretation and its by-then notorious application to the probability of causes, although he took a more temperate, if not sanguine, view of the probabilities of testimony and judgments. Boole, like Mill, defined mathematical

[8] Mill, *Logic*, 8th edition, p. 382.

probabilities in terms of statistical frequencies, or rather "as the limit toward which the ratio of the favourable to the whole number of observed cases approaches (the uniformity of nature being presupposed), as the observations are indefinitely continued."[9] Although he admitted "some analogy between *opinion* and *sensation*," Boole hesitated to interpret mathematical probabilities as "an emotion of the mind."[10] Despite his respect for De Morgan, foremost British advocate of the subjective interpretation,[11] Boole concluded that the law of succession and other uses of Bayes' theorem in conjunction with the Principle of Indifference had illegitimately extended the empire of numbers "even beyond their ancient claim to rule the World," and condemned the assumption of an a priori "equal distribution of our knowledge, or rather of our ignorance" as an "arbitrary method of procedure."[12]

Boole did not reproach the probability of judgments directly with affronts to common sense or bizarre hypotheses. All applications of mathematical probability perforce invoked hypotheses. The key question was whether there was sufficient data available to formulate reliable hypotheses. In some cases, Boole warned that "an undue readiness to form hypotheses in subjects which from their very nature are placed beyond the human ken" could only reflect badly on the mathematical credentials of probability theory.[13] Yet on the whole, Boole believed that hypotheses concerning "the conduct of our own species" in the probability of judgments were more legitimate than hypotheses concerning the natural order

[9] George Boole, *An Investigation of the Laws of Thought* (1854) (New York: Dover, 1958), p. 14.

[10] Boole, *Laws*, p. 244.

[11] In many ways, De Morgan was more representative of what Mill and Boole conceived to be a Laplacean interpretation than was Laplace himself. De Morgan's approach to probabilities was consistently and explicitly subjective, with few of the objective elements that tempered Laplace's work. In a particularly revealing letter to Herschell, dated 2 September 1849, Royal Society, Herschel Correspondence, 6.255, De Morgan wrote:

My dear Sir John

I have no objection to any one word in your letter.

But I think you are—as I was for a long time—apt to escape from the subjective into the objective in viewing a question of probabilities.

When you look at my judge—the ego of the problem—you are no ways bound by his opinion of himself—To you he is what he is in the mind of another ego.

My question is not whether he is right in his preconception—but what the theory of probabilities binds him to if he has—and cannot help having—preconceptions.

[12] Boole, *Laws*, pp. 368, 370.   [13] Boole, *Laws*, p. 20.

in the probability of causes, since our daily experience of human nature supplemented the meager statistics.[14] Despite this charitable provision, Boole argued that no amount of statistical documentation could yield certain values for either the individual probability of a correct decision or of the guilt of the accused prior to trial without either some hypothetical assumption regarding the absolute independence of individual judgments or some substitution of mean values for these probabilities, both of which Mill had rejected as irreconcilable with common sense.[15]

There were French as well as British critics of the classical interpretation of probability and its applications, including Antoine Destutt de Tracy and Auguste Comte. Among mathematicians, perhaps the most prominent was Joseph Bertrand, Perpetual Secretary of the Académie des Sciences and author of a text, *Calcul des probabilités* (1883), which presented several paradoxes later discussed by Henri Poincaré and Paul Lévy.[16] Bertrand rang the familiar changes on the earlier critiques of the probability of causes, of testimony, and of judgments, dismissing the urn model of the classical probabilists as a totally inadequate description of both natural causes and human decision making.[17] Like Mill, Bertrand ridiculed the probability of decisions for neglecting important features of individual cases and for assuming the independence of each judge's or juror's decision.[18] Bertrand marveled that mathematicians as astute as M.J.A.N. Condorcet, Cournot, Poisson, and especially Laplace should have trusted results derived from such outlandish assumptions. Although he admitted that none of the probabilists had committed any mathematical errors— that is, their results followed rigorously from their hypotheses—Bertrand found their readiness to assimilate both natural and moral causes to the ubiquitous urn model almost incomprehensible. If we believe that the sun will rise tomorrow, it was because of "the discovery of astronomical laws and not by renewed success in the same game of chance."[19] If a judge erred, it was for a particular reason: he had not "put his hand in an urn" and made an unlucky draw.[20] Bertrand branded all such speculations the "scan-

---

[14] Boole, *Laws*, p. 376.   [15] Boole, *Laws*, p. 386.

[16] Henri Poincaré, *Cours de calcul des probabilités* (Paris, 1896); Paul Lévy, *Calcul des probabilités* (Paris: Gauthier-Villars, 1925).

[17] Joseph Bertrand, *Calcul des probabilités* (Paris, 1889), p. xliv.

[18] Bertrand, *Calcul*, pp. 319–326.

[19] Bertrand, *Calcul*, p. 174.

[20] Bertrand, *Calcul*, p. 326.

dal of mathematics," ironically quoting Condorcet's appeal for *"truly* enlightened men."[21]

Bertrand and Mill, and to a lesser extent Boole, regarded the most characteristic applications of the classical theory of probability as inimical to good sense. By 1850, the theory once advertised by its most prominent practitioners as "good sense reduced to a calculus" struck mathematicians and philosophers alike as "an aberration of the intellect." The classical interpretation itself drew sharp criticism from mathematicians who, like Boole and Cournot, regarded it as suspiciously subjective. These shifts in attitude owed chiefly to changing conceptions of good sense, of the psychology of belief, and of the methodological relations between the natural and moral sciences.

The mid-nineteenth century conception of "good" or "common sense" had diverged significantly from the late seventeenth- and eighteenth-century notion of "reasonableness." As the history of the concept of probabilistic expectation shows, "reasonableness" itself was not a static notion. It evolved from a legal sense of fairness to a more commerical sense of prudently cultivated self interest. The common thread connecting all of these various shades of "reasonableness" was an emphasis on comparison and calculation. Reasonableness consisted of making informed decisions to act or to believe on the basis of past experience. Such decisions implied a choice among alternatives and a system of converting all of the possible outcomes to a homogeneous medium that made comparisons of degree possible. Reasonableness was therefore abstract, deliberate, and analytical—and quantifiable.

In contrast, nineteenth-century conceptions of good sense stressed the intuitive, synthetic, and spontaneous aspects of understanding. The *homme éclairé* of the post-Revolutionary era was an *homme de sensibilité* first and an *homme d'esprit* second. The transition from the formal, hierarchical system of legal proofs to the discretionary, intuitive proofs of "intimate conviction" epitomized this contrast. Critics of the probabilistic approach queried whether good sense, far from being a kind of implicit calculus, might not actually be antithetical to calculation, at least as far as human affairs were concerned. Risueño d'Amador, an opponent of medical statistics, summarized these doubts in an address to the Académie de Médecine:

[21] Bertrand, *Calcul*, p. 327. Emphasis is Bertrand's.

The *calculus of probabilities* has been described as good sense reduced to calculation; but with a little of what one wants to calculate, one might well wonder whether good sense is calculable, the same as human intelligence, emotions, mental states, etc.—in short, all that concerns the moral, intellectual, and emotional life of man.[22]

Eighteenth-century theories of associationist psychology supported the alliance of mathematical probability and reasonableness by describing mental processes in combinatorial, probabilistic terms. If Bernoulli's theorem was "level to the common understanding," it was because the "common understanding" operated implicitly by Bernoulli's theorem, tallying, correlating, and comparing frequencies of events. Associationism joined the subjective and objective sides of the classical interpretation of probability: subjective belief mirrored objective frequencies. However, as associationist theories came to place increasing emphasis upon the ways in which habit, passion, prejudice, and self-interest distorted this correspondence, the bonds between subjective and objective probabilities loosened. By 1840, probabilists recognized two distinct and opposed interpretations of their theory, and categorized the classical interpretation as subjective. The classical probabilists had viewed the law of succession (and the probability of causes in general) as a mathematicized induction. Yet to Boole and Mill, the law of succession stemmed from a contempt for empirical observation and an unseemly eagerness to substitute calculation for facts.

But the single most important factor in the decline of the classical interpretation of probability during the first half of the nineteenth century was the emergence of a new program for the moral sciences. Although some of the older assumptions persisted in the work of Quetelet, several of the most prominent spokesmen for the ascendant moral sciences, such as Auguste Comte, proclaimed the methodological and ontological autonomy of their discipline. They sanctioned judicious borrowing from the natural sciences and believed in the existence of determinate (if not deterministic) social laws, but attacked the reductionist model propounded by the probabilists as a kind of physico-mathematical imperialism. In their opinion, the "social mathematics" of Condorcet and his school abounded in ill-chosen,

---

[22] Risueño d'Amador, *Mémoire sur le calcul des probabilités appliqué à la médecine*, Lu à l'Académie royale de Médecine dans la séance du 25 avril 1837 (Paris, 1837), pp. 56–57.

unjustified or even bizarre assumptions, vastly oversimplified hypotheses, and a pernicious tendency to obscure the characteristic complexity of the moral realm by taking averages. Although there was little consensus as to alternative methods or models, the philosophers of the moral sciences were generally united in rejecting both the problems and the assumptions of the probabilistic model.[23]

In his 1842 presidential address to the recently reinstituted Académie des Sciences Morales et Politiques, Victor Cousin congratulated the "science of man" on having come of age, recalling the dark days when "The most liberal academies of Europe would not admit it for its own sake: they demanded that it speak in fine language, or ally itself with either a profound erudition or the genius of the mathematical sciences."[24] Cousin's rhetoric concerning the emancipation of the moral sciences from their joint bondage to belles-lettres and mathematics was already a familiar theme by 1842. Auguste Comte had declared the independence of the science of society in his *Course de philosophie positive* (1830–42). Although Comte strongly endorsed the study of mathematics (as well as that of the natural sciences, which logically and historically preceded sociology in his classification of the sciences), he contended that the complexity of biological and a fortiori social phenomena precluded the application of mathematical methods to such subjects. Comte singled out the attempts to apply probability theory to the moral realm as particularly misguided, and rebuked Condorcet, Laplace, and their followers for "repeating the fancy, in heavy algebraic language, without adding anything new, abusing the credit which justly belongs to the true mathematical spirit."[25]

John Stuart Mill warned against the same misplaced abstraction in his discussion of the relation of the "geometrical or abstract method" to the moral sciences. The geometrical method deduced conclusions from a "sup-

---

[23] These criticisms of the probabilistic approach were not unique to the moral sciences. Many of the same themes recurred in a debate over the "méthode numérique" which took place in the Académie de Médecine only a week after the animated exchange between Poisson, Poinsot, Navier, and Dupin at the Académie des Sciences (the two academies shared several members), and dominated meetings for a month thereafter. See *Bulletin de l'Académie royale de Médecine* 1 (October 1836–September 1837): 622–805.

[24] Cousin, [Public Address], *Séances et travaux de l'Académie des Sciences Morales et Politiques* 1 (1842), pp. 4–9, on p. 8.

[25] Auguste Comte, *The Positive Philosophy of Auguste Comte*, Harriet Martineau, trans., 2nd edition (London, 1875), vol. 2, p. 100.

positious set of circumstances," and was valid only to the extent that those suppositions were both true and comprehensive:

> If the set of circumstances supposed have been copied from those of any existing society, the conclusions will be true of that society, provided, and in as far as, the effect of those circumstances shall not be modified by others which have not been taken into the account. If we desire a nearer approach to concrete truth, we can only aim at it by taking, or endeavoring to take, a greater number of individualizing circumstances into the computation.[26]

The critics of the probabilistic approach to the moral sciences attacked its hypotheses as at best crude oversimplifications of an intricate social reality and at worst bizarre caricatures of social phenomena. With the collapse of the eighteenth-century notion of reasonableness and the psychological theories that had rendered it plausible, the urn analogy had lost much of its verisimilitude. Viewed within the nineteenth-century framework, even natural causes resisted comparison with the independent drawings of colored balls from Bernoulli's urn. First, philosophers and scientists were by and large more confident in the certainty of natural laws, if not in the accessibility of hidden causes: Bertrand's Adam believed that the sun would rise tomorrow because his astronomical theory predicted the event. Second, some natural causes seemed too complex to be readily assimilated to Bernoulli's assumptions. Statisticians like Bienaymé questioned the accuracy of the assumption of independent trials for phenomena such as rainfall.[27] The introduction of inverse probabilities committed further sins against complexity by automatically assuming the simplest— that is, uniform—prior distribution of probabilities, often flying in the face of information that suggested that this was not even an acceptable approximation of the true distribution. If the natural world proved so intractable, how could the tangled causes of the moral realm be forced into the procrustean mold of classical probability theory?

However, the criticisms of the probabilistic approach to the moral sciences penetrated below the level of hypothesis and technique. They also

---

[26] Mill, *Logic*, p. 622.

[27] I. J. Bienaymé, "Communication sur un principe que M. Poisson avait cru découvrir et qu'il avait appelé Loi des Grands Nombres," *Séances et travaux de l'Académie des Sciences Morales et Politiques*, 3rd series, 1 (1855): 379–389, on pp. 386–387.

challenged its choice of problems. For the most part, the perspective of the eighteenth-century moral sciences had been individualistic and psychological. The study of societies as coherent units, rather than as aggregates of individuals, was a nineteenth-century innovation. (Comte, perhaps the most original and distinguished proponent of this truly sociological approach, went so far as to deny psychology any independent scientific status.) Classical probabilists incorporated this individualistic outlook into their treatments of moral phenomena such as judgment and testimony. Although some tenets of associationist psychology were linked to mathematical probability, there was no inherent reason why probabilists should have preferred an individualistic to a collective approach. Indeed, tools such as Bernoulli's and Bayes' theorems obliged them to awkwardly consider individual phenomena from the standpoint of averages. Nonetheless, the probabilities of judgment and testimony, with their assumptions about the *lumières* and veracity of individuals belonged to the psychological tradition that Comte and his followers regarded as obsolete.

Also, the legal cast of the probabilistic applications was alien to nineteenth-century social scientists who did not share the *philosophes'* avid interest in legal reforms. Jurists, for their part, wanted nothing to do with a system that either subverted the time-honored standards for trustworthy testimony or recalled the discarded formalism of the hierarchy of legal proofs.[28] If the hypotheses of the probabilists appeared oversimplified to the point of absurdity, their interest in testimony and judgments seemed archaic and irrelevant.

Finally, the advocates of an autonomous science of society questioned the ontology of probabilistic approaches to the moral sciences. It was not so much abstraction per se—all social theories resorted to some abstraction—nor the implied determinism—Comte affirmed that the laws of society were as inexorable as those of nature—which repelled these writers, but rather the *degree* of abstraction and the *kind* of determinism. Probabilists had presented at least two models of social causation, represented by

[28] Bertrand recounted an anecdote about the physicist and statesman Arago, a protégé of Laplace and co-editor of Condorcet's *Oeuvres*, who reportedly upheld Laplace's views on the required majority of jurors for an acceptably low probability of an erroneous conviction in a legislative debate over judicial reform. When a fellow deputy expressed his reservations, Arago curtly replied that "when he spoke in the name of science, it was not for ignoramuses to contradict him"; see Bertrand, *Calcul*, p. 371. Arago's arrogance and bad manners apparently won few friends for the probability of judgments.

the views of Laplace and Poisson. Laplace constructed an elaborate analogy between the physical and moral realms. Not only were the methods of the natural and social sciences identical; Laplace hinted that the very causes which governed the two spheres were in some sense the same. He envisioned the science of man as a kind of societal mechanics, in which abrupt changes squandered the social equivalent of *vis viva*, and spoke of "sympathetic vibrations" which touched off collective emotional reactions. Although he admitted that moral causes were "far more complicated" than their physical analogues, Laplace nonetheless contended that his mechanical and dynamical metaphors were more than just metaphors: "Hesitation between opposed motives is an equilibrium of equal forces. . . . An intense, continuous effort of attention exhausts the sensorium, as a long series of shocks exhausts a voltaic pile, or the electrical organ of a fish. Almost all of the comparisons which we draw from material objects to render intellectual things palpable are at bottom identities."[29]

Poisson's conception of moral causes preserved Laplace's emphasis on the methodological model provided by the natural sciences but abandoned Laplace's detailed analogies. Poisson adopted Laplace's view that in the long run accidental, variable causes would cancel out, revealing the uniform, constant cause, but declined to speculate on the nature of these causes. Poisson's understanding of causes, both natural and moral, was totally agnostic. He repeatedly emphasized that his law of large numbers applied indifferently to weather patterns, birthrates, judicial decisions, and urn drawings because it made absolutely no hypotheses as to the *nature* of the cause. Whereas Laplace had assimilated the moral to the natural sciences by making mechanics the universal science of reference, Poisson had united the two by abstracting out any specific feature aside from uniformity and independence. Social phenomena were simply a subset of the vast class of phenomena described by the law of large numbers. Patient observation of anything would eventually reveal regular causes at work, but observations alone tell us nothing about the nature of these causes.

Quetelet's "social physics" united Laplace's vision of a societal mechanics with the featureless regularity of Poisson's law of large numbers. Quetelet's conceptual innovations were few; his real contributions lay in his untiring campaign for massive compilations of statistical data on all sub-

---

[29] Pierre Simon Laplace, *Essai philosophique sur les probabilités* (1814), in *Oeuvres complètes*, Académie des Sciences (Paris, 1886), vol. 7, p. cxxxviii.

jects and in his unshakable faith that regular patterns invisible at the individual level would emerge at the societal level. Although he is best known for his statistical studies of social phenomena, he also participated in international projects to collect data on terrestrial magnetic fluctuations and meteorology. In every domain, Quetelet saw the archetype of the normal curve behind the great welter of data. He professed Laplace's and Poisson's faith that all phenomena conformed to some version of Bernoulli's theorem. Quetelet's understanding of the moral sciences was societal rather than individualistic, and he sought regularities at the macroscopic level. But unlike Poisson, he was not content with the regularities themselves. Quetelet, perhaps the most faithful of Laplace's disciples, translated these regularities into physicalist terms. Perhaps it was the literalness with which Quetelet construed his "social physics" which persuaded Comte to retitle his own social physics "sociology."

Quetelet made the law of large numbers the "fundamental principle" in all science, and in his review of Poisson's *Recherches sur la probabilité des jugements* predicted that the evolution of the moral sciences would retrace the path blazed by the natural sciences: the steady observation of phenomena, study of attendant circumstances, and numerical evaluation of the data would ultimately reveal the laws that "govern the moral and intellectual, as well as the material, worlds." In the first edition of his *Sur l'homme et le développement de ses facultés, ou Essai de physique sociale*, Quetelet proclaimed the hegemony of the law of large numbers:

> Thus, moral phenomena observed in the mass come to resemble the order of physical phenomena, and we will be obliged to admit as the fundamental principle of these sorts of investigations that the greater the number of individuals observed, the more individual peculiarities, be they physical or moral, disappear, leaving the series of general facts by which society exists and endures to predominate.[30]

In Quetelet's social physics, individuals were the equivalent of variable, accidental causes. Although Quetelet believed that social laws could be gradually modified by deliberate human action, he emphasized the reality and regularity of social causes.[31]

[30] Adolphe Quetelet, *Sur l'homme et le développement de ses facultés, ou Essai de physique sociale* (Paris, 1835), vol. 1, p. 12. On Quetelet's work and influence, see Theodore M. Porter, *The Rise of Statistical Thinking, 1820–1900* (Princeton: Princeton University Press, 1986), parts 1–2.

[31] Quetelet, *Sur l'homme*, vol. 1, p. 13.

As the alternative title, *Essai de physique sociale*, suggests, Quetelet's treatise abounded in Laplacean analogies to mechanics. Society was governed by "moral forces"; the fictitious *l'homme moyen* functioned as the center of gravity in society; population growth encountered resistance in proportion to the square of its growth rate; revolutions equaled in violence the severity of past abuses, just as equal and opposite reactions accompanied every physical action. The "perturbational forces" of individual action fluctuated within ever-tighter limits "defined by nature" around the mean value determined by the uniform forces of social laws, just as the variable causes of error oscillated around the mean defined by the true observational value,[32] or as a physical system oscillated around its equilibrium state. Social phenomena, like the complex physical phenomena of terrestrial magnetism or weather patterns, followed regular periodic cycles: diurnal, monthly, seasonal, and annual. Quetelet extended this physicalist perspective in full confidence that not only the methods but also the laws of the moral sciences would ultimately mirror those of the physical sciences: "This extension of a physical law, so happily confirmed when applied to the records furnished by society, offers yet another example of the analogies one finds in so many cases between the laws which govern material phenomena and those which concern man."[33]

Although moral scientists recognized the importance of statistics, few shared Quetelet's conviction that statistical data analyzed by probabilistic techniques would reveal the laws of social organization. The reconstituted Académie des Sciences Morales et Politiques included a section for Political Economy and Statistics, but its members generally viewed statistical data as a diagnostic tool for social ills and as an aid to legislators and administrators. For example, Poisson's *Recherches sur la probabilité des jugements* went unremarked by the Académie, although it appointed a commission to review the Ministry of Justice's statistical supplement on the administration of civil justice.[34] The reporter, Alphonse Bérenger, characteristically dwelled upon the practical legal implications of the supple-

[32] See Victor Hilts, "Statistics and social science," in *Foundations of Scientific Method: The Nineteenth Century*, Ronald N. Giere and Richard S. Westfall, eds. (Bloomington: Indiana University Press, 1973), pp. 206–233.

[33] Quetelet, *Sur l'homme*, vol. 1, p. 277. Quetelet wrote a brief, unpublished treatise that develops these physicalist ideas in striking detail; see Fonds Quetelet, MS. 110, Bibliothèque Albert 1er, Brussels.

[34] Garde des Sceaux, Ministère de la Justice, *Compte général de l'administration de la justice civile en France* (Paris, 1831).

ment rather than the regularities that had struck Quetelet, André Guerry, and Poisson.[35] The French statisticians Louis-René Villermé and Louis-François Benoiston de Chateauneuf concentrated their research on the traditional areas of demography and mortality rates, in contrast to the moral phenomena investigated by Quetelet.[36] Quetelet's belief that the laws of society paralleled those of the physical universe, and that Bernoulli's theorem described both physical and moral realms, simultaneously allied him with the classical probabilists and isolated him from the mainstream moral sciences of his day.

Quetelet marks both the end of the classical interpretation and the beginning of a new statistical view of probability. Although Quetelet adhered to the Laplacean view of probability as a measure of ignorance and did not hesitate to invoke the law of succession, his preoccupation with statistics and distributions was shared by the frequentists. He abandoned the individualist, psychological framework or the probabilities of judgments and of testimony for a collective, sociological perspective; yet he persisted in the belief that the calculus of probabilities provided a mathematical model for the moral sciences.

The famous *l'homme moyen* epitomizes the Janus-like position of Quetelet with respect to the classical interpretation of probability. Quetelet's *l'homme moyen* was a statistical composite of the physical, moral, and intellectual traits of the entire society. All that was merely individual, particular, or specific was obliterated in *l'homme moyen*, who represented the average in all things. At first glance, nothing seems further removed from the elite of reasonable men *(hommes éclairés)* who featured so prominently in the writings of the eighteenth-century probabilists. The *hommes éclairés* belonged to a select minority distinguished by an unusual ability to intuitively approximate probabilities and expectations; *l'homme moyen* possessed no distinction of any kind, unless it was his dead-center averageness. Yet *l'homme éclairé* and *l'homme moyen* both served as standards to be described

[35] *Mémoires de l'Académie royale des Sciences Morales et Politiques*, 2nd series, 1 (1837), pp. clxxviii–clxxxiv.

[36] See, for example, the memoirs by Villermé and Benoiston de Chateauneuf in the *Mémoires de l'Académie royale des Sciences Morales et Politiques*. The typical recipient of the Académie des Sciences Prix de Statistique, established in 1818, was a descriptive statistical profile of a *département* or colony, and after 1829, surveys of medical and social topics such as cholera, foundlings, and prostitution; see Archives de l'Académie des Sciences, *Concours 1818 à 1837*.

by the calculus of probabilities in order that they might be emulated by others. Quetelet held up *l'homme moyen* as a literary, artistic, moral, and intellectual ideal, literally the golden mean for a given society: the great man was he who subsumed his individuality in "all of humanity, nature, and the universal order."[37]

The ideals represented by *l'homme éclairé* and *l'homme moyen* were indeed very different, and reflected an altered understanding of both the moral sciences and the calculus of probabilities. The *homme éclairé* exemplified the criteria of rational belief and action, codified in the definition and comparison of probabilistic expectations. *L'homme moyen* personified the uniform, constant laws of society, as opposed to the accidental, variable conduct of individuals, revealed by the normal curve. However, *l'homme moyen* and *l'homme éclairé* were still brothers under the skin: both embodied social standards to be mathematically codified and mathematically enforced. Translated into the universal terms of mathematical probability, the good sense of *l'homme éclairé* could be taught to all; defined by the peak of the binomial distribution, *l'homme moyen* was the "center of gravity" of the society.

Rationality, whether defined by *l'homme éclairé* or *l'homme moyen*, remained a matter of calculation, and calculation compelled agreement. Of course, in one case rationality lay at the level of individual action and belief; in the other, at that of the operation of whole societies, *despite* the irrationality of their individual members. Consequently, the targets of persuasion also differed: Quetelet wanted governments to change their ways on the basis of his figures, not individuals. But both sorts of probabilistic rationality presupposed the stable, orderly phenomena that made calculation possible, even if they singled out different *kinds* of phenomena as quantifiable. Classical probabilists believed that judicial decisions, but not traffic accidents, were regular; their successors believed just the reverse. Although their calculus may not have been the Universal Characteristic Leibniz had envisioned, the probabilists remained true to the spirit of Leibniz's Characteristic for almost two centuries:

> For all inquires which depend on reasoning would be performed by the transposition of characters and by a kind of calculus, which would immediately facilitate the discovery of beautiful results. For we should not have to break our heads as much as is necessary today, and

[37] Quetelet, *Sur l'homme*, vol. 2, p. 277.

yet we should be sure of accomplishing everything the given facts allow.

Moreover, we should be able to convince the world what we should have found or concluded, since it would be easy to verify the calculation either by doing it over or by trying tests similar to that of casting out nines in arithmetic. And if someone would doubt my results, I should say to him: "Let us calculate Sir": and thus by taking to pen and ink, we should soon settle the question.[38]

[38] Gottfried Wilhelm Leibniz, "Preface to the Universal Science" (1677), in *Leibniz Selections*, Philip P. Wiener, ed. (New York: Charles Scribner's Sons, 1951), p. 15.

BIBLIOGRAPHY

Archival and Manuscript Sources

ARCHIVES DE L'ACADÉMIE DES SCIENCES, PARIS
    Dossiers for sessions (by date)
    MS. *Procès-verbaux* ("plumatifs")

ARCHIVES NATIONALES, PARIS
    F^4 1011–1012. Loterie royale, 1741–89
    F I 148. Loteries et maisons de jeu, An XIII–1817
    F^{12} 795–798^d. Loteries et assurance, 1701–1814
    F^{30} 101. Registre de la Loterie Nationale
    F^{12} 795. Loteries, banques, etc. XVIIIe siècle
    F^{10} 220. Poudres et Loterie, 1791–An II

BIBLIOTHÈQUE DE L'INSTITUT, PARIS
    MS. 1793. D'Alembert (unpub. vol. 9 of *Opuscules mathématiques*)
    MSS. 874–875. Condorcet (mathematical works, including draft of 1785 *Essai*)
    MSS. 883–884. Condorcet (includes several works on probability)
    MS. 853. Condorcet (letters to Turgot)
    MS. 857. Condorcet (includes pieces on probability theory, annuities, political arithmetic)
    MS. 2396. Lacroix (letters to Condorcet and Quetelet)
    MS. 2242. Laplace (includes draft of paper on population of France)
    MS. 955. Poinsot (lecture notes on probability theory)

BIBLIOTHÈQUE ALBERT Ier, BRUSSELS
    Quetelet MSS. (Salle des Cartes et Plans)
        See Lillian Wellens-De Donder, "Inventaire de la correspondance d'Adolphe Quetelet deposée à l'Académie royale de Belgique." *Académie royale de Belgique. Classe des Sciences. Memoires* 37 (1966)

EQUITABLE LIFE ASSURANCE SOCIETY, LONDON
    Office Copies of Proposals, 1758–71
    Rough Minutes of the Weekly Courts

Claims Books, 1806–38
Committee Minutes, 1765–
Annual Accounts, 1767–1900
James Dodson, MS. "Lectures on Insurances"
Orders of the Court of Directors, 1774–1848

LONDON GUILDHALL LIBRARY
MS. 8740. Royal Assurance Company. Assurance Book on Lives
MS. 18,847. Charles Povey. "Proposals" (1706)

ROYAL SOCIETY OF LONDON
Correspondence of John Herschel: vol. 6 (De Morgan); vol. 14 (Quetelet)

AMEV LIBRARY, UTRECHT
MS. G610. Johannes Hudde. Stades-finantie geredresfeert in den jaare
1679. Reckoning sheets on annuities

Primary Sources

[D'Alembert, Jean]. "Croix ou pile." In d'Alembert and Diderot, eds., *Encyclo-pédie, ou Dictionnaire raisonné des sciences, des arts et des métiers.* Vol. 4, p. 513.
———. *Mélanges de littérature, d'histoire et de philosophie.* 5 vols. Amsterdam, 1770.
———. *Oeuvres de d'Alembert.* A. Belin, ed. 5 vols. Paris, 1821–22.
———. *Oeuvres et correspondance inédites de d'Alembert.* Charles Henry, ed. Paris, 1887.
———. *Opuscules mathématiques.* 8 vols. Paris, 1761–80.
———, C. Bossut et al., eds. *Dictionnaire encyclopédique des mathématiques.* 3 vols. Paris, 1784–89.
———. "Fortuit (Metaphys.)." In d'Alembert and Diderot, eds., *Encyclopédie, ou Dictionnaire raisonné des sciences, des arts et des métiers.* Vol. 7, pp. 204–205.
———. "Fortune (Morale)." In d'Alembert and Diderot, eds., *Encyclopédie, ou Dictionnaire raisonné des sciences, des arts et des métiers.* Vol. 7, pp. 205–206.
D'Alembert, Jean, and Denis Diderot, eds. *Encyclopédie, ou Dictionnaire raisonné des sciences, des arts et des métiers.* 17 vols. of text. Paris and Neuchâtel, 1751–65.
Anon. *Directions for Breeding Game Cocks.* London, 1780.
———. *A Modest Defence of Gambling.* London, 1754.
———. *Reflexions on Gaming.* London, n.d.
Arbuthnot, John. "An argument for divine providence, taken from the regularity

observ'd in the birth of both sexes." *Philosophical Transactions of the Royal Society of London* 27 (1710–12): 186–190.

Aristotle. *The Art of Rhetoric.* Tr. John Henry Freese. Loeb Classical Library. Cambridge, Mass., 1959.

———. *Eudemian Ethics.* Tr. Michael Woods. Oxford, 1982.

———. *Nicomachean Ethics.* Tr. H. Rackham. Loeb Classical Library. London, 1962.

Arnauld, Antoine, and Pierre Nicole. *La logique, ou l'Art de penser* (1662). Pierre Clair and François Girbal, eds. Paris, 1965.

Bacon, Francis. *The Works of Francis Bacon.* Basil Montagu, ed. 16 vols. London, 1825–36.

———. *The Works of Francis Bacon.* James Spedding, Douglas Heath, and Robert Leslie Ellis, eds. 15 vols. Boston, 1860–64.

Balmford, James. *A Modest Reply to Certain Answeres.* N.p., 1623.

Barbeyrac, Jean. *Traité du jeu.* 2nd revised edition. 3 vols. Amsterdam, 1737.

Bauny, R.P.E. *Somme des pechez qui se commettent en tous les états.* Lyons, 1646.

Bayes, Thomas. *Divine Benevolence, or an Attempt to Prove that the Principal End of the Divine Providence and Government Is The Happiness of His Creatures.* London, 1731.

———. "An essay towards solving a problem in the doctrine of chances." *Philosophical Transactions of the Royal Society of London* 53 (1763): 370–418.

———. *Facsimiles of Two Papers by Bayes.* W. Edwards Deming, ed. Washington, D.C., 1940.

Bayle, Pierre. *Oeuvres diverses* (1727). Elisabeth Labrousse, ed. Hildesheim, 1965.

———. *Pensées diverses sur la comète* (1681). A. Prat, ed. 2 vols. Paris, 1911.

Beccaria, Cesare Bonesana. *Des délits et des peines.* J.A.S. Collin de Plancy, ed. Paris, 1823.

———. *Traité des délits et des peines* (1764). Tr. Abbé Morellet. Lausanne, 1766.

[Bellin, Jacques-Nicolas]. "Assurance." In d'Alembert and Diderot, eds., *Encyclopédie, ou Dictionnaire raisonné des sciences, des arts et des métiers.* Vol. 1, p. 774.

Benoiston de Chateauneuf, Louis-François. "Sur les résultats des *Comptes de l'administration de la justice criminelle en France, de 1825 à 1839.*" *Séances et travaux de l'Académie des Sciences Morales et Politiques* 1 (1842): 324–341.

Bentham, Jeremy. *Traité des preuves judiciaires.* 2 vols. Paris, 1823.

Bernoulli, Daniel. "Essai d'une nouvelle analyse de la mortalité causée par la petite vérole et des avantages de l'inoculation pour la prévenir." *Histoires et Mémoires de l'Académie des Sciences* (1760; publ. 1766), part 2, pp. 1–79.

———. "Recherches physiques et astronomiques, sur le problème proposé. . . . Quelle est la cause physique de l'inclinaison des plans des orbites des planètes par rapport au plan de l'équateur de la révolution du soleil autour de son axe; et d'ou vient que les inclinaisons de ces orbites sont différentes entre elles?" *Recueil des pièces qui ont remportés les prix de l'Académie royale des Sciences* 3 (1752): 93–122 (French text); 123–145 (Latin original).

———. "Specimen theoriae novae de mensura sortis." *Commentarii academiae scientarum imperialis Petropolitanae* 5 (1730–33; pub. 1738): 175–192. Tr. L. Sommer. "Exposition of a new theory on the measurement of risk." *Econometrica* 22 (1954): 23–36.

Bernoulli, Jakob. *Ars conjectandi* (1713). In *Werke*. Vol. 3, pp. 107–259.

———. *Ars conjectandi: Translations from James Bernoulli*. Tr. Bing Sung, with a preface by A. P. Dempster. Harvard University Department of Statistics Technical Report No. 2 (12 February 1966).

———. *Die Werke von Jakob Bernoulli*. Basel Naturforschende Gesellschaft. 3 vols. Basel, 1969–75.

Bernoulli, Nicholas. *De usu artis conjectandi in iure* (1709). In *Die Werke von Jakob Bernoulli*. Vol. 3, pp. 287–326.

Bertrand, Joseph. *Calcul des probabilités*. Paris, 1889.

Bicquelley, Charles-François. *De calcul des probabilités*. Toulouse, 1783.

Bienaymé, I. J. "Communication sur un principe que M. Poisson avait cru découvrir et qu'il avait appelé Loi des Grands Nombres." *Séances et Travaux de l'Académie des Sciences Morales et Politiques*, 3rd series, 11 (1855): 379–389.

Boethius. *Philosophiae consolationis*. In *Boethius: The Theological Tractates*. H. F. Stewart, E. K. Rand, and S. J. Tester, eds. Loeb Classical Library. Cambridge, Mass., 1973.

Boole, George. *An Investigation of the Laws of Thought* (1854). New York, 1958.

Borda, Jean-Charles. "Sur la forme des élections." *Mémoires de l'Académie royale des Sciences. Année 1781* (1784): 657–665.

Borély, Joseph. *Discours prononcé devant la Cour royale d'Aix*. Aix, 1836.

Boyle, Robert. *The Works of the Honourable Robert Boyle*. 6 vols. London, 1772.

[Brissot de Warville, J. P.]. *Dénonciation au public d'un nouveau projet d'agiotage, ou Lettre à M. le Comte de S * * ***. London, 1786.

———. *Seconde lettre contre la compagnie d'assurance pour les incendies à Paris, & contre l'agiotage en général*. London, 1786.

Browne, Thomas. *The Works of Sir Thomas Browne*. Geoffrey Keynes, ed. 4 vols. London, 1964.

Buffon, George Leclerc. "Essais d'arithmétique morale." *Histoire naturelle. Supplément*. Vol. 4, pp. 46–148. Paris, 1777.

———. "Des probabilités de la durée de la vie." *Histoire naturelle. Supplément*. Vol. 4, pp. 149–323. Paris, 1777.

Butler, Joseph. *The Analogy of Religion, Natural and Revealed, to the Constitution and Course of Nature*. London, 1736.

Cardano, Gerolamo [Cardanus, Hieronymus]. *The Book on Games of Chance*. Tr. Sydney Henry Gould. New York, 1961.

————. *De ludo aleae* (comp. ca. 1520). In his *Opera omnia*. Vol. 1, pp. 262–276.

————. *Opera omnia*. 10 vols. Stuttgart-Bad Cannstatt, 1966; facsimile of Lyons, 1663, edition.

Charleton, Walter. *Physiologia Epicuro-Gassendo-Charltoniana* (1654). New York and London, 1966.

Cicero, Marcus Tullius. *De inventione. De optimo genere oratorum. Topica* (ca. 85 B.C.). Tr. H. M. Hubbell. Loeb Classical Library. Cambridge, Mass., 1960.

Clark, Samuel. *Considerations Upon Lottery Schemes In General*. London, 1775.

————. *The Laws of Chance*. London, 1758.

————. *A Letter to Richard Price, D.D. and F.R.S.* London, 1777.

Cleirac, Estienne. *Les us, et coutumes de la mer*. Rouen, 1671.

Collier, Jeremy. *An Essay Upon Gaming*. London, 1713. (Reprinted as no. 8 of *Collectanea Adamantaea*, Edinburgh, 1885.

Compagnie Royale d'Assurances. *Prospectus de l'établissement des assurances sur la vie*. Paris, 1788.

Company of Parish Clerks of London. *London's Dreadful Visitation: Or, A Collection of All the Bills of Mortality for This Present Year*. London, 1665.

Comte, Auguste. *Cours de philosophie positive*. 6 vols. Paris, 1830–42.

————. *The Positive Philosophy of Auguste Comte*. Tr. Harriet Martineau. 2nd edition. 2 vols. London, 1875.

————. "Considérations sur les tentatives qui ont été faites pour fonder la sciences sociale sur la physiologie et sur quelques autres sciences." *Revue occidentale philosophique, sociale et politique* 8 (1882): 386–399.

————. "Sur les travaux politiques de Condorcet." *Revue occidentale philosophique, sociale et politique* 8 (1882): 400–409.

Condillac, Étienne. *Oeuvres*. 23 vols. Paris, An VI/1798.

Condorcet, Marie-Jean-Antoine-Nicholas Caritat. *Eléments du calcul des probabilités*. Paris, An XIII/1805.

————. *Essai sur l'application de l'analyse à la probabilité des décisions rendues à la pluralité des voix*. Paris, 1785.

————. "Mémoire sur le calcul des probabilités."

"Première partie. Réflexions sur le règle générale qui préscrit de prendre pour valeur d'un événement incertain, la probabilité de cet événement, multipliée par la valeur de l'événement en lui-même." *Mémoires de l'Académie royale des Sciences*. Année 1781 (1784): 707–720.

"Deuxième partie. Application de l'analyse à cette question: Déterminer la probabilité qu'un arrangement régulier est l'effet d'une intention de le produire." *Mémoires de l'Académie royale des Sciences*. Année 1781 (1784): 720–728.

"Troisième partie. Sur l'évaluation des droits éventuels." *Mémoires de l'Académie royale des Sciences*. Année 1782 (1785): 674–691.

"Quatrième partie. Réflexions sur la méthode de déterminer la probabilité des événements futurs d'après l'observation des événements passés." *Mémoires de l'Académie royale des Sciences. Année 1783* (1786): 539–553.

"Cinquième partie. Sur la probabilité des faits extraordinaires." *Mémoires de l'Académie royale des Sciences. Année 1783* (1786): 553–559.

"Sixième partie. Applications des principes de l'article précédent à quelques questions de critique." *Mémoires de l'Académie royale des Sciences. Année 1784* (1787): 454–468.

———. *Oeuvres de Condorcet.* F. Arago and A. Condorcet-O'Connor, eds. 12 vols. Paris, 1847–49.

———. "Probabilité." *Dictionnaire encyclopédique des mathématiques.* D'Alembert, Bossut et al., eds. Vol. 2, pp. 649–663.

Condorcet, M.-J.-A.-N. Caritat; Jean d'Alembert; and C. Bossut. *Nouvelles expériences sur la résistance des fluides.* Paris, 1777.

Condorcet, M.-J.-A.-N. Caritat; Du Sejour; and Laplace. "Essai pour connaître la population du royaume, et le nombre des habitants de la campagne, en adaptant sur chacune des cartes de M. Cassini, l'année commune des naissances, tant des villes que des bourgs et des villages dont il est fait mention sur chaque carte." *Mémoires de l'Académie royale des Sciences. Année 1783* (1786): 703–718; *Année 1784* (1787): 577–593; *Année 1785* (1788): 661–689; *Année 1786* (1788): 703–717; *Année 1787* (1789): 601–610; *Année 1788* (1791): 755–767.

[Cotton, Charles]. *The Compleat Gambler.* 2nd edition. London, 1680.

Coudrette, Abbé. *Dissertation théologique sur les loteries.* N.p., 1742.

Cournot, Antoine Augustin. *Exposition de la théorie des chances et des probabilités.* Paris, 1843.

———. "Mémoire sur les applications du calcul des chances à la statistique judiciaire." *Journal de mathématiques pures et appliquées* 3 (1838): 257–334.

———. *Recherches sur les principes mathématiques de la théorie des richesses.* Paris, 1838.

Cousin, Victor. *Cours d'histoire de la philosophie moderne.* 1st series, 4 vols. Paris, 1846.

———. [Public Address]. *Séances et travaux de l'Académie des Sciences Morales et Politiques* 1 (1842): 4–9.

Craig, John. *Theologiae christianae principia mathematica.* London, 1699. Tr. excerpts in "Craig's rules of historical evidence." *History and Theory.* Beiheft 4. The Hague, 1964.

Daneau, Lambert. *Trve and Christian Friendshippe.* London, 1586.

Defoe, Daniel. *An Essay on Projects.* Menston, 1969; reprint of original 1697 edition.

———. *The Gamester.* London, 1719.

———. *A Journal of the Plague Year.* London, 1722.

De La Roche, Estienne. *L'Arismethique*. Lyon, 1520.

De Moivre, Abraham. *Annuities for Life*. London, 1725.

———. *Annuities on Lives*. 3rd edition. London, 1750.

———. *The Doctrine of Chances*. 3rd edition. London, 1756.

De Morgan, Augustus. *An Essay on Probabilities*. In Dionysius Lardner, ed., *The Cabinet Cyclopaedia*. London, 1838.

———. "Theory of Probabilities." *Encyclopaedia Metropolitana*. Vol. 2, pp. 393–490. London, 1845.

———. "Review of *Théorie Analytique des Probabilités*." *Dublin Review* 2 (1837): 338–354.

Deparcieux, A. *Essai sur les probabilités de la durée de la vie humaine*. Paris, 1746.

Deparcieux, A. (nephew). *Traité des annuités*. Paris, 1781.

[De Prades, Abbé.] "Certitude." In d'Alembert and Diderot, eds., *Encyclopédie, ou Dictionnaire raisonné des sciences, des arts et des métiers*. Vol. 2, pp. 845–862.

Derham, William. *Physico-Theology, or, A Demonstration of the Being and Attributes of God from His Works of Creation*. 4th revised edition. New York, 1977; reprint of 1716 edition.

Descartes, René. *Oeuvres de Descartes*. Charles Adam and Paul Tannery, eds. 11 vols. Paris, 1964-74.

Destutt de Tracy, Antoine Louis Claude. *Commentaire sur l'Esprit des lois de Montesquieu, suivis d'observations inédites de Condorcet sur le vingt-neuvième livre du même ouvrage*. Paris, 1828.

———. *Élémens d'idéologie*. 2nd edition. 4 vols. Paris, 1804–18.

De Witt, Johann. *Waerdye van Lyf-Renten* (1671). In *Die Werke von Jakob Bernoulli*. Vol. 3, pp. 328–350. Tr. F. Hendriks. In Robert G. Barnwell, *A Sketch of the Life and Times of John De Witt*. New York, 1856.

Diannyère, Antoine. *Essais d'arithmétique politique*. Paris, An VIII/1800.

[Diderot, Denis]. "Fortuit (Gramm.)." In d'Alembert and Diderot, eds., *Encyclopédie, ou Dictionnaire raisonné des sciences, des arts et des métiers*. Vol. 7, p. 204.

———. "Luxe." In d'Alembert and Diderot, eds., *Encyclopédie, ou Dictionnaire raisonné des sciences, des arts et des métiers*. Vol. 9, pp. 763–771.

———. "Probabilité." In d'Alembert and Diderot, eds., *Encyclopédie, ou Dictionnaire raisonné des sciences, des arts et des métiers*. Vol. 13, pp. 393–400.

———. "Risque"/"Risquer." In d'Alembert and Diderot, eds., *Encyclopédie, ou Dictionnaire raisonné des sciences, des arts et des métiers*. Vol. 14, pp. 301–302.

Dodson, James. *The Mathematical Repository*. 3 vols. London, 1753–55; 2nd edition, London, 1775.

Domat, Jean. *Les loix civiles dans leur ordre naturel* (1689–94). Nouvelle edition . . . augmenteé des Troisième et Quatrième Livres du Droit Public, par M. de Héricourt. Paris, 1777.

Du Moulin, Charles. *Summaire du livre analytique des contractz usures, rentes constituees, interestz & monnoyes*. Paris, 1554.

Dupont de Nemours, Pierre. *Opinion de Du Pont (de Nemours), sur les projets de loterie, & sur l'état des revenus ordinaires de la République*. Paris, An V/1797.

Dusaulx, Jean. *Lettre et réflexions sur la fureur du jeu*. Paris, 1775.

Duvillard de Durand, Emmanuel. *Plan d'une association de prévoyance, dans laquelle ses membres feront entr'eux et pour eux, de la manière la plus avantageuse possible, tous les arrangements connus sous la dénomination d'assurances sur la vie*. Paris, [1790].

Ellis, R. L. "On the foundations of the theory of probabilities." *Transactions of the Cambridge Philosophical Society* 8 (1849): 1–6.

[Erskine, Thomas]. *Reflections on Gaming, Annuities, and Usurious Contracts*. London, 1776.

Euler, Leonhard. *Opera omnia*. Series 1 (Opera mathematica). 29 vols. Leipzig and Berlin, 1911–56.

Ferriere, Claude de. *Corps et compilation de tous les commentateurs anciens et modernes sur la coutume de Paris*. 3 vols. Paris, 1685.

Fielding, Henry. *The Lottery*. London, 1732.

Fontenelle, Bernard de. "Éloge de M. Bernoulli." *Histoire de l'Académie royale des Sciences*. Année 1705. Amsterdam, 1746.

Forcadel, Pierre. *L'Arithmeticque*. Paris, 1557.

Fuss, Nicolas. *Éclairissements sur les établissemens publics en faveur tant des veuves que des morts*. St. Petersburg, 1776.

Fuss, Paul Heinrich, ed. *Correspondance mathématique et physique de quelques célèbres géomètres du XVIIIème siècle*. New York and London, 1968; reprint of 1843 edition.

Galilei, Galileo. *Opere*. Antonio Favaro, ed. 20 vols. Florence, 1968.

Gard, T. *The Odds and Chances of Cocking, and Other Games*. London, n.d.

Garde des Sceaux, Ministère de la Justice. *Compte général de l'administration de la justice criminelle en France*. Paris, 1827.

———. *Compte général de l'administration de justice civile en France*. Paris, 1831.

Gataker, Thomas. *Of the Natvre and Vse of Lots*. London, 1619.

Glanvill, Joseph. *Essays on Several Important Subjects in Philosophy and Religion*. London, 1676.

———. *The Vanity of Dogmatizing*. London, 1661.

Graunt, John. *Natural and Political Observations Mentioned in a Following Index and Made Upon the Bills of Mortality*. London, 1662.

'sGravesande, Willem. *Mathematical Elements of Natural Philosophy*. Tr. J. T. Desaguliers. 5th edition, 2 vols. London, 1737.

———. *Oeuvres philosophiques et mathématiques*. 2 vols. Amsterdam, 1774.

Great Britain. House of Commons. *Report from the Committee, Appointed to Examine the Book, Containing an Account of the Contributors to the Lottery 1753: And The Proceedings of the House Thereupon.* London, 1754.

Gregory IX, Pope. *Decretales Gregorii Noni Pontificis cum epitomis, divisionibus, et glosis ordinariis.* Lugduni, 1558.

Grimaudet, François. *Paraphrase des droicts des usures pignoratifs.* Paris, 1583.

Grotius, Hugo. *The Rights of War and Peace* (1625). Tr. A. C. Campbell. Washington and London, 1901.

———. *The Truth of the Christian Religion* (1627). Tr. John Clarke. London, 1800.

Guerry, André Michel. *Essai sur la statistique morale.* Paris, 1833.

Guerry, André Michel, and Adriano Balbi. *Statistique comparée de l'état de l'instruction et du nombre des crimes dans les divers arrondissements des académies et des cours royales de France.* Paris, 1829.

Halley, Edmund. "An estimate on the degrees of mortality of mankind, drawn from curious tables of the birth and funerals at the city of Breslaw; with an attempt to ascertain the price of annuities upon lives." *Philosophical Transactions of the Royal Society of London* 17 (1693): 596–610.

Hartley, David. *Observations on Man, His Frame, His Duty, and His Expectations.* 2 vols. London, 1749.

———. *The Progress of Happiness Deduced from Reason.* London, 1734.

Henry, Charles, ed. *Correspondance inédite de Condorcet et de Turgot, 1770–1779.* Paris, 1883.

Herschel, John. "Review of the *Lettres à S.A.R. le Duc regnant de Saxe-Coburg et Gotha sur la théorie des probabilités appliquée aux sciences morales et politiques*, par M. A. Quetelet. . . ." *Edinburgh Review* 96 (1850): 1–57.

[Hooper, George]. "A calculation of the credibility of human testimony." *Philosophical Transactions of the Royal Society of London* 21 (1699): 359–365.

Hoyle, Edmund. *The Polite Gamester.* Dublin, 1761.

Hume, David. *An Enquiry Concerning Human Understanding* (1758). Charles W. Hendel, ed. Indianapolis, 1955.

———. *Essays Moral, Political and Literary* (1741–42). London, 1963.

———. *The Philosophical Works.* Thomas H. Green and T. H. Grose, eds. 4 vols. London, 1882.

———. *A Treatise of Human Nature* (1739). L. A. Selby-Bigge, ed. Oxford, 1975.

Huygens, Christiaan. *Oeuvres complètes.* Société Hollandaise des Sciences. 22 vols. The Hague, 1888–1967.

———. *De ratiociniis in ludo aleae* (1657). In *Oeuvres.* Vol. 14, pp. 50–91.

Jados, Stanley, ed. *Consulate of the Sea and Related Documents.* University, Alabama, 1975.

Kersey, John, ed. *Mr. Wingate's Arithmetick.* 5th edition. London, 1670.

Lacroix, Silvestre-François. *Traité élémentaire du calcul des probabilités*. Paris, 1816.

La Fontaine, [Louis?] de. *Précis d'un projet d'opération de finance, par forme de loterie, &c. &c.* London, 1775.

La Harpe, Jean-François. *Lycée, ou cours de littérature ancienne et moderne*. 16 vols. Paris, 1813.

La Mothe le Vayer, François de. *Des anciens et principavs historiens grecs et latins*. Paris, 1646.

——. *Discours de l'histoire*. Paris, 1638.

Laplace, Pierre Simon. *Essai philosophique sur les probabilités* (1814). In *Oeuvres*. Vol. 7, pp. i–cliii. Tr. Frederick Wilson Truscott and Frederick Lincoln Emory, *A Philosophical Essay on Probabilities*. New York, 1951.

——. "Mémoire sur la probabilité des causes par les événements" (1774). In *Oeuvres*. Vol. 8, pp. 27–65.

——. "Mémoire sur les probabilités" (1781). In *Oeuvres*. Vol. 9, pp. 383–485.

——. *Oeuvres complètes*. Académie des Sciences. 14 vols. Paris, 1878–1912.

——. "Recherches sur l'intégration des équations différentielles aux différences finies et sur leur usage dans la théorie des hasards" (1773). In *Oeuvres*. Vol. 8, pp. 69–200.

——. "Sur les naissances, les mariages et les morts à Paris, depuis 1771 jusqu'en 1784. . . ." *Mémoires de l'Académie royale des Sciences. Année 1783*, pp. 693–702. Paris, 1786.

——. *Théorie analytique des probabilités* (1812). In *Oeuvres*. Vol. 7.

La Placette, Jean. *Divers traités sur des matières de conscience*. Amsterdam, 1697.

Laurent, H. *Traité du calcul des probabilités*. Paris, 1873.

Le Clerc, Jean. *Réflexions sur ce que l'on appelle bonheur et malheur en matière de loteries, et sur le bon usage qu'on en peut faire*. Amsterdam, 1696.

Leibniz, Gottfried Wilhelm. *G. W. Leibniz: Mathematische Schriften*. C. I. Gerhardt, ed. Hildesheim, 1962; reprint of 1855 edition.

——. *Leibniz Selections*. Philip Wiener, ed. New York, 1951.

——. *Nouveaux essais sur l'entendement humain* (comp. 1703–1705; publ. 1765). In *Sämtliche Schriften und Briefe*. Sechste Reihe, vol. 6.

——. *Opuscules et fragments inédits de Leibniz*. Louis Couturat, ed. Hildesheim, 1961; reprint of 1903 edition.

——. *Sämtliche Schriften und Briefe*. 6th edition. Akademie der Wissenschaften zu Berlin, Sechste Reihe: Philosophische Schriften. 4 vols. (1, 2, 3, 6). Berlin, 1962–80.

Leti, G. *Critique historique, politique, morale, économique, & comique sur les lotteries*. 2 vols. Amsterdam, 1697.

Locke, John. *Discourse on Miracles*. London, 1702.

——. *An Essay Concerning Human Understanding* (1689; all editions dated 1690). Alexander Campbell Fraser, ed. 2 vols. New York, 1959.

Lucas, Theophilus. *Memoirs of the Lives, Intrigues, and Comical Adventures of the Most Famous Gamesters and Celebrated Sharpers.* London, 1714.

Machiavelli, Niccolò. *The Prince* (1532). Tr. Edward Dacres (1640). Reprinted with introduction by Henry Cust. London, 1905.

Magens, Nicolas. *An Essay on Insurances.* 2 vols. London, 1755.

[Mallet, Abbé]. "Assûrer." In d'Alembert and Diderot, eds., *Encyclopédie, ou Dictionnaire raisonné des sciences, des arts et des métiers.* Vol. 1, p. 775.

Mallet, E. "[Review of] *Recherches sur la probabilité des jugements* . . . , par M. Poisson." *Bibliothèque universelle de Genève*, 2nd series, 10 (1837): 125–132.

Marseille, J. B. *La pierre, et la vraie pierre philosophale des loteries impériales de France.* 2nd edition. Paris, 1807.

Marshall, Samuel. *Treatise on the Law of Insurance.* Boston, 1805.

Masius, E. A. *Lehre der Versicherung und statistische Nachweisung aller Versicherungs-Anstalten in Deutschland.* Leipzig, 1846.

Massé de la Rudelière. *Défense de la doctrine des combinaisons.* Paris, 1763.

Masterson, Thomas. *His First Books of Arithmeticke.* London, 1652.

Melon, Jean-François. *Essai politique sur le commerce* (1734; revised edition 1736). In Eugène Daire, ed., *Economistes–financiers du XVIIIe siècle.* Paris, 1843.

[Menestrier, Claude François]. *Dissertation des lotteries.* Lyons, 1700.

Menochius, Jacobus. *De praesumptionibus conjecturis, signis et indicis commentaria, in VI distincta libros.* Cologne, 1595.

Mersenne, Marin. *De la vérité des sciences.* Paris, 1625.

Messance. *Nouvelles recherches sur la population de la France, avec des rémarques importantes sur divers objets d'administration.* Lyons, 1788.

Mill, John Stuart. *A System of Logic.* London, 1843.

———. *A System of Logic* (1843). 8th edition. New York, 1881.

Montaigne, Michel de. *Essais.* Maurice Rat, ed. 2 vols. Paris, 1962.

Montesquieu, Charles de Secondat de. *De l'esprit des lois* (1748). G. Truc, ed. 2 vols. Paris, 1961.

———. *The Spirit of the Laws* (1748). Tr. Thomas Nugent. London, 1900.

Montmort, Pierre de. *Essay d'analyse sur les jeux de hazard.* Paris, 1708.

———. *Essai d'analyse sur les jeux de hazard.* 2nd edition. Paris, 1713.

Montucla, J. F. *Histoire des mathématiques.* 2nd edition. Completed by J. Lalande. 4 vols. Paris, An VII–X/1798–1802.

Morris, Corbyn. *Observations on the Past Growth and Present State of the City of London.* London, 1750.

Mumford, Erasmus. *A Letter to the Club at White's.* London, 1750.

Newton, Isaac. *Philosophiae naturalis principia mathematica.* 2nd edition. Cambridge, Eng., 1713.

———. *Mathematical Principles of Natural Philosophy.* Tr. Andrew Motte (1729), rev. Florian Cajori. Berkeley and Los Angeles, 1962.

Newton, Isaac. *Opticks*. Based on the 4th edition of 1730. New York, 1952.

Ordonnance, statut et police. Nouvellement Faicte par le Roy Nostre Sire, svr le faict des contractz des assevrances es Pays-Bas. Anvers, 1571.

Parisot, Sebastien-Antoine. *Traité du calcul conjectural, ou l'Art de raisonner*. Paris, 1810.

Pascal, Blaise. *Lettres provinciales* (1657). J. J. Pauvert, ed. Holland, 1964.

―――. *Monsieur Pascall's Thoughts, Meditations, and Prayers, Touching Matters Moral and Divine*. Tr. Joseph Walker. London, 1688.

―――. *Oeuvres complètes*. Jean Mesnard, ed. Vol. 1, parts 1 and 2. Paris, 1964, 1970.

―――. *Pensées*. Louis Lafuma, ed. Paris, 1962.

―――. "Traité du triangle arithmétique" (1665). In *Oeuvres complètes*. Vol. 1, part 2, pp. 1166–1195.

Payen, Antoine-François. *Extrait d'une lettre . . . contenant l'observation de l'éclipse de soleil*. N.p., n.d.

―――. *Senelion, ou Apparition luni-solaire en l'Isle de Gergonne*. Paris, 1666.

Pelseneer, Jean. "Lettres inédites de Condorcet." *Osiris* 10 (1952): 322–327.

Petty, William. *An Essay Concerning the Multiplication of Mankind*. London, 1682.

―――. "[Review of] Natural and Political Observations Made Upon the Bills of Mortality." *Journal des Sçavans*. Amsterdam edition 1 (1665–66): 585–590.

―――. *Several Essays in Political Arithmetic*. London, 1699.

Poisson, Siméon-Denis. "Note sur la loi des grands nombres." *Comptes rendus hebdomadaires des séances de l'Académie des Sciences* 2 (1836): 377–380.

―――. "Recherches sur la probabilité des jugements." *Comptes rendus hebdomadaires des séances de l'Académie des Sciences* 1 (1835): 473–494.

―――. *Recherches sur la probabilité des jugements en matière criminelle et en matière civile*. Paris, 1837.

Pothier, Robert Joseph. *Traité des contrats aléatoires*. Paris and Orléans, 1775.

―――. *Traité du contrat de constitution de rente*. 2 vols. Paris, 1774.

Prestet, Jean. *Elemens des mathématiques*. Paris, 1675.

Price, Richard. "A demonstration of the second rule in the Essay towards the solution of a problem in the doctrine of chances." *Philosophical Transactions of the Royal Society of London* 54 (1764): 296–325.

―――. *Four Dissertations* (1767). 5th edition. London, 1811.

―――. *Observations on Reversionary Payments*. London, 1769; 3rd enlarged edition, London, 1773; 6th edition revised by William Morgan. 2 vols. London, 1803.

―――. *Review of the Principal Questions in Morals* (1758). D. D. Raphael, ed. Oxford, 1974.

Pufendorf, Samuel. *Le droit de la nature et des gens* (1682). Tr. Jean Barbeyrac. 2 vols. London, 1740.

Purser, William. *Compound Interest and Annuities*. London, 1634.

Quesnay, François. "Evidence." In d'Alembert and Diderot, eds., *Encyclopédie, ou Dictionnaire raisonné des sciences, des arts et des métiers*. Vol. 6, pp. 146–157.

———. *Oeuvres économiques et philosophiques de Quesnay*. Auguste Oncken, ed. Paris, 1888.

———, and Victor de Riquetti Mirabeau. *Philosophie rurale*. 3 vols. Amsterdam, 1764.

Quetelet, Adolphe. *Instructions populaires sur le calcul des probabilités*. Brussels, 1828.

———. *Lettres à S.A.R. le Duc regnant de Saxe-Coburg et Gotha, sur la théorie des probabilités, appliquée aux sciences morales et politiques*. Brussels, 1846.

———. "[Review of] *Recherches sur la probabilité des jugements* par M. Poisson." *Correspondance mathématique et physique de l'Observatoire de Bruxelles* 9 (1837): 485–486.

———. *Sur l'homme et le développement de ses facultés, ou Essai de physique sociale*. 2 vols. Paris, 1835.

———. *Théorie des probabilités*. Brussels, 1843.

Raithby, John, ed. *The Statutes at Large, of England and of Great Britain*. London, 1811.

Risueño d'Amador. *Mémoire sur le calcul des probabilités appliqué à la médecine*. Paris, 1837.

Roederer, Pierre-Louis, ed. *Collection de divers ouvrages d'arithmétique politique par Lavoisier, de Lagrange et autres*. Paris, An IV/1796.

Saint-Cyran, Paul-Edmé. *Calcul des rentes viagères*. Paris, 1779.

Scott, S. P., tr. *The Civil Law, including the Twelve Tables, the Institutes of Gaius, the Rules of Ulpian, the Opinions of Paulus, the Enactments of Justinian, and the Constitution of Leo*. Cincinnati, 1932.

Sévigné, Marie de Rabutin Chantal. *Correspondance*. Roger Duchêne, ed. 3 vols. Paris, 1974.

Short, Thomas. *A Comparative History of the Increase and Decrease of Mankind in England and Several Countries Abroad According to the Different Soils, Business of Life, Use of the* NON-NATURALS *&c.* London, 1867.

———. *New Observations, Natural, Moral, Civil, Political, and Medical on City, Town, and Country Bills of Mortality*. London, 1750.

Simon, Richard. *Histoire critique du Vieux Testament*. Nouvelle édition. Rotterdam, 1685.

Simpson, Thomas. *The Doctrine of Annuities and Reversions*. London, 1742.

———. *The Nature and Laws of Chance*. London, 1740.

Smollett, Tobias. *The Adventures of Ferdinand Count Fathom*. 2 vols. Dublin, 1753.

Society for Equitable Assurances. *The Plan of the Society for Equitable Assurances on Lives and Survivorships; Established by Deed*. London, 1766.

Society for Equitable Assurances. *A Short Account of the Society for Equitable Assurances on Lives and Survivorships*. London, 1764.

———. *A Short Account of the Society for Equitable Assurances on Lives and Survivorships*. London, 1781.

———. *Tables Showing the Total Number of Persons Assured in the Equitable Society*. London, 1834.

Spirito, Lorenzo. *Le passetemps de la fortune des dez* (1508). Lyons, 1583.

[Steele, Richard]. *The Guardian*. 2 vols. London, 1714.

Stevin, Simon. *L'Arithmetique*. Leyden, 1585.

Struyck, Nicolas. *Les oeuvres de Nicholas Struyck (1687–1769)*. Tr. J. A. Vollgraff. Amsterdam, 1912.

Süssmilch, Johann Peter. *Die göttliche Ordnung in den Veränderungen des menschlichen Geschlechts, besonders im Tode*. Berlin, 1756; 3rd revised edition. 3 vols. Berlin, 1775.

Taillefer, P., ed. *Methodiques institutions de la vraye et parfaite arithmetique de Iacques Chauvet*. Paris, 1615.

[Talleyrand-Perigord, Charles de]. *Des loteries*. Paris, 1789.

Tetens, Johann Nicholas. *Einleitung zur Berechnung der Leibrenten und Anwartschaften die vom Leben und Tode einer oder mehrerer Personen abhängen*. Leipzig, 1785.

Tillotson, John. *The Works of the Most Reverend Dr. John Tillotson*. 10 vols. Edinburgh and Glasgow, 1748.

Turgot, Ann-Robert-Jacques. *Oeuvres de Turgot et documents le concernant*. Gustave Schelle, ed. 5 vols. Tauners, F.R.G., 1972; reprint of 1913–23 edition.

———. *Les réflexions sur la formation et la distribution des richesses* (1766). *Oeuvres*. Vol. 2, pp. 534–601.

Underwriters' Agency of New York. *The Agent*. July, 1872.

Voltaire, François Marie Arouet de. *Lettres philosophiques* (1734). Raymond Naves, ed. Paris, 1964.

———. *Oeuvres complètes de Voltaire*. 70 vols. Paris, 1785–89.

Wallis, John. *A Discourse of Combinations, Alternations and Aliquot Parts* (1685). In Francis Masères, ed., *The Doctrine of Chances*. London, 1795.

Wilkins, John. *Of the Principles and Duties of Natural Religion* (1675) 4th edition. London, 1699.

Secondary Sources

Ackerknecht, Erwin. "Villermé and Quetelet." *Bulletin of the History of Medicine* 26 (1952): 317–329.

Alauzet, Isidore. *Traité général des assurances*. 2 vols. Paris, 1843.

Allais, M. "Le comportement de l'homme rationel devant le risque: Critique des postulats et axiomes de l'École Américaine." *Econometrica* 21 (1953): 503–546.

Allard, Albéric. *Histoire de la justice criminelle au seizième siècle.* Aalen, 1970; facsimile of the Ghent, Paris, Leipzig, 1868 edition.

Amzalak, Moses Benedict. *Trois précurseurs portugais.* Paris, 1935.

Appleby, Joyce Oldham. *Economic Thought and Ideology in Seventeenth-Century England.* Princeton, 1978.

Archibald, R. C. "A rare pamphlet of Moivre and some of his discoveries." *Isis* 8 (1926): 671–684.

Archibald, R. C. and Karl Pearson. "Abraham De Moivre." *Nature* 117 (1926): 551–552.

Arrow, Kenneth. "Alternative approaches to the theory of choice in risk-taking situations." *Econometrica* 19 (1951): 404–437.

———. "Formal theories of social welfare." In *Dictionary of the History of Ideas.* Philip Wiener, editor-in-chief. Vol. 4, pp. 276–284.

Ashton, John. *A History of English Lotteries.* London, 1893.

———. *The History of Gambling in England.* London, 1898.

Ayer, A. J. *Probability and Evidence.* London, 1972.

Baker, Keith M. *Condorcet: From Natural Philosophy to Social Mathematics.* Chicago, 1975.

Ball, J. N. *Merchants and Merchandise.* New York, 1977.

Barker, John. *Strange Contrarieties: Pascal in England During the Age of Reason.* Montreal and London, 1975.

Biermann, Kurt-Reinhard. "Eine Untersuchung von G. W. Leibniz über die jährliche Sterblichkeitsrate." *Forschungen und Fortschritte* 28 (1955): 205–208.

———. "G. W. Leibniz und die Berechnung der Sterbewahrscheinlichkeit bei J. de Witt." *Forschungen und Fortschritte* 33 (1959): 168–173.

Black, Duncan. *The Theory of Committees and Elections.* Cambridge, Eng., 1958.

Black, Max. "Induction and probability." In Raymond Klibansky, ed., *Philosophy in the Mid-Century*, pp. 154–163. Florence, 1958.

———. "Probability." In *Encyclopedia of Philosophy.* Paul Edwards, editor-in-chief. Vol. 6, pp. 464–479.

Blum, Edgar. "Les assurances terrestres en France sous l'Ancien Regime." *Revue d'Histoire Économique et Sociale* 8 (1920): 95–104.

Boas, George. "Primitivism." In *Dictionary of the History of Ideas*, Philip Wiener, editor-in-chief. Vol. 3, pp. 577–598.

Boiteux, L. A. *La fortune de la mer.* Paris, 1968.

Bonar, James. *Theories of Population from Raleigh to Arthur Young.* London, 1931.

Borel, Emile. *Probabilité et certitude.* Paris, 1950.

Bouchary, Jean. *Les manieurs d'argent à Paris à la fin du XVIIIe siècle.* 3 vols. Paris, 1939.

Braun, Heinrich. *Geschichte der Lebensversicherung und der Lebensversicherungtechnik.* Nuremberg, 1925.

Briggs, Morton. "D'Alembert." In *Dictionary of Scientific Biography.* Charles C. Gillispie, editor-in-chief. Vol. 1, pp. 110–117.

Brugmans, Henri. *Le séjour de Christian Huygens à Paris et ses relations avec les milieux scientifiques français.* Paris, 1935.

Brumfitt, J. H. *Voltaire Historian.* London, 1958.

Buck, Peter. "Seventeenth-century political arithmetic: Civil strife and vital statistics." *Isis* 68 (1977): 67–84.

Burke, Peter. *Popular Culture in Early Modern Europe.* London, 1978.

Burns, R. M. *The Great Debate on Miracles.* Lewisburg, Pa.: 1981.

Byrne, Edmund F. *Probability and Opinion: A Study in the Medieval Pre-suppositions of Post-Medieval Theories of Probability.* The Hague, 1968.

Caillois, Roger. *Man, Play, and Games.* Tr. Meyer Barash. New York, 1961.

Campbell, Sybil. "The economic and social effect of the usury laws in the eighteenth century." *Transactions of the Royal Historical Society,* 4th series 16 (1933): 197–210.

———. "Usury and annuities of the eighteenth century." *Law Quarterly Review* 44 (1928): 473–491.

Carnap, Rudolf. *Logical Foundations of Probability.* Chicago, 1950.

Carswell, John. *The South Sea Bubble.* London, 1960.

Cassirer, Ernst. *The Philosophy of the Enlightenment.* Tr. F.C.A. Koelln and J. P. Pettegrove. Princeton, 1951.

Charpaux, Marcel. *Almanach de la Loterie Nationale 1539–1949.* Paris, 1949.

Chaufton, Albert. *Les assurances.* 2 vols. Paris, 1884.

Cioffari, Vincenzo. "Fortune, fate, and chance." In *Dictionary of the History of Ideas.* Philip P. Wiener, editor-in-chief. Vol. 2, pp. 225–236.

Cockerell, H.A.L., and Edwin Green. *The British Insurance Business, 1547–1970.* London, 1976.

Cohen, L. Jonathan. "Some historical remarks on the Baconian conception of probability." *Journal of the History of Ideas* 41 (1980): 219–231.

Comp, B. H. "Definitions of probability." *American Mathematical Monthly* 39 (1932): 285–288.

Corblet, J. *Étude historique sur les loteries.* Paris, 1861.

Coste, Pierre. *Les loteries d'état en Europe et la Loterie Nationale.* Paris, 1933.

Coumet, Ernest. "La théorie du hasard est-elle née par hasard?" *Annales: Économies, Sociétés, Civilisations* (May-June, 1970): 574–598.

Couturat, Louis. *La logique de Leibniz d'après des documents inédits.* Paris, 1901.

Crosland, Maurice. "Nature and measurement in eighteenth-century France." *Studies on Voltaire and the Eighteenth Century* 87 (1972): 277–309.

————. *The Society of Arcueil*. Cambridge, Mass., 1967.

Dahm, John J. "Science and apologetics in the early Boyle Lectures." *Church History* 39 (1970): 172–186.

Dainville, François. "Un dénombrement inédit du XVIIIe siècle: L'enquête du contrôleur general Ouy." *Population* 7 (1952): 49–68.

Daston, Lorraine J. "D'Alembert's critique of probability theory." *Historia Mathematica* 6 (1979): 259–279.

————. "The domestication of risk: Mathematical probability and insurance, 1650–1830." In L. Krüger et al., eds., *The Probabilistic Revolution*. Vol. I, *Ideas in History*, pp. 237–260.

————. "Folklore and natural history." *Harvard Advocate* 107 (Autumn 1983): 35–38.

————. "Probabilistic expectation and rationality in classical probability theory." *Historia Mathematica* 7 (1980): 234–260.

————. "Rational individuals versus laws of society: From probability to statistics." In L. Krüger et al., eds. *The Probabilistic Revolution*. Vol. I, *Ideas in History*, pp. 100–125.

David, F. N. *Games, Gods and Gambling*. London, 1962.

Davis, Natalie Zemon. *Society and Popular Culture in Early Modern France*. London, 1975.

Defert, D.; J. Donzelot; F. Ewald; G. Maillet; and C. Mevel. *Socialisation du risque et pouvoir dans l'entreprise. Histoire des transformations politiques et juridiques qui ont permis la légalisation du risque professionel*. [Paris], 1977. Typescript from Ministère du Travail.

Delumeau, Jean. *La peur en Occident (XIVe–XVIIIe siècles)*. Paris, 1978.

Demain, Thomas. "Probabilisme." In *Dictionnaire de théologie catholique*. A. Vacant and E. Mangenot, eds. Vol. 13, cols. 417–619. Paris, 1935.

Dickson, P.G.M. *The Sun Insurance Office 1710–1960*. London, 1960.

Doren, A. "Fortuna im Mittelalter und in der Renaissance." *Vorträge der Bibliothek Warburg* 2 (1922): 71–144.

Downes, D. M.; B. P. Davies; M. E. David; and P. Stone. *Gambling, Work and Leisure: A Study Across Three Areas*. London, 1976.

Dupaquier, Jacques. "Sur une table (prétendument) florentine d'espérance de vie." *Annales. Économies, Sociétés, Civilisations* (July-August 1973): 1066–1070.

Eadington, William R., ed. *Gambling and Society*. Springfield, Ill., 1976.

Edwards, Paul, editor-in-chief. *Encyclopedia of Philosophy*. 8 vols. New York, 1967.

Ehrard, Jean. *L'Idée de la nature en France dans la première moitié du XVIIIe siècle*. 2 vols. Paris, 1970.

Elias, Norbert. *The Court Society.* Tr. Edmund Jephcott. New York, 1983.

Esmein, A. *Histoire de la procédure criminelle en France.* Paris, 1882.

Favre, Robert. *La mort dans la littérature et la pensée françaises au siècle des lumières.* Lyon, 1978.

Feller, William. *An Introduction to Probability Theory and Its Applications.* 2nd edition. New York, 1957.

Fine, Terence L. *Theories of Probability: An Examination of the Foundations.* New York and London, 1973.

Florange, Charles. *Curiosités financières sur les emprunts et loteries en France depuis les origines jusqu'à 1783.* Paris, 1928.

Foirers, Paul. "La conception de la preuve dans l'école de droit naturel." In *La Preuve. Deuxième partie: Moyen Age et temps modernes. Recueils de la Société Jean Bodin pour l'histoire comparative des institutions* 17 (1965): 169–192.

Fox-Genovese, Elizabeth. *The Origins of Physiocracy.* Ithaca, N.Y., 1976.

Francis, John. *Annals, Anecdotes and Legends: A Chronicle of Life Assurance.* London, 1853.

Fraser-Harris, D. F. "Smallpox in non-medical literature." *Medical Life* 37 (1930): 522–567.

Fry, Thornton C. *Probability and Its Engineering Uses.* New York, 1928.

Fulton, John F. *Bibliography of the Honourable Robert Boyle.* Oxford, 1961.

Garber, Daniel, and Sandy Zabell. "On the emergence of probability." *Archive for History of Exact Sciences* 21 (1979): 33–53.

Gay, Peter. *The Enlightenment: An Interpretation.* 2 vols. New York, 1966–69.

Georgescu-Roegen, Nicholas. "Utility and value in economic thought." In *Dictionary of the History of Ideas.* Philip Wiener, editor-in-chief. Vol. 4, pp. 450–458.

Gierke, Otto. *Natural Law and the Theory of Society 1500–1800.* Tr. with an introduction by Ernest Barker. Boston, 1957.

Gilissen, John. "La preuve en Europe du XVIe au debut du XIXe siècle. Rapport de synthèse." In *La Preuve. Deuxième partie: Moyen Age et temps modernes. Recueils de la Société Jean Bodin pour l'histoire comparative des institutions,* Vol. 17 (1965), pp. 755–833.

Gillispie, Charles C., editor-in-chief. *Dictionary of Scientific Biography.* 14 vols. plus Supplement. New York, 1970– .

———. "Intellectual factors in the background of analysis by probabilities." In A. C. Crombic, ed., *Scientific Change,* pp. 431–453. London, 1963.

———. "Laplace." In *Dictionary of Scientific Biography.* Charles C. Gillispie, editor-in-chief. Supplement 1, pp. 273–403.

———. "Probability and politics: Laplace, Condorcet, and Turgot." *Proceedings of the American Philosophical Society* 16 (1972): 1–20.

———. *Science and Polity in France At the End of the Old Regime.* Princeton, 1980.

Goffman, Erving. "Where the action is." In his *Interaction Ritual*, pp. 149–270. Garden City, N.Y., 1967.

Granger, Gilles-Gaston. *La mathématique sociale du Marquis de Condorcet*. Paris, 1956.

Greenwood, Major. *Medical Statistics from Graunt to Farr*. Cambridge, Eng., 1948.

Grimsley, Ronald. *Jean d'Alembert (1717–1783)*. Oxford, 1963.

Gouraud, Charles. *Histoire du calcul des probabilités depuis ses origines jusqu'à nos jours*. Paris, 1848.

Gusdorf, George. *Les sciences humaines et la pensée occidentale*. 7 vols. Paris, 1966–76.

Hacking, Ian. "Equipossibility theories of probability." *British Journal for the Philosophy of Science* 22 (1971): 339–355.

———. *The Emergence of Probability*. Cambridge, Eng., 1975.

———. "Jacques Bernoulli's *Art of Conjecturing*." *British Journal for the Philosophy of Science* 22 (1971): 209–229.

———. "The Leibniz-Carnap program for inductive logic." *Journal of Philosophy* 68 (1971): 597–610.

———. "The logic of Pascal's wager." *American Philosophical Quarterly* 9 (1972): 186–192.

Hahn, Roger. *The Anatomy of a Scientific Institution: The Paris Academy of Sciences, 1666–1803*. Berkeley, Los Angeles, and London, 1971.

———. "Laplace's first formulation of scientific determinism in 1773." *Actes du XIe Congrès International d'Histoire des Sciences*. Vol. 2, pp. 167–171. Cracow, 1968.

Hald, A. "Nicholas Bernoulli's theorem." *International Statistical Review* 52 (1984): 93–99.

———. "A. De Moivre: 'De Mensura Sortis' or 'On the Measurement of Chance'." *International Statistical Review* 52 (1984): 229–262.

Halliday, Jon, and Peter Fuller, eds. *The Psychology of Gambling*. New York, 1974.

Halperin, Jean. *Les assurances en Suisse et dans le monde*. Neuchâtel, 1946.

Hankins, Thomas L. *Jean d'Alembert: Science and the Enlightenment*. Oxford, 1970.

Hara, Kokiti. "Pascal et l'induction mathématique." *Revue d'Histoire des Sciences* 15 (1962): 287–302.

Harrison, Brian. "Religion and recreation in nineteenth-century England." *Past & Present*, no. 38 (December 1967): 98–125.

Heath, James. *Eighteenth-century Penal Theory*. Oxford, 1963.

Heckshaw, Eli F. *Mercantilism*. Tr. Muriel Shapiro. 2 vols. London, 1935.

Heidelberger, Michael. "Fechner's indeterminism: From freedom to laws of chance." In L. Krüger et al., eds. *The Probabilistic Revolution*. Vol. I, *Ideas in History*, pp. 117–156.

Henry, Charles. "Correspondance inédite de d'Alembert avec Cramer." *Bulletino di bibliografia e di storia delle scienze matematiche e fisiche* 18 (1885): 507–570, 605–649.

Heyde, C. C., and E. Seneta. *I. J. Bienaymé: Statistical Theory Anticipated*. New York, 1977.

Hill, Christopher. "The uses of Sabbatarianism." In his *Society and Puritanism in Pre-Revolutionary England*, pp. 145–218. London, 1964.

Hilts, Victor. "Statistics and social science." In Ronald N. Giere and Richard S. Westfall, eds., *Foundations of Scientific Method: The Nineteenth Century*. Bloomington, Ind., 1973.

Hirschman, Albert O. *The Passions and the Interests*. Princeton, 1977.

Holland, J. D. "The Reverend Thomas Bayes, F.R.S. (1702–1761)." *Journal of the Royal Statistical Society*, Series A, 125 (1962): 451–461.

Horvath, Robert. "*L'Ordre Divin* de Süssmilch. Bicentenaire du premier traité spécifique de démographie (1741–1761)." *Population* 17 (1962): 267–288.

Imbert, Jean, ed. *Quelques procès criminels des XVIIe et XVIIIe siècles*. Paris, 1964.

Jacob, James. *Robert Boyle and the English Revolution*. New York, 1977.

Jacob, Margaret C. *The Newtonians and the English Revolution, 1689–1720*. Ithaca, N.Y., 1976.

Jacobson, David. "Trial by jury and criticism of the Old Regime." *Studies on Voltaire and the Eighteenth Century* 153 (1976): 1099–1111.

Jarret, Derek. *England in the Age of Hogarth*. New York, 1974.

Jorland, Gérard. "The St. Petersburg Paradox (1713–1937)." In L. Krüger et al., eds., *The Probabilistic Revolution*. Vol I, *Ideas in History*, pp. 157–190.

Kahneman, Daniel; Paul Slovic; and Amos Tversky, eds. *Judgment Under Uncertainty: Heuristics and Biases*. Cambridge, Eng., 1982.

Kargon, Robert. "Atomism in the seventeenth century." In *Dictionary of the History of Ideas*. Philip Wiener, editor-in-chief. Vol. 1, pp. 132–141.

Kauder, Emil. *A History of Marginal Utility Theory*. Princeton, 1965.

Kendall, M. G. "The beginnings of a probability calculus." *Biometrika* 43 (1956): 1–14. Reprinted in E. S. Pearson and M. G. Kendall, eds., *Studies in the History of Statistics and Probability*. Vol. 1, pp. 19–34.

Kendall, M., and R. L. Plackett, eds. *Studies in the History of Statistics and Probability*. Vol. 2. London, 1977.

Keynes, John Maynard. *A Treatise on Probability*. London, 1943.

Knobloch, Eberhard. "The mathematical studies of G. W. Leibniz on combinatorics." *Historia Mathematica* 1 (1974): 409–430.

———. "Musurgia universalis: Unknown combinatorial studies in the age of Baroque absolutism." *History of Science* 17 (1979): 258–275.

Kolmogorov, Andrei. *Foundations of the Theory of Probability*. Tr. Nathan Morrison. New York, 1950.

Koyré, Alexandre. "Pascal savant." In *Blaise Pascal, l'homme et l'oeuvre*, pp. 259–295. Cahiers de Royaumont, Paris, 1956.

Krakeur, L. G., and R. L. Krueger, "The mathematical writings of Diderot." *Isis* 33 (1941): 219–232.

Krieger, Leonard. *The Politics of Discretion: Pufendorf and the Acceptance of Natural Law.* Chicago and London, 1965.

Krüger, Lorenz; Lorraine J. Daston; and Michael Heidelberger, eds. *The Probabilistic Revolution.* Vol. I, *Ideas in History.* Cambridge, Mass., 1987.

Langbein, John H. *Prosecuting Crime in the Renaissance: England, Germany, France.* Cambridge, Mass., 1974.

———. *Torture and the Law of Proof.* Chicago and London, 1976.

Lazarsfeld, Paul. "Notes on the history of quantification in sociology: Trends, sources and problems." *Isis* 52 (1961): 277–333.

Le Cam, Lucien, and Jerzy Neymann, eds. *Bernoulli 1713; Bayes (1763); Laplace (1813). Anniversary Volume.* Proceedings of an International Research Seminar. Statistical Laboratory, University of California, Berkeley. Berlin, Heidelberg, and New York, 1965.

Leeson, Francis. *A Guide to the Records of the British State Tontines and Life Annuities of the 17th and 18th Centuries.* Shalfleet Manor, Eng., 1968.

Leonnet, Jean. *Les loteries d'état en France aux XVIIIe et XIXe siècles.* Paris, 1963.

Lévy, Jean Philippe. *La hiérarchie de preuves dans le droit savant du Moyen Age.* In *Annales de l'Université de Lyon,* 3rd series: Droit. Fascicule 5. Paris, 1939.

Lévy, Paul. *Calcul des probabilités.* Paris, 1925.

Lopes, Lola L. "Doing the impossible: A note on induction and the experience of randomness." *Journal of Experimental Psychology: Learning, Memory, and Cognition* 8 (1982): 626–636.

Lottin, Joseph. *Quetelet, statisticien et sociologue.* Louvain and Paris, 1912.

McGuire, P. M. "Atoms and the 'analogy of nature': Newton's Third Rule of Philosophizing." *Studies in the History and Philosophy of Science* 1 (1970–71): 3–58.

McKibbin, Ross. "Working-class gambling in Britain 1880–1939." *Past & Present,* no. 82 (February 1979): 147–178.

McManners, John. *Death and the Enlightenment.* Oxford and New York, 1985.

MacPherson, Colin. *The Political Theory of Possessive Individualism: Hobbes to Locke.* Oxford, 1972.

Maire, Albert. *Pascal philosophe.* Vol. 4 of *Les Pensées-Editions originales, ré-impressions successives.* Paris, 1926.

Maistrov, L. E. *Probability Theory: A Historical Sketch.* Tr. Samuel Kotz. New York and London, 1974.

Malcomson, Robert W. *Popular Recreations in English Society 1700–1850.* Cambridge, Eng., 1973.

Manuel, Frank. *Isaac Newton, Historian.* Cambridge, Mass., 1963.

Mauzi, Robert. "Écrivains et moralistes du XVIIIe siècle devant les jeux de hasard." *Revue des sciences humaines* 26 (1958): 219–256.

Mertz, Rudolf. "Les amités françaises de Hume et le mouvement des idées." *Revue de littérature comparée* 9 (1929): 644–713.

Meuvret, Jean. "Les données démographiques et statistiques en histoire moderne et contemporaine." In his *Études d'histoire économique, Cahiers des Annales 32* (Paris, 1971), pp. 313–340.

Molina, E. C. "The Theory of probability: Some comments on Laplace's *Théorie analytique*." *Bulletin of the American Mathematical Society* 36 (1930): 369–392.

Morgenstern, Oskar, and John von Neumann. *The Theory of Games and Economic Behavior*. 2nd edition, Princeton, 1947.

Morize, André. *L'Apologie du luxe au XVIIIe siècle et "Le Mondain" de Voltaire*. Geneva, 1970; reprint of 1909 edition.

Murphy, James J. *Rhetoric in the Middle Ages*. Berkeley, Los Angeles, and London, 1947.

Murray, David. *Chapters in the History of Bookkeeping, Accountancy and Commercial Arithmetic*. Glasgow, 1930.

Nagel, Ernest. *Principles of the Theory of Probability*. In Otto Neurath, Charles Morris, and Rudolf Carnap, eds., *International Encyclopedia of Unified Science*. 2 vols. Vol. 1, part 2. Chicago, 1955.

Nisbet, Robert. "The French Revolution and the rise of sociology in France." *American Journal of Sociology* 49 (1943–44): 156–164.

Noonan, John T., Jr. *The Scholastic Analysis of Usury*. Cambridge, Mass., 1957.

Noxon, James. *Hume's Philosophical Development*. Oxford, 1973.

Ogborn, Maurice Edward. *Equitable Assurances*. London, 1962.

Ore, Oystein. *Cardano: The Gambling Scholar*. Princeton, 1953.

Oschilewski, Walther Georg. *Lotto-Toto Lotterie 1763–1963*. Berlin, 1963.

Owen, G.E.L. "Tithenai ta phainomena." In S. Mansion, ed., *Aristote et les problèmes de méthode*, pp. 83–92. Louvain, 1961.

Park, Katharine, and Lorraine J. Daston. "Unnatural conceptions: The study of monsters in sixteenth- and seventeenth-century France and England." *Past & Present*, no. 92 (August 1981): 20–54.

Pearson, E. S., and M. G. Kendall, eds. *Studies in the History of Statistics and Probability*. Vol. 1. Darien, Conn., 1970.

Pearson, Karl. *The History of Statistics in the 17th and 18th Centuries*. E. S. Pearson, ed. London and High Wycombe, 1978.

———. "Historical note on the origin of the normal curve of errors." *Biometrika* 16 (1924): 402–404.

———. "James Bernoulli's Theorem." *Biometrika* 17 (1925): 201–210.

Peller, Sigismund. "Studies in mortality since the Renaissance." *Bulletin of the History of Medicine* 13 (1943): 427–461; 16 (1944): 362–381; 21 (1947): 51–101.

Poincaré, Henri. *Cours de calcul des probabilités*. Paris, 1896.

Popkin, Richard. *The History of Scepticism from Erasmus to Descartes.* Assen, The Netherlands, 1964.

Popper, Karl. "The propensity interpretation of probability." *British Journal for the Philosophy of Science* 10 (1959): 25–42.

Porter, Theodore M. "A statistical survey of gases: Maxwell's social physics." *Historical Studies in the Physical Sciences* 12 (1981): 77–116.

———. *The Rise of Statistical Thinking, 1830–1900.* Princeton, 1986.

Proust, Jacques. *Diderot et l'Encyclopédie.* Paris, 1962.

Prudential Insurance Company of America. *The Documentary History of Insurance, 1000 B.C.–1875 A.D.* Newark, N.J., 1915.

Rabinovitch, Nachum L. *Probability and Statistical Inference in Ancient and Medieval Jewish Literature.* Toronto, 1973.

Rankin, Bayard. "The history of probability and the changing concept of the individual." *Journal for the History of Ideas* 27 (1966): 483–504.

Rashed, Roshdi. *Condorcet: Mathématique et société.* Paris, 1974.

Raymond, Pierre. *De la combinatoire aux probabilités. Le combinatoire de Cardan à Jacques Bernoulli.* Paris, 1975.

Reinhard, Marcel. "La statistique de la population sous le Consulat et l'Empire. Le Bureau de Statistique." *Population* 5 (1950): 103–120.

Reiser, Stanley J. *Medicine and the Reign of Technology.* Cambridge, Eng., 1978.

Rosin, Albert. *Lebensversicherung und ihre geistesgeschichtlichen Grundlagen.* Leipzig, 1932.

Roth, Eugen. *Das grosse Los.* Nordrhein-Westfalen, 1965.

Rouault de la Vigne, René. *La loterie à travers les ages et plus particulièrement en France.* Paris, 1934.

Rowbotham, Arnold. "The *philosophes* and the propaganda for inoculation of smallpox in eighteenth-century France." *University of California Publications in Modern Philology* 8 (1935): 265–290.

Saint-Aubert, Gaston de. *L'Assurance contre l'invalidité et la vieillese en Allemagne.* Paris, 1900.

Sambursky, S. "On the possible and the probable in ancient Greece." *Osiris* 12 (1956): 35–48.

Samuelson, Paul. "Probability, utility, and the independence axiom." *Econometrica* 20 (1952): 670–678.

Savage, L. J. *The Foundations of Statistics.* 2nd edition. New York, 1972.

Schmitt-Lermann, Hans. *Der Versicherungsgedanke im deutschen Geistesleben des Barock und der Aufklärung.* Munich, 1954.

Schneider, Ivo. *Die Entwicklung des Wahrscheinlichkeitsbegriff in der Mathematik von Pascal bis Laplace.* Munich, 1972.

———. "Leibniz on the probable." In Joseph Dauben, ed., *Mathematical Perspectives*, pp. 201–219. New York, 1981.

Schneider, Ivo. "Der Mathematiker Abraham De Moivre (1667–1754)." *Archive for History of Exact Sciences* 5 (1968–69): 177–317.

———. "Why do we find the origin of a calculus of probabilities in the seventeenth century?" In J. Hintikka, D. Gruender, and E. Agazzi, eds., *Pisa Conference Proceedings*, vol. 2, pp. 3–24. Dordrecht and Boston, 1980.

———. "Laplace and thereafter: The status of the probability calculus in the nineteenth century." In L. Krüger et al., eds., *The Probabilistic Revolution.* Vol. I, *Ideas in History*, pp. 191–214.

Schöpfer, Gerald. *Sozialer Schutz im 16.–18. Jahrhundert.* Graz, Austria, 1976.

Scott, J. F. "John Craig." In *Dictionary of Scientific Biography.* Charles C. Gillispie, editor-in-chief. Vol. 2, pp. 458–459.

———. *The Mathematical Work of John Wallis.* London, 1938.

Sergescu, Pierre. "La contribution de Condorcet à l'*Encyclopédie*." *Revue d'Histoire des Sciences* 4 (1951): 233–237.

Sewall, Hannah Robie. "The theory of value before Adam Smith." *Publications of the American Economic Association.* Series 3, vol. 2, no. 3 (August 1901).

Shafer, Glenn. "Bayes's two arguments for the rule of conditioning." *Annals of Statistics* 10 (1982): 1075–1089.

———. "The combination of evidence." Typescript, 1984.

———. "Non-additive probabilities in the work of Bernoulli and Lambert." *Archive for History of Exact Sciences* 19 (1978): 309–370.

Shapiro, Barbara J. "Law and science in seventeenth-century England." *Stanford Law Review* 21 (1969): 727–766.

———. *Probability and Certainty in Seventeenth-Century England.* Princeton, 1983.

———. *John Wilkins, 1614–1672.* Berkeley and Los Angeles, 1969.

Sheynin, O. B. "Daniel Bernoulli's work on probability." *Rete* 1 (1972): 273–300.

———. "R. J. Boscovich's work on probability." *Archive for History of Exact Sciences* 9 (1973): 306–324.

———. "J. H. Lambert's work on probability." *Archive for History of Exact Sciences* 7 (1970–71): 244–256.

———. "Laplace's theory of error." *Archive for History of Exact Sciences* 17 (1977): 1–61.

———. "P. S. Laplace's work on probability." *Archive for History of Exact Sciences* 16 (1976): 137–187.

———. "Mathematical treatment of astronomical observations." *Archive for History of Exact Sciences* 11 (1973): 97–126.

———. "Newton and the classical theory of probability." *Archive for History of Exact Sciences* 7 (1970–71): 217–243.

———. "On the early history of the law of large numbers." In E. S. Pearson and M. G. Kendall, eds., *Studies in the History of Statistics and Probability.* Vol. 1, pp. 231–239.

———. "On the mathematical treatment of observations by L. Euler." *Archive for History of Exact Sciences* 9 (1972): 45–56.

———. "On the prehistory of the theory of probability." *Archive for History of Exact Sciences* 12 (1974): 97–141.

Société Générale Néerlandaise d'Assurances sur la Vie et de Rentes Viagères. *Mémoires pour servir à l'histoire des assurances sur la vie et des rentes viagères au Pays-Bas.* Amsterdam, 1898.

Sorabji, Richard. *Necessity, Cause and Blame: Perspectives on Aristotle's Theory.* London, 1980.

Stefani, Giuseppe. *Insurance in Venice from the Origins to the End of the Serenissima.* Tr. Arturo Dawson. Trieste, 1958.

Steinmetz, Andrew. *The Gaming Table: Its Votaries and Victims.* 2 vols. London, 1870.

Stigler, Stephen M. "John Craig and the probability of history: From the death of Christ to the birth of Laplace." Technical Report No. 165, Department of Statistics, University of Chicago (September 1984; revised February 1985).

———. "Napoleonic Statistics: The work of Laplace." University of Wisconsin Department of Statistics Technical Report No. 376 (June 1974).

———. "Who discovered Bayes' theorem?" *The American Statistician* 37 (1983): 296–325.

———. *The History of Statistics: The Measurement of Uncertainty Before 1900.* Cambridge, Mass., 1986.

Stone, Lawrence. *The Family, Sex and Marriage in England 1500–1800.* London, 1977.

Stove, David C. *Probability and Hume's Inductive Scepticism.* Oxford, 1973.

Supple, Barry. *The Royal Exchange Assurance.* Cambridge, Eng., 1970.

Swijtink, Zeno G. "D'Alembert and the maturity of chances." *Studies in History and Philosophy of Science* 17 (1986): 327–349.

———. "The objectification of observation: Measurement and statistical methods in the nineteenth century." In L. Krüger et al., eds., *The Probabilistic Revolution.* Vol. I, *Ideas in History,* pp. 261–285.

Tenenti, Alberto. *Naufrages, corsaires et assurances maritimes à Venise 1592–1909.* Paris, 1959.

Tenenti, Branislava and Alberto. "L'assurance en Méditerranée." *Annales. Économies, Sociétés, Civilisations* (March-April, 1976): 411–413.

Thomas, Keith. *Religion and the Decline of Magic.* New York, 1971.

———. "Work and Leisure in pre-industrial society." *Past & Present,* no. 29 (December 1964): 50–62.

Thompson, E. P. "Time, work-discipline, and industrial capitalism." *Past & Present,* no. 38 (December 1967): 56–97.

Todhunter, Isaac. *A History of the Mathematical Theory of Probability from the Time of Pascal to that of Laplace* (1865). New York, 1949.

Trennery, C. F. *The Origin and Early History of Insurance*. London, 1926.

Trexler, Richard C. "Une table florentine d'espérance de vie." *Annales. Économies, Sociétés, Civilisations* (January-February 1971): 137–139.

Tribe, Lawrence. "Trial by mathematics: Precision and ritual in the legal process." *Harvard Law Review* 84 (1970–71): 1329–1393.

Vacant, A., and E. Mangenot, eds. *Dictionnaire de théologie catholique*. 16 vols. Paris, 1903–51.

Van Leeuwen, Henry. *The Problem of Certainty in English Thought, 1630–1690*. The Hague, 1963.

Van Rooijen, J. P. *La notion de probabilité et la science actuarielle*. Amsterdam, 1935.

Verdon, Jean. *Les loisirs en France au Moyen Age*. Paris, 1980.

Von Bar, Carl Ludwig. *A History of Continental Criminal Law*. Tr. Thomas S. Bell. Boston, 1916.

Von Mises, Richard. *Probability, Statistics and Truth*. 2nd English edition. London, 1957.

Waldman, Theodore. "Origins of the legal doctrine of reasonable doubt." *Journal of the History of Ideas* 20 (1959): 299–316.

Walford, Cornelius. *The Insurance Guide and Hand-Book*. New York, 1868.

Walker, Helen. *Studies in the History of the Statistical Method*. Baltimore, 1929.

Weber, Jean D. *Historical Aspects of the Bayesian Controversy*. Division of Economic and Business Research, University of Arizona, Tuscon (January 1973).

Westergaard, Harald. *Contributions to the History of Statistics*. London, 1932.

Wiener, Philip P., editor-in-chief. *Dictionary of the History of Ideas*. 4 vols. New York, 1973.

Yamazaki, E. "D'Alembert et Condorcet: Quelques aspects de l'histoire du calcul des probabilités." *Japanese Studies in the History of Science* 10 (1971): 60–93.

Young, Robert M. "David Hartley." *Dictionary of Scientific Biography*. Charles C. Gillispie, editor-in-chief. Vol. 6, pp. 138–140.

Zelizer, Viviana A. Rotman. *Morals and Markets: The Development of Life Insurance in the United States*. New York and London, 1979.

INDEX

Académie des Sciences, Paris, 84–85, 96, 98, 99, 106, 145, 172n, 277, 347, 364–365, 375
Académie des Sciences Politiques et Morales, 366, 378, 383
actuaries, 116, 168–169, 180
Albertus Magnus, 153
aleatory contracts. *See* contracts
d'Alembert, Jean, xxi, xxii, 17, 206, 368; and Beccaria, 344–345; and Condorcet, 78, 96–97, 211–212, 278, 283; criticisms of probabilistic arguments for order, 252–253; exception among classical probabilists, 106, 111; and Laplace, 78, 103–104, 105, 191, 275, 283; on mixed mathematics, 53–55; on risk taking, 81, 88–89; on St. Petersburg problem, 76–81, 88, 95; on smallpox inoculation, 83–89
Allais, M., 109
Allen, Sir Thomas, 170
American School, of economics, 109
Amicable Society, 164, 167, 170–171, 174, 176, 177
analogy, 193–194, 204, 205–206, 208, 220, 229, 244–245, 254, 307, 381, 383
annuities, xix, 14, 19, 23, 50, 112, 166, 167, 173, 174, 176, 178, 348; history of, 116, 121–122, 138–139, 142; mathematically based, 27–30, 55, 69–70, 104, 136, 168–169, 172n; obstacles to application of statistics to, 122, 124–125, 133, 169–170, 172–174; prices, 121–122, 124–125
Appleby, Joyce, 51

applications of probability theory, xx–xxiv, 13, 33, 48, 49, 229–230, 238, 369, 372
applied mathematics, 3–6, 53–54, 354–355, 364
Aquinas, Saint Thomas, 37–38, 151, 153
Arago, François, 380n
Arbuthnot, John, 127, 131, 138, 248, 252, 276, 286
Archimedes, 236, 239
argument from design, 187, 248, 251–253, 267. *See also* natural theology
Aristotle, 5, 8, 38, 51, 61, 151, 153, 154
Arnauld, Antoine, 17, 59
Arrow, Kenneth, 56
assemblies, 215, 343
associationism, xxi, xxiii, 47, 191, 196–225, 228, 255, 261, 290–291, 368, 370–371, 377, 380
astronomy, 35, 192, 216, 228, 271, 364, 379
atomism, 246–249
Augustine, Saint, 151, 153

Babbage, Charles, 181
Bacon, Francis, 51, 65, 127, 228, 241–242, 307
Baker, Keith, xvi, 96, 104, 283
Barbeyrac, Jean, 159
Barrow, Isaac, 59
Bayes, Thomas: assumption of equal prior probabilities, 223, 258–262, 266; expectation approach to probability theory, 31–33, 258, 261–262; and Hartley, 204, 255–256; on inverse probabilities, 208n, 221n, 230, 233,

413

Bayes, Thomas (*cont.*)
    257–262; and Price, 256–257, 258,
    262–264
Bayes' theorem, Bayes' version, 257–264;
    Laplace's version, 268–269
Bayle, Pierre, 159, 304, 309–311, 321,
    323
Beaumarchais, Eustache, 148
Beaumont, Charles d'Eon de, 165
Beccaria, Cesare, 304–305, 343–345,
    352, 353
Benoiston de Chateauneuf, Louis-François,
    360n, 384
Bentham, Jeremy, 297n, 355
Bentley, Richard, 252
Bérenger, Alphonse, 383–384
Bernoulli, Daniel, xxii, 100, 147, 253n
    on expectation, 70–77, 93–95, 104,
    105, 274; on smallpox inoculation, 53,
    83–89, 96
Bernoulli, Jakob, xix, xxi, xxvi, 13, 55,
    64, 68, 125, 126, 128, 137, 138; *Ars
    conjectandi*, 7, 10, 24, 32, 48, 98, 189,
    236–237, 296, 316, 318; correspond-
    ence with Leibniz, 28n, 129–130, 232–
    233, 237–240, 286; on degrees of cer-
    tainty, 5, 33, 34–35, 38; on determin-
    ism, 35, 37, 281, 284; on estimation of
    evidence, 40–44, 306, 311; on frequen-
    cies and statistics, 26, 28n, 301; model
    of causation, 238–241, 246, 249–250,
    262, 268, 285; on objective/subjective
    probabilities, 188, 189–190; and prob-
    ability of causes, 227–253, 294; rules
    for conjecture, 49–50
Bernoulli, Nicholas, xxvi, 7, 13, 172,
    312, 347; criticism of probabilistic ar-
    guments for order, 252–253; on expec-
    tation, 25, 48, 69–70, 76; on insur-
    ance, 136–137; on life expectancy, 87,
    128, 135, 136; on probability of testi-
    mony, 46, 193, 306, 311–312, 319,
    334
Bernoulli's theorem: formulation of, 231–

257; and law of large numbers, 268,
    285
Bertrand, Joseph, 375–376, 379, 380n
Bible, 92, 123, 131, 133, 154, 306,
    308–309, 313–316
Bicquelley, Charles-François, 339
Bienaymé, I. J., 288n, 379
Binomial expansion, 235, 256, 264, 285,
    350
Biot, J. B., 291, 293
Boethius, Anicius Manlius, 151–152, 153
Boltzmann, Ludwig, 246
Boole, George, 223, 373–376
Borda, Jean-Charles, 145
Bossuet, Jacques Bénigne, 321, 323
Bossut, Charles, 278, 352
bottomry, 20, 116–117. *See also* insur-
    ance, maritime
Bourchier, Dick, 149
Boyle, Robert: corpuscularism, 241–242,
    246–249; on empiricist modesty, 226–
    227; on evaluation of testimony, 63–
    307; on kinds of certainty, 38, 56–57,
    63–64; and natural theology, 63–65,
    205, 267; on reasonableness, 59–66;
    and skepticism, xix, 283
Boyle lectures, 59, 131
Brissot de Warville, J. P., 167, 179n,
    343–344
British Museum, 143
Buffon, George Leclerc, 172n, 344; coin-
    tossing experiment, 95n, 287; on credi-
    bility of testimony, 330–331; on expec-
    tation, 76, 90–95, 147; on induction,
    206–208; on life expectancy and mor-
    tality, 86, 184, 301, 302; on measure-
    ment of risk, 91–92, 185, 348
Burnet, Thomas, 310, 313
Butler, Joseph, 62, 204–205, 267

Caillois, Roger, 148
Calas, Jean, 304, 343, 349, 353
calculation, 152, 157–159, 161–163,

173, 187, 215–218, 221, 369, 376, 385–386. *See also* mental calculations; self-interest
Calzabigi, Florentin, 143
Cambridge Platonists, 248
Caminade de Castres, Marc-Alexandre, 145n
Canning, Elizabeth, 343
Canton, John, 256
Carcavy, P., 24
Cardano, Gerolamo, 13, 15, 23, 36–37, 51, 124, 158, 231, 234
Carnap, Rudolf, 189
Cartesianism, xx, 156, 300
Casanova, Giovanni-Jacopo de Seingalt, 143
causes: complexity of, 79–80, 275, 281; hidden, 208, 226, 237, 240–245, 253, 379; in Hume, 199–201, 228, 265–266, 290–291; manifest, 207–208, 226, 241; perturbing, 263, 270–272; reasoning from effects to, 226, 232, 240, 268; regular and irregular, 270–272, 280–281, 286–287, 290, 381, 383; urn model of, 230–231, 237–249, 268–270, 279–280, 294, 375
certainty: degrees of, 14, 33–47, 57, 58–59, 126, 138, 192; moral, 38, 39, 57, 60, 63–64, 91, 205, 206, 207–208, 213, 227, 235, 315, 318, 322, 334n, 344, 348
chance: as absence of cause, 10, 25, 36, 38, 79, 199, 288; and urn model of causation, 237, 245–246. *See also* fortune, randomness
charity, 175, 177, 179
Charleton, Walter, 246–249
Chillingworth, William, 60
Cicero, Marcus Tullius, 38, 153
Civiale, Jean, 367
Clark, Samuel, 136n, 156
classical probability theory, xix–xxvi, 3–6, 52, 67, 116; decline of, 210–211, 223–225, 284, 370–386; description and

prescription in, 52–53, 88, 102, 107–108, 198–199, 211–212, 299; good sense interpretation of, xix, xxii, xxvi, 50, 67, 108, 198, 211, 218, 352, 357–359, 360; objective/subjective meanings of probability in, 189–191, 261–262, 370–371, 377; urn model in, 229–231, 238, 379
Clausius, Rudolf, 246
Clavière, Étienne, 166n
Cleirac, Estienne, 50, 118–120
cockfighting, 124, 157, 160, 163
combinatorics, 9, 15–17, 31, 34, 47, 235, 240–241, 245–246
common sense. *See* good sense
Compagnie Royale d'Assurance, 178–179
Comte, Auguste, 298, 354, 366, 375, 377–378, 380, 382
Condillac, Étienne, 93, 110, 191, 198, 208–212, 214, 221, 302
Condorcet, M.J.A.N., 78, 90, 145n, 172n, 198, 206, 318n; and Beccaria, 344–345; educational projects, 98, 216–218; on expectation, 76, 96–103, 349; and Laplace, 104, 268; on meaning of probability, 191, 211–219, 224, 281–282; on measurement, 92–93, 97–98, 216–217; on natural rights, 303, 348–349; praise of printing, 322; on probability of causes, 208, 223, 230, 233, 267, 278–283, 285; on probability of judgments, 107, 215, 296–297, 299, 342–343, 345–352, 354, 356–357, 361–363, 375; on probability of testimony, 323, 324, 329–335, 339–340; on social mathematics, 67, 104, 211, 217–218, 304, 355, 377, 378; and statistics, 302
consensus, 66–67, 108, 193, 197, 217–218, 352, 359
contracts, xxii, 7, 14–15, 21–22, 102, 347; aleatory, 14–15, 18–33, 47–48 69–70, 112, 115, 117, 121, 125, 137–138, 140, 155, 166–167, 169, 172–

contracts (*cont.*)
174, 178; equity in, 19–23, 25–33, 66, 69, 71, 94, 99–100, 136
corpuscles, 242, 244, 246–247, 249. *See also* atomism
Coulomb, Charles-Auguste, 352
Coumet, Ernest, 7, 18
Council of Trent, 126
Counter-Reformation, xix, 308, 320
Cournot, A. A., 7, 189, 190–192, 223–224, 233–234, 284, 359, 370–371, 375
Cousin, Victor, 371, 378
Cox, James, 144
Craig, John, 46, 64, 92, 312–316, 320, 323, 333, 342
Cramer, Gabriel, 95
credibility. *See* testimony

Darwin, Erasmus
decision making, xix, 44, 48, 60, 208, 213, 224, 227
Defoe, Daniel, 140, 147, 164, 165, 166, 170–171
De Moivre, Abraham, xxvi, 268; on annuities and life expectancy, 135, 136n, 137–138, 169; on argument from design, 187, 205–206, 248, 253–257, 267, 288; on Bernoulli's theorem, 198, 204, 228–229, 250, 253–254, 257, 263, 267, 286; on definition of probability, 31–32; on inverse of Bernoulli's theorem, 233, 251–252, 256–257, 262; on mortality statistics, 126; on nonexistence of chance, 10–11, 37, 266
De Morgan, Augustus, 181, 370, 374, 374n
Deparcieux, Antoine, 138, 172n
De Prades, Jean-Martin, 330–331, 337
Derham, William, 131–132, 252
Descartes, René, 49, 50, 93, 226, 241, 243
Destutt de Tracy, Antoine Louis Claude, 354–355, 366, 375

determinism: linked with classical probability theory, 10, 34–37, 51; and moral sciences, 300, 380; opposed to chance, 153, 154–156, 159; and probability of causes, 241, 245, 279; social context of, 113; and statistical regularities, 133, 183
De Witt, Johann, 7, 13, 17, 23, 27–29, 31, 45, 125, 128, 129, 137, 139
Diderot, Denis, 73, 318, 321–322, 344
distributions, 108, 183, 258, 363, 379. *See also* normal curve
Dodson, James, 114, 169, 176, 179
Domat, Jean, 22, 306, 344
doubt, xx, 43, 57, 59, 194, 201, 241
Dover/Calais packetboat (as standard of acceptable risk), 92–93, 348
Du Moulin, Charles, 19, 22, 50
Dupin, Charles, 106, 365–367, 378n
Dupont de Nemours, Pierre, 157
Durkheim, Émile, 298
Duvillard, Étienne, 179n

economics, 109. *See also* political economy
Elizabeth I (queen of England), 141, 143, 322
Ellenborough, Lord, 176n
Ellis, Robert Leslie, 191, 223, 284
empiricism, 226–229, 241–245, 373
Encyclopédie, 54, 78, 97, 318–321, 323, 330, 337, 343
Enlightenment, xxi–xxv, 51, 55, 109–110, 150, 153, 156, 193, 199, 298–299, 302, 321
equiprobability, 23, 29–33, 77, 79, 80, 124, 125, 137, 199, 260, 270, 273, 306, 331. *See also* Principle of Indifference
Equitable Society (est. 1712), 167
Equitable Society for the Assurance of Lives (est. 1762), 169, 174–182
equity, 19, 21–23, 25–26, 28, 53, 57, 67, 69, 76, 100, 110, 133. *See also* contracts

error law, 271–272, 383
Erskine, Thomas, 166–167
Euler, Leonhard, 144
evidence, xix, 61; historical, 320–323, 369; and induction, 193–196, 228; legal, xxiii, 43–44, 45–47, 48, 192–196, 228, 306, 356, 369; to persuade reasonable man, 62–63, 202, 300; of signs, 11–12
expectation, xxii, xxiv, 6, 47, 48, 53, 57, 133, 140, 174; applied to probability of testimony, 315; average, 96, 98–99; and contracts, 15–30, 117; mathematical, 57, 58, 70 (definition), 76–78, 81, 104; moral, 57, 58, 70, 71 (definition)– 77, 94–95, 102, 104–105, 110; origins of concept, 14–33, 138; and probabilites, 16–18, 24–34, 69, 135; and reasonableness, 58–110, 205, 296

Fermat, Pierre, 3, 5, 7, 9, 13, 15–18, 23, 31
Fielding, Henry, 156
Finetti, Bruno de, 189
foenus nauticum. See bottomry
Fontenelle, Bernard de, 316–317
Fortuna, 151–152
fortune, 36–37, 143, 147, 148, 149, 150, 151–153, 158–163, 186
Francis I (king of France), 126, 141
Frederick II (king of Prussia), 143, 144, 346, 347, 349, 352
free will, 300–304, 317, 342
French Revolution, xxii, 52, 58, 106–107, 147, 211, 217, 298, 342–343, 352– 354, 356, 366
frequency. *See* statistics
Fréret, Nicolas, 333–334
friendly societies, 170, 170n

Galiani, Ferdinand, 71, 110
Galilei, Galileo, 5, 13, 37, 124
Galton, Francis, 295
gambling, 8, 13–14, 18, 19, 23, 34, 61,

102, 114, 115–116, 125, 139; and life insurance, 122–123, 135, 137, 140– 141, 163–168, 171, 173–176, 182, 184–185; mathematical critique of, 75, 94–95, 104, 105, 147, 156; moral and religious critique of, 104, 117, 123, 146, 147–151, 154–156, 161–163, 173; practices, 124, 157–163; universality of, 123, 141–144, 160–161. *See also* lotteries
Gambling Act, 123, 175
Garber, Daniel, 12–13
Gassendi, Pierre, xix, 56, 59, 243, 246– 247
Gataker, Thomas, 155
Gay, John, 203
Gentile, Benedetto, 143
geometry, xx, 3, 4, 54, 226
Glanvill, Joseph, 38, 56, 59, 63, 248, 253, 322
Goffman, Erving, 153, 162
Gonzaga, Louis (duke of Nevers), 155n
good sense: ambiguity in, 110–111, 297; classical probability theory as, xix, xxii, xxvi, 50, 67, 108, 198, 211, 218, 352, 357–359, 360; conflicts with classical probability theory, 52–53, 68, 81, 341, 355, 368–369, 370–371, 376, 377; defined by moral sciences, 106; probability theory altered to fit, 95, 99, 102–103. *See also* St. Petersburg problem
Goodwin, William, 203
Grattan-Guinness, Ivor, 368n
Graunt, John: on bills of mortality, 12, 26, 127–129; influence of, 131, 132, 133, 137, 237; knowledge of probability theory, 128; mortality table, 31, 128, 135, 136, 138
's Gravesande, Willem, 65, 253n
Gregory IX (pope), 117
Grotius, Hugo, xix, 20, 22, 56, 60, 205
Guerry, André Michel, 287n, 359, 360n, 384

Hacking, Ian, xxiv, 7, 11–13, 30, 34, 188, 189–190, 234
Halley, Edmund, 13, 31, 51, 87, 126, 128, 129, 130, 131, 132, 136, 137, 139, 234, 317
Hartley, David, 191, 198, 203–206, 210, 212, 214, 224, 245, 254–256, 302, 370
Helvetius, Claude Adrien, 302, 344
Henry VIII (king of England), 126
Herodotus, 321
Herschel, John, 370, 374n
Hewit, Beau, 158
Hilbert, David, 3
Hirschman, Albert, 51
history, xxiv; probability of testimony applied to, 304, 313–316, 333–334; written sources in, 310–322
Hobbes, Thomas, 10, 74
Hogarth, William, 165
d'Holbach, P.H.D., 10, 302, 344
homme moyen, 298, 383, 384–385
hommes éclairés, 90, 211, 216, 298, 376, 384–385. See also reasonable man
Hooper, George, 315–316, 341
Houdini, Harry, 186
Hudde, Johannes, 27, 125, 145
L'Huillier, Simon, 339
human nature, predictability of, 51–52, 162–163, 302–303
Hume, David: on associationism, 191, 198–203, 210, 212, 214, 222, 224, 293; and Beccaria, 344; and Bishop Butler, 205n; on causation and induction, 228, 264–267, 280, 283, 325–326; on hope and fear under uncertainty, 185–186; on luxury, 73; on miracles, 154, 321, 323, 324–330, 334–335, 340–341
Hutchenson, Francis, 371n
Huygens, Christiaan, 5, 7, 9–10, 13, 17, 21, 23, 45, 97, 114, 126, 261; De ratiociniis in aleae ludo, 24–28, 30, 31–32, 34, 128, 134, 158; on life expectancy, 128, 133–135
Huygens, Ludwig, 128, 129, 133–135

idéologues, 354
induction: in Hartley, 204, 205–206; in Hume, 202, 264–267; limitations of, 243, 245; in Locke, 193–195; quantification of, xxiii, 228–229, 237, 262, 267–293, 294–295, 377; reliability of, 254–256
insurance, 8, 19–21, 112, 139–141; fire, 114, 164, 166, 168, 172, 174; life, 114–116, 137, 163–182, 184–185 (see also gambling); maritime, 50, 55, 116–120, 136–137, 165, 168, 174; mathematically based, 136–137, 164, 167, 170–171, 175–182; and middle-class values, 174–178, 184–185, 187; obstacles to application of statistics in, 124–125, 133, 172–174, 192; policies, 118–120, 171, 181; premiums, 117–120, 124, 136–137, 171, 177–182, 185
interest rates, 20, 112, 121–122, 169, 179, 181

Jacob, James, 60
Jacob, Margaret, 60
Jeffreys, Harold, 189
Jesuits, 7, 20, 124, 308
jurisprudence: influence on probability theory, xxiii, 6–7, 14–15, 47, 99 (see also contracts, proofs, testimony); probability theory applied to, xxiv, 296, 342 (see also probability of judgments, probability of testimony); Roman-canon, 15, 19, 41–43, 46, 76, 124, 154n, 353; as model of reasonableness, 57, 63, 66, 106. See also expectation
Justinian, 117

Keynes, John Maynard, 38, 189
Kolmogorov, A. N., 3

Koyré, Alexandre, 7
Kramp, Chrétien, 172n

La Barre, Jean-François LeFèvre de, 304, 343, 346, 349
La Condamine, Charles Marie, 84
Lacroix, Silvestre-François, 339–341
Lagrange, Joseph Louis, 278
La Harpe, Jean-François, 354, 366
Lalande, Joseph-Jérôme, 90
Lambert, Johann Heinrich, 338
La Mothe le Vayer, François de, 322
Laplace, Pierre Simon, xix, xxi, xxvi, 253n, 378; on annuities and insurance, 172n, 182; and associationism, 206, 219–222; and Bayes' theorem, 208, 268–270, 274; on connection of natural to moral sciences, 220, 275–277, 299, 381–383; on determinism, 10, 274–275, 284; on expectation, 76, 103–106, 274–275; influence of d'Alembert on, 78, 90, 275, 283; on meaning of probability, 189–191, 210, 218–223, 224, 273–274, 281–282, 371, 374n; and Principle of Indifference, 29–30, 269, 273, 275, 279, 289, 372; on probability of causes, 13, 208, 223, 230, 233, 267–278, 294–295, 299; on probability of judgments, 107, 304, 350, 356–361, 364, 367, 375, 380n; on probability of testimony, 13, 323, 324, 335–338, 339, 340; on probability theory as good sense, 50, 68, 111, 297, 337, 357–359, 360; and statistics, 195, 222, 275–276, 284, 302
latitudinarianism, 59–60
Law, John, 164
law. See jurisprudence
law of large numbers, 181, 204, 205, 285–289, 299, 362, 365, 366–367, 381–382
law of succession, 270, 277–279, 295, 334, 370, 372–374, 377, 386
Le Clerc, Jean, 154, 156, 166

legal reform, 298, 304–305, 342–347, 352–354, 356, 360n, 380
Legendre, Adrien Marie, 145n, 172n
Leibniz, Gottfried Wilhelm, xxiii, 6, 7, 26, 126, 188; combinatorics, 9–10, 235; correspondence with Jakob Bernoulli, 28n, 129–130, 232–233, 237–240, 250, 255, 285–286; on legal probabilities, 44–46, 308, 310–311; and mortality data, 128, 132; on probability theory as new logic, 45, 266; on universal characteristic, 66, 385, 386
Leibniz-Clarke correspondence, 252
Leti, Gregorio, 141, 142, 146
life expectancy: methods of computing, 83, 86–87, 126, 128, 133–136, 172; tables, 112, 169, 318
Livy, 323
Lloyds of London, 120, 164, 165, 168
Locke, John: on associationism, 191, 196–199, 200, 201, 203, 208, 210–212, 222, 261; on evidence, 193–196, 291, 323; king of Siam parable, 195, 228, 280, 307, 328; and possessive individualism, 74; on probability, 45, 193–194, 201, 244, 282, 307, 324; and skepticism, xix; theory of knowledge, 55, 227, 241–242
London Assurance, 164, 170–171, 177
lotteries, xix, 78, 88, 91, 144, 141–163, 168, 173; blank, 142–143, 156; Genoan-style, 143; mathematical basis, 144–145, 157; players, 145–163; popularity, 141–144, 148; profits, 144–145, 157
Louis XV (king of France), 165
Louis XVI (king of France), 144
Lucas, Theophilus, 158
luck, 36–37, 124, 152, 153, 156–158. See also fortune
luxury, xxii, 72–74, 94–95

Machiavelli, Niccolo, 152–153, 157
MacPherson, C. B., 74

Magens, Nicolas, 166
Malebranche, Nicolas, 310
Malesherbes, Chrétien-Guillaume de La-
    moignon de, 345
Mandeville, Bernard, 72–73, 94
Martianus Capella, 151
Marx, Karl, 298
Massé de la Rudelière, 96
mathematical model, xx, 56, 110–111
mathematical theory, xx, 4, 110
Maxwell, James Clerk, 246
measurement: common units of, 97–98;
    psychological units of, 91–94
Melon, Jean-François, 72–73, 94
mental calculations, 93–95, 98, 108, 197,
    201, 305. See also associationism
Méré, Antoine Gombault, 15, 17
Merlin de Douai, Philippe Antoine, 352
Mersenne, Marin, xix, 56, 59, 243, 283
Michodière, 302
Mill, James, 203
Mill, John Stuart, 203, 297, 297n, 372–
    373, 375, 376, 377, 378–379
miracles, 155, 199, 202, 210, 267, 289,
    304, 307, 310, 322–327, 329, 332–
    335, 337–338, 342
Mises, Richard von, 189
mixed mathematics, xxi, 53–56, 79, 88,
    110, 369
Molesworth, John, 156
Montesquieu, Charles de Secondat de, 73,
    97, 300, 303, 343
Montmort, Pierre de, 10, 32, 69, 253n,
    312, 316–318, 341, 342
moral sciences: Enlightenment character
    of, 52, 55–56, 110, 298–305; influence
    on probability theory, 102, 106–107,
    110–111; and natural sciences, 219,
    275, 294, 296, 299–304, 347, 381–
    383; not subject to mathematical treat-
    ment, 354–355, 366, 368, 377–379;
    probability theory applied to, xxiv, 93,
    107, 213, 219, 277, 288, 294, 296–
    369, 377–379; and social sciences,

106–108, 110, 298–299, 369, 377–
    382, 384–385. See also jurisprudence;
    political economy; psychology
Morand, 302
Morellet, André, 304, 343–344
Morgan, William, 180–181
Morgenstern, O., 109
Morris, Corbyn, 180
mortality: bills of, 12, 126–127, 192;
    rates of, 113, 121, 122, 123, 171, 334,
    362, 384; regularity of, 127–133, 135,
    137, 138, 171, 172, 175, 181, 183–
    184, 301; tables of, 31, 89, 93, 127–
    129, 136, 137, 172–173, 176, 180,
    181, 185, 301, 317. See also statistics,
    demographic
Mulgrave, Earl of, 149
Mylon, Claude, 23

Nagel, Ernest, 6, 29
Napoleon I, 106, 356
natural laws, 55, 202, 208, 270, 280–
    281, 299–304, 310, 324–325, 337–
    338, 379
natural philosophy, 56–57, 59, 227, 240,
    242–243, 246–248, 254–255, 267
natural theology, 56–57, 59, 62, 65,
    131–132, 137, 203, 205, 253, 267
Navier, Claude, 265, 367, 378n
Neumann, John von, 109
Neumann, Kaspar, 132
Newton, Isaac, 5, 51, 131, 203, 220,
    226, 241, 243–244, 250, 255, 299,
    300, 313, 333
Nicole, François, 172n
Nicole, Pierre, 17
normal curve, 250, 382, 385

Occam's razor, 250
Oersted, H. C., 291–293

Pacioli, Luca, 9, 18
Paracelsus, Phillipus, 12
parish records, 126–127

Pascal, Blaise, xix, 5, 7, 9, 13, 45, 81, 235; correspondence with Fermat, 3, 15–18, 23, 31; *Lettres provinciales*, 308; *Pensées*, 60–61; wager, 57, 60–63, 67, 99, 221

passions, 51–52, 106, 161–162, 209, 214, 216, 221, 356, 364

Paul V (pope), 126

Payen, Antoine-François, 346

Pearson, Karl, 131

Pepys, Samuel, 160, 161

Petty, William, 127, 128, 171, 232

Peyronnet, Charles-Ignace de, 286, 359–360

physiocrats, 55, 100, 110, 151, 300–301

Pitt, William, 175n

Poinsot, Louis, 106, 365–369, 378n

Poisson, Siméon-Denis, 46, 57, 58, 111, 187, 253, 302, 304; on connection of natural to moral sciences, 287–288, 296, 365–366, 381–382; on expectation, 76, 105–106; on meaning of probability, 191, 219, 222–223, 371; on probability of causes, 227, 267, 284–294, 299; on probability of judgments, 107, 284, 289, 296–297, 304, 343, 350, 359–368, 370, 372, 375, 378n; on probability of testimony, 306, 323, 324, 338–339, 370

political arithmetic, 26, 67, 130, 192, 232

political economy, xxiii, 68, 70–76, 94–95, 99–101, 106, 110–111, 296

Popkin, Richard, xix

popular errors, xxii, 153–154, 321

Port Royal *Logique*, 17, 21, 23, 39–40, 50, 61, 64, 65, 212, 234, 243, 244, 307, 323

Porter, Theodore, xxiv

Pothier, Robert, 172

Povey, Charles, 114

Prestet, Jean, 9–10, 235

Price, Richard: on annuities and insurance, 136n, 169, 179–181; and Bayes'

essay, 32–33, 205, 206, 208n, 212n, 230, 256–257, 258, 262–267; criticism of Hume, 203, 228, 248, 264–267, 278, 283, 323, 326–330

Priestly, Joseph, 203

Principle of Indifference, 29–30, 68, 103, 260, 273, 275, 279, 289, 290, 329, 372, 374

probabilism, 308

probability: inverse, 103, 204, 205, 206, 208, 218, 223, 230–231, 236–237, 256–263, 267–293; objective, xxi, 11–13, 30, 37, 103, 188–225, 229, 257, 261–262, 282, 283–284, 289–290, 370, 374n, 377; prehistory of, 3–47, 308; qualitative, 11–12, 14, 38–40, 48, 63, 194–195, 201, 205, 305, 330; quantitative, 7–8, 17–18, 24, 38–40, 194–195, 201, 205, 305; subjective, xxi, 11-13, 14, 30, 33–41, 46, 188–225, 229, 257, 261–262, 282, 289–290, 370, 374n, 377

probability of causes, 13, 229, 318; Bayes' theorem, 253–267, 274, 275, 277, 279, 293; Bernoulli's theorem, 230–253, 274, 275, 282, 285, 288n; critique of, 294–295, 372–375; in moral sciences, 277, 294–295, 296, 299, 302–303, 325, 338–339, 349, 351, 358, 359, 361–362, 365–366; and scientific reasoning, 267–293

probability of judgments, 6, 284, 289; critique and decline of, 106, 107, 229, 297, 305, 354–355, 359, 364, 365–370, 372–373; defended, 215, 358, 364, 367; and legal reform, 298, 304–305, 342–347, 349, 356, 380; political content of, 296, 304–305, 348–349, 357–358, 361–362, 364, 366

probability of testimony, 6; applied to historical sources, 304, 313–316, 333–334; applied to miracles, 323–332, 334–335; critique and decline of, 13, 193, 297, 305, 316–318, 355, 359,

probability of testimony (*cont.*)
370, 372–376; defended, 337; in *Encyclopédie*, 318–320; extrinsic and intrinsic components of, 306, 310, 311, 313, 316, 318–319, 323, 327–329, 331, 335, 339; historical context of, 308–310, 320–323, 341, 380; link with early probability theory, 304–305, 306, 308; and proto-quantification, 133, 305
problem of points, 9, 15–18, 24
proofs, legal, 14–15, 40–47, 192–193, 306, 353–354, 368–369, 376, 380
providence, 131–132, 137, 151, 154–156, 248–253, 286, 288, 310
psychology, 92–93, 109, 191, 199, 202, 203, 209–210, 220–221, 225, 305, 371, 380. *See also* associationism
Pufendorf, Samuel, 22, 28
pure mathematics, 53–54
pyrrhonism. *See* skepticism

quantification, xxiii–xxiv, 4–6, 14, 23, 38, 42, 47–48, 52, 131–133, 305, 355, 369
Quesnay, François, 100, 300
Quetelet, Adolphe, 106–108, 182, 286–287, 294, 298, 359, 369, 370–372, 377, 381–385

randomness, 11, 35, 157n, 240, 248, 251–252, 254. *See also* chance
rationalism, 240–241, 373
rationality, xix, xx, xxii–xxvi, 48, 50, 52, 57, 59, 61, 89, 108–111, 116, 152, 187, 196, 202, 211, 385. *See also* reasonableness
reasonable calculus, xix, 211
reasonable man, xx, xxii, 52–53, 56–111, 187, 197, 222, 293, 322, 342
reasonableness, xix, 48, 50, 53, 59, 90, 211, 262, 267, 369; and associationism, 200, 209, 212; and expectation, 57–82, 93–111; and miracles, 326; nineteenth-century attacks on, 376. *See*

also good sense
Reformation, xix, 308, 322
risk, 18–23, 50, 53, 76; averse, 163, 173, 178, 182, 185–186; before probability theory, 116–125; comparison of, 90, 213, 345, 348–349; mathematical theory of, 125–138; perception, 90–95, 103, 112–116, 163, 171, 185–186; quantification of, 90–95, 120–121, 124–133, 173; relation to values and beliefs, 112–116, 140–141, 172–174; seeking, 163, 165, 167–168, 173, 178; of smallpox inoculation, 82–89. *See also* annuities; gambling; insurance; lotteries
Risueño d'Amador, 376–377
Roberval, Giles, 5, 17, 23
Royal Exchange Assurance, 164, 169, 170–171, 177, 181–182
Royal Society of London, 38, 60, 61, 62, 64, 131–132, 176, 196, 205, 248, 256, 265, 315
Royer-Collard, Pierre Paul, 366
rule of fellowship, 20
rules of reasoning, xxiii, 40, 49, 202, 241, 243–244

Saint-Cyran, Paul-Edmé, 301
St. Petersburg problem, xxi, 53, 56, 68–71, 75–82, 88, 95, 97, 99, 101, 103, 104–105, 106, 126n, 198
Schneider, Ivo, xxiv, xxv
scholasticism, xx, 57
Schooten, F. van, 9–10, 24, 235
scientific method, 222, 227, 240, 267, 272, 279–281, 290–293
self-interest, 51, 57, 59, 60, 62, 161, 173, 184, 208–210, 216, 376
Shapiro, Barbara, xxiv, 59
Sheynin, B. O., xxiv, xxv
signs, 11–12, 40–41, 43
Simon, Richard, 309
simplicity in nature, 239–240, 249–252, 281
Simpson, Thomas, 136, 169

Sixtus V (pope), 117n
skepticism: Condorcet's, 215, 283; constructive or moderate, 57–58, 65, 243, 247, 283, 307; historical, 322–323; Hume's, 199, 203, 228, 264, 267; probability theory used against, 228, 253–254, 267, 315, 329; sixteenth-century revival of, xix, 308, 322
smallpox inoculation, xxii, 33, 82–89
Smollet, Tobias, 157
social sciences, 106–108, 110, 298–299, 369, 372, 379–380. See also moral sciences
South Sea Bubble, 167
Spencer, Herbert, 298
Spinoza, B. de, 10
statistical inference, 190, 232–234, 295
statistical regularities, 107–108, 130–133, 137, 182–184, 187, 192, 288, 294, 301, 359, 372
statistics: demographic, 27, 91, 126–129, 138, 276–277, 301–302, 372, 384; lack of, 26–27, 30, 67, 83, 85, 120, 126, 136, 286; medical, 367, 376–377, 378n, 384n; relation to probability, 47, 49, 98, 103, 125–126, 135, 138, 230–234; social, 107–108, 286–287, 294, 302, 359, 371–372, 382, 384n; and stable conditions, 120, 124–125, 174, 183–185. See also mortality
Stigler, Stephen, xxiv, xxv, 256n
Stone, Lawrence, 184
Struyck, Nicholas, 131
Sun Fire Office, 114
Süssmilch, Johann, 130–131, 301

Talleyrand-Périgord, Charles-de, 146, 156, 157
Tartaglia, Nicolo, 18
testimony, 39–40, 63, 193, 195, 308–310, 320–323, 344, 352, 380. See also probability of testimony
Tetens, Johann, 166
Thomas, Keith, 164–165

Thucydides, 323
Tillotson, John, xix, 59, 60, 61–62, 66, 205, 242, 267
time, 115, 161–163, 173, 186
Tonti, Lorenzo, 167n
tontines, 122, 166, 167, 167n, 168, 170n, 178
tribunals, 6, 215, 238, 345, 366, 369. See also probability of judgments
Turgot, Ann-Robert Jacques, 96, 97, 100–101, 102, 104, 110, 278, 304, 345, 351

Ulpian, 30, 116, 121, 122
underwriters, 119–120, 165
uniformity of nature, 202, 208, 210, 237, 244, 262–263, 266, 283, 285, 338, 374
Ussher, Henry (bishop of Armagh), 207
usury, 18, 19–21, 115, 117, 121, 124, 160, 167, 169, 173
utility, 70–76, 95, 109. See also expectation, moral

value, 71, 74, 76, 77, 93, 99–101, 102, 105, 108, 110. See also utility
Vernier, Théodore, 166
Villermé, Louis-René, 384
Voltaire, François Marie Arouet de, 72–73, 83, 85, 94, 144, 304, 321, 323, 334, 343–344, 348, 353–354

Walford, Cornelius, 139
Walker, Joseph, 61
Wallis, John, 9, 235, 249
Walpole, Robert, 165
Westholme, William Dorghtie de, 143
White's of London, 160, 165, 167
widows' funds, 169, 174
Wilkins, John, 5, 38, 56, 59, 60, 62–63, 66–67, 242, 267, 283, 322

Zabell, Sandy, 12–13

CPSIA information can be obtained
at www.ICGtesting.com
Printed in the USA
JSHW022059160623
43393JS00001B/1